ADVANCES IN CHEMICAL PHYSICS

VOLUME XXXIV

Advances in

CHEMICAL

PHYSICS

EDITED BY

I. PRIGOGINE

University of Brussels,
Brussels, Belgium
and
University of Texas,
Austin, Texas

AND

STUART A. RICE

Department of Chemistry
and
The James Franck Institute
The University of Chicago,
Chicago, Illinois

VOLUME XXXIV

AN INTERSCIENCE® PUBLICATION

JOHN WILEY AND SONS

NEW YORK · LONDON · SYDNEY · TORONTO

CONTRIBUTORS TO VOLUME XXXIV

HANS C. ANDERSEN, Department of Chemistry, Stanford University, Stanford, California

L. BLUM, Department of Physics, University of Puerto Rico, Rio Piedras, Puerto Rico

DAVID CHANDLER, School of Chemical Sciences, University of Illinois, Urbana, Illinois

N. G. VAN KAMPEN, Institute for Theoretical Physics of the University, Utrecht, Netherlands

J. F. KARNICKY, Division of Chemistry and Chemical Engineering, California Institute of Technology, Pasadena, California

A. H. NARTEN, Chemistry Division, Oak Ridge National Laboratory, Oak Ridge, Tennessee

C. J. PINGS, Division of Chemistry and Chemical Engineering, California Institute of Technology, Pasadena, California

WILLIAM A. STEELE, Department of Chemistry, The Pennsylvania State University, University Park, Pennsylvania

JOHN D. WEEKS, Bell Laboratories, Murray Hill, New Jersey

INTRODUCTION

In the last decades chemical physics has attracted an ever-increasing amount of interest. The variety of problems, such as those of chemical kinetics, molecular physics, molecular spectroscopy, transport processes, thermodynamics, the study of the state of matter, and the variety of experimental methods used, makes the great development of this field understandable. But the consequence of this breadth of subject matter has been the scattering of the relevant literature in a great number of publications.

Despite this variety and the implicit difficulty of exactly defining the topic of chemical physics, there are a certain number of basic problems that concern the properties of individual molecules and atoms as well as the behavior of statistical ensembles of molecules and atoms. This new series is devoted to this group of problems which are characteristic of modern chemical physics.

As a consequence of the enormous growth in the amount of information to be transmitted, the original papers, as published in the leading scientific journals, have of necessity been made as short as is compatible with a minimum of scientific clarity. They have, therefore, become increasingly difficult to follow for anyone who is not an expert in this specific field. In order to alleviate this situation, numerous publications have recently appeared which are devoted to review articles and which contain a more or less critical survey of the literature in a specific field.

An alternative way to improve the situation, however, is to ask an expert to write a comprehensive article in which he explains his view on a subject freely and without limitation of space. The emphasis in this case would be on the personal ideas of the author. This is the approach that has been attempted in this new series. We hope that as a consequence of this approach, the series may become especially stimulating for new research.

Finally, we hope that the style of this series will develop into something more personal and less academic than what has become the standard scientific style. Such a hope, however, is not likely to be completely realized until a certain degree of maturity has been attained—a process which normally requires a few years.

At present, we intend to publish one volume a year, and occasionally several volumes, but this schedule may be revised in the future.

In order to proceed to a more effective coverage of the different aspects of chemical physics, it has seemed appropriate to form an editorial board. I want to express to them my thanks for their cooperation.

I. PRIGOGINE

CONTENTS

ix

ADVANCES IN CHEMICAL PHYSICS

VOLUME XXXIV

THE ROTATION OF MOLECULES IN DENSE PHASES

WILLIAM A. STEELE

Department of Chemistry,
The Pennsylvania State University,
University Park, Pennsylvania

CONTENTS

1

I. INTRODUCTION

A. General

Interest in molecular rotation or libration in dense phases has increased considerably over the past few years. Progress has been both experimental and theoretical, and in large part has been due to the introduction and wide acceptance of the time-correlation function formalism as a representation of experimental results and as the primary object of theoretical speculation.

At present, the theory of molecular rotation in dense phase is at the model-building level. That is, the usual approach is to produce a plausible physical description of the dynamics of the system; one then computes correlation functions for this model as a function of one or more adjustable parameters, and compares the results with selected data as a test of the model. It is now known that the range of observed rotational behavior for molecules in dense phases is extremely wide, varying from almost free motion to nearly complete hindrance to reorientation. A correspondingly wide range of models has been suggested. Since it is almost always possible to find a system that will at least superficially agree with predictions based on a particular model, one must be particularly cautious in defining the limits of applicability of the model. This is not always a straightforward procedure, especially when one has a disposable parameter or two to improve the fit between theory and experiment and a limited number of experimental data to deal with. It is to be hoped that progress will soon occur at a more fundamental level where molecular shapes and interaction laws play fundamental roles rather than parameters such as "collision frequency" or "average jump angle" that so often appear in present theories. Nevertheless, a critical review and discussion of current practice should be useful, especially if it includes discussion of some specific examples that have been successfully treated.

The relationships between measurable quantities and the variables involving molecular reorientation can be conveniently exhibited via time-correlation functions;[1,2] a listing of the most important of these should include rotational band shapes in infrared and Raman spectra, nuclear magnetic relaxation times, dielectric relaxation, and inelastic neutron scattering by polyatomic molecules. Although this article is primarily concerned with descriptions of molecular reorientation and the calculation of time-correlation functions from these theories, we will at least list some of the best-known expressions that relate observables to time-correlation functions, and comment upon the complications that often arise in applying them to actual experimental situations.

We will approach the problem by assuming that the best way to express an arbitrary function of the Eulerian orientation angles $\alpha, \beta, \gamma \equiv \Omega$ of a molecule is to expand it in the complete orthogonal set of functions $D_{km}{}^j(\Omega)$. Rose's definitions[3] for the generalized spherical harmonics will be used; thus,

$$D_{km}{}^j(\Omega) = e^{-ik\alpha}\, d_{km}{}^j(\beta)e^{-im\gamma} \tag{1.1}$$

where the $d_{km}{}^j(\beta)$ can be expressed in terms of the Jacobi polynomials, or one can use one of the explicit equations given by Rose (among others[4, 5]). A few of the most useful properties of these functions are

$$\frac{1}{8\pi^2}\int D^{*j}_{km}(\Omega)D^{j'}_{k'm'}(\Omega)\,d\Omega \equiv \langle D^{*j}_{km}(\Omega)D^{j'}_{k'm'}(\Omega)\rangle = \left(\frac{1}{2j+1}\right)\delta_{j,\,j'}\delta_{k,\,k'}\delta_{m,\,m'} \tag{1.2}$$

$$D_{k0}{}^j(\alpha\beta\gamma) = (-1)^k D_{0k}{}^j(\gamma\beta\alpha) = \left(\frac{4\pi}{2j+1}\right)^{1/2}Y^*_{jk}(\alpha\beta) \tag{1.3}$$

where Y_{lm} is a normalized spherical harmonic. If one rotates first by $\Omega(0)$, and subsequently by $\delta\Omega$, to a final orientation $\Omega(t)$, one has

$$D_{km}{}^j(\Omega(0)) = \sum_{r=-j}^{j} D_{kr}{}^j(\Omega(t))D^{*j}_{mr}(\delta\Omega) \tag{1.4}$$

$$D_{km}{}^j(\Omega(t)) = \sum_{r=-j}^{j} D_{kr}{}^j(\Omega(0))D_{rm}{}^j(\delta\Omega) \tag{1.5}$$

$$D_{km}{}^j(\delta\Omega) = \sum_{r=-j}^{j} D_{rm}{}^j(\Omega(t))D^{*j}_{rk}(\Omega(0)) \tag{1.6}$$

These expressions can be shown to be internally consistent by using the completeness property:

$$\sum_{r} D^{*j}_{m'r}(\Omega)D^{j}_{m''r}(\Omega) = \delta_{m',\,m''} \tag{1.7}$$

$$\sum_{r} D^{*}_{rm'}(\Omega)D^{j}_{rm''}(\Omega) = \delta_{m',\,m''} \tag{1.8}$$

B. Magnetic Relaxation

One of the earliest explicit uses of time-correlation functions was in connection with the theory of nuclear magnetic relaxation due to magnetic dipole-dipole interactions in a fluid or solid. For a pair of spins with angular momentum operators $\mathbf{I}_1$ and $\mathbf{I}_2$, the interaction appears as a perturbation term $\mathscr{H}_p$ in the spin Hamiltonian for the system; the explicit form of $\mathscr{H}_p$ is

$$\mathscr{H}_p = \gamma_1\gamma_2\,\hbar^2\mathbf{I}_1\cdot\mathbf{T}_{12}\cdot\mathbf{I}_2 \tag{1.9}$$

where γ_i is the gyromagnetic ratio for spin i and the tensor $\mathbf{T}_{12}$ is

$$\mathbf{T}_{12} = \frac{1}{r_{12}^3}\left(1 - 3\frac{\mathbf{r}_{12}\mathbf{r}_{12}}{r_{12}^2}\right) \tag{1.10}$$

If $\mathcal{H}_p$ is small, a straightforward time-dependent perturbation theory calculation yields expressions for the relaxation times. It is important to realize that the relaxation is due to fluctuations in $\mathbf{T}_{12}$ that occur at the nuclear Larmor frequency ω_I (or a simple multiple of it). These fluctuations are due to changes in the length and orientation of $\mathbf{r}_{12}$. The total dipole-dipole perturbation of a relaxing spin is the sum of the pairwise $\mathcal{H}_p$ terms over all other spins in the system. Consequently, changes in $\mathbf{r}_{12}$ can be due to relative translational and rotational motions of spins on two different molecules, or they can be due to the time dependence of the vector for a pair of spins on the same molecule. We assume that contributions to the relaxation from spins on different molecules can be eliminated, either experimentally by dilution in a nonmagnetic solvent or by choosing the relaxing spin such that the entire relaxation is due to a few nearby spins in the same molecule (for example, relaxation of the ^{13}C nucleus in a hydrocarbon is generally dominated by the protons that are bonded to the carbon atom). The contribution to the spin–lattice relaxation time T_1^{DD} and the spin-spin relaxation time T_2^{DD} due to the interaction of a pair of identical spins with quantum number I are written as[6]

$$\left(\frac{1}{T_1^{DD}}\right) = \gamma^4 \hbar^2 I(I + 1)\{\mathcal{I}_{2,1}(\omega_I) + 4\mathcal{I}_{2,2}(2\omega_I)\} \tag{1.11}$$

$$\left(\frac{1}{T_2^{DD}}\right) = \gamma^4 \hbar^2 I(I + 1)\{\tfrac{3}{2}\mathcal{I}_{2,0}(0) + \tfrac{5}{2}\mathcal{I}_{2,1}(\omega_I) + \mathcal{I}_{2,2}(2\omega_I)\} \tag{1.12}$$

where $\mathcal{I}_{l,m}(\omega_I)$ is a spectral density defined as

$$\mathcal{I}_{l,m}(\omega) = \int_{-\infty}^{\infty} c_{l,m}(t) e^{i\omega t} \, dt \tag{1.13}$$

Focusing on the rotational contributions to the spin relaxation, one has $r_{12}(0) = r_{12}(t) = b$, where b is the separation of a pair of spins in the same molecule (assumed to be rigid, for simplicity). In this case,

$$c_{l,m}(t) = \frac{1}{b^6(2l + 1)} C_{l,m}^{(v)}(t) \tag{1.14}$$

$$C_{l,m}^{(v)}(t) = (2l + 1)\langle D_{m0}^l(\theta(0), \phi(0)) D_{m,0}^{*l}(\theta(t), \phi(t))\rangle \tag{1.15}$$

where the superscript (v) denotes a correlation function for the (12) spin-spin vector which is fixed in the molecular frame. However, it is almost always more convenient to calculate correlation functions for reorientation of the principal axes of a molecule than it is for a particular vector in the molecule. Therefore, we let η, ζ be the polar and azimuthal angles of the spin-spin vector relative to a set of axes fixed in the molecule; after adopting the notation of generalized spherical harmonic addition theorem shown in (1.5), one finds

$$C_{l,m}^{(v)}(t) = (2l + 1) \sum_{r,r'} D_{r0}{}^l(\eta, \zeta)D_{r'0}^{*l}(\eta, \zeta) \langle D_{mr}{}^l(\Omega(0))D_{mr'}^{*l}(\Omega(t)) \rangle \quad (1.15a)$$

Equation (1.15) can be simplified somewhat (without loss in generality) by remembering that these correlation functions depend on the molecular reorientation $\delta\Omega$ in time interval t and not upon the initial orientation $\Omega(0)$, if the system is isotropic. One again uses (1.5), and averages over the random distribution of initial orientations in an isotropic fluid using (1.2) to find

$$\langle D_{mr}{}^l(\Omega(0))D_{m'r'}^{*l}(\Omega(t)) \rangle_{\Omega(0)} = \frac{1}{2l + 1} \langle D_{rr'}^{*l}(\delta\Omega) \rangle \delta_{mm'} \quad (1.16)$$

and thus

$$C_{l,m}^{(v)}(t) = \sum_{r,r'} D_{r0}{}^l(\eta, \zeta)D_{r'0}^{*l}(\eta, \zeta) \langle D_{rr'}^{*l}(\delta\Omega) \rangle \quad (1.17)$$

Often, internal rotations in a nonrigid molecule will keep b constant, but will cause η, ζ to depend on time (H—H dipolar interaction in a rotating methyl group is the best-known example); clearly, these problems can be treated by making a trivial extension of (1.17). We now see that these correlation functions actually do not depend upon the index m; we thus omit it and refer to $C_l^{(v)}(t)$. In addition, it is almost invariably true that the time required for the attainment of a random distribution of reorientations in fluids (and thus for $\langle D_{rr'}^l(\delta\Omega) \rangle$ to become zero) is short compared to ω_I^{-1} for a nuclear spin; in this case, we can write

$$\mathscr{I}_{2,m}(2\omega_I) = \mathscr{I}_{2,m}(\omega_I) = \mathscr{I}_{2,m}(0) = \frac{2}{5}b^{-6}\int_0^\infty C_2^{(v)}(t)\,dt \quad (1.18)$$

$$= \frac{2}{5b^6}\tau_2^{(v)} \quad (1.19)$$

The first of these equations expresses the fact that correlation functions of the classical rotors that will be considered here depend upon $|t|$ only. $\tau_2^{(v)}$ is the correlation time for the spin-spin vector associated with spherical harmonics with $l = 2$. We now see that (1.11) and (1.12) become

$$\left(\frac{1}{T_1^{DD}}\right)_{rot} = \left(\frac{1}{T_2^{DD}}\right)_{rot} = \frac{2\gamma^4\hbar^2 I(I + 1)}{b^6}\tau_2^{(v)} \quad (1.20)$$

$$\tau_2^{(v)} = \sum_{r,r'=-2}^{2} D_{r0}{}^2(\eta, \zeta)D_{r'0}^{*2}(\eta, \zeta)\int_0^\infty \langle D_{rr'}^{*2}(\delta\Omega) \rangle\,dt \quad (1.21)$$

Equations (1.20) and (1.21) still must be summed over all spin pairs that contribute to the relaxation of a single spin. The summation over other spin pairs in a single molecule is far from trivial because of strong correlations in the relative motions of spins attached to a rigid frame.[7] Even in the simplest case where the dipole-dipole interaction of a single spin pair dominates the relaxation, it should be emphasized that the magnitude of $\tau_2^{(v)}$ depends upon the orientation of the spin-spin vector relative to the molecular frame. Except for molecules with tetrahedral (or higher) symmetry, different rates of reorientation are generally found for the different axes in a molecule, with the most rapid motions occurring about axes of high symmetry.

One final comment is in order here: if it happens that the spin-spin vector is parallel to the molecular z-axis in the principal axis system, $\eta = 0$. Since

$$D_{km}{}^{j}(0) = \delta_{k,m} \tag{1.22}$$

(1.17) and (1.21) reduce to

$$C_2(t) = \langle D_{00}{}^{2}(\delta\Omega) \rangle \tag{1.23}$$

$$\tau_2 = \int_0^\infty C_2(t)\, dt \tag{1.24}$$

where we omit the (v) when the vector of interest happens to be the molecular z-axis.

Of course, we have not yet discovered the criteria for choice of principal axes; although one might at first suppose that these should be the inertial axes, this is not always the case. As we will see later, the principal axes for a molecule undergoing rotational random walk are determined by the symmetry of the intermolecular torques rather than by its mass distribution. In the worst case, it can happen that the body-fixed orientation of the principal axes for a particular system will vary with the density, being inertial at low density and torque-fixed at high; it is conceivable that it will be temperature dependent as well.

Another relaxation mechanism frequently encountered in nmr studies arises from the electric quadrupolar interaction of a nuclear spin with $I > \frac{1}{2}$ with an electric field gradient at the nucleus. The details of the argument are considerably more complicated than in the case of dipolar relaxation; however, if one assumes isotropy in the initial orientations and a relaxation time that is short compared to $\omega_I{}^{-1}$, one can eventually show[8,9] that

$$\left(\frac{1}{T_1{}^Q}\right) = \left(\frac{1}{T_2{}^Q}\right) = \frac{3(2I + 3)}{40I^2(2I - 1)} \left(\frac{e^2qQ}{\hbar}\right)^2 \tau_2^{(v)} \tag{1.25}$$

where Q is the nuclear quadrupole moment. For simplicity, we have assumed that the electric field has cylindrical symmetry at the nucleus so that q is the magnitude of the only nonzero component of the field gradient; the direction of this gradient relative to the molecular axes is given by the vector $\mathbf{v}$, and $\tau_2^{(v)}$ in (1.25) is identical to the correlation time defined in (1.21) except that η, ζ now denote the orientation angles of the field gradient vector.

A third type of interaction that can give an important contribution to nuclear magnetic relaxation is caused by the coupling of the nuclear moment with the magnetic moment due to rotation of the molecule with its associated electric charges. This spin-rotation interaction is dependent upon molecular angular velocity as well as a spin-rotation tensor $\mathbf{C}$ (units of frequency). Explicitly, the spin-rotation energy $\mathscr{H}_{SR}$ has the form

$$\mathscr{H}_{SR}(t) = -\gamma\hbar\mathbf{H}(t)\cdot\mathbf{I} = \mathbf{L}(t)\cdot\mathbf{C}(t)\cdot\mathbf{I} \qquad (1.26)$$

where $\mathbf{H}(t)$ is the rotational magnetic field, $\mathbf{L}(t)$ is the angular momentum of the molecule at time t, and $\mathbf{C}(t)$ depends upon time because it is fixed to the rotating molecular frame. Again making the assumptions that the decay of the correlations involved here is rapid compared to ω_I^{-1} and that the fluid is isotropic, one finds that the spin-rotation contribution to the nuclear magnetic relaxation times becomes

$$\frac{1}{T_1^{SR}} = \frac{1}{T_2^{SR}} = \frac{8\gamma^2 I(I+1)\langle H^2\rangle}{9h^2}\tau_H \qquad (1.27)$$

where τ_H is defined by:

$$\tau_H = \frac{1}{\langle H^2\rangle}\int_0^\infty \langle\mathbf{H}(0)\cdot\mathbf{H}(t)\rangle\,dt \qquad (1.28)$$

Inasmuch as the magnetic field $\mathbf{H}(t)$ is $\gamma^{-1}\mathbf{L}(t)\cdot\mathbf{C}(t)$, it is now apparent that the spin-rotational correlation time τ_H involves not only the reorientational motions that give rise to the time dependence of the spin-rotation tensor $\mathbf{C}(t)$, but also upon the change in angular momentum $\mathbf{L}$ due to the torques exerted on a molecule in a dense phase.

For simplicity, we assume that the spin-rotation tensor can be diagonalized in some set of molecule-fixed axes; it will then possess nonzero components C_{xx}, C_{yy}, C_{zz}. It does not appear that an expression for τ_H has been derived for the case where the principal axes of $\mathbf{C}$ do not coincide with the principal inertial axes; however, when the two sets do coincide, it can be shown that[10]

$$\langle\mathbf{H}(0)\cdot\mathbf{H}(t)\rangle = \sum_{k,k'=-1}^{1} C_{kk}C_{k'k'}\langle D_{k'k}^{*1}(\delta\Omega)L_k(t)L_{k'}^*(0)\rangle$$

$$+ (\Delta C)^2 \sum_{k,k'=1,-1}^{1}\langle D_{k'k}^{*1}(\delta\Omega)L_{-k}^*(t)L_{-k}^*(0)\rangle \qquad (1.29)$$

The components of **C** and **L** are expressed here in a spherical basis so that

$$
\mathbf{C} = \begin{vmatrix} C_{\perp} & 0 & \Delta C \\ 0 & C_{zz} & 0 \\ \Delta C & 0 & C_{\perp} \end{vmatrix}
\tag{1.30}
$$

where

$$
C_{\perp} = \tfrac{1}{2}(C_{xx} + C_{yy})
\tag{1.31}
$$

$$
\Delta C = \tfrac{1}{2}(C_{yy} - C_{xx})
\tag{1.32}
$$

One also has

$$
L_{\mu} = L D_{\mu,0}^{*1}(\phi_L, \theta_L)
\tag{1.33}
$$

where ϕ_L, θ_L are the orientation angles of **L** in the body-fixed frame.

Calculations of mixed angular momentum-orientation correlations of (1.29) are considerably more difficult than pure orientational correlations. Although the angular momenta in (1.29) are expressed in the body-fixed frame, it is not difficult to switch to the space-fixed frame for these variables, if desired. In fact, (1.29) is only one of several equivalent mathematical formulations of this problem.[11-14]

The time-correlation function formalism is also used in the calculation of electron spin relaxation times. However, the most important interactions differ from those for nuclear spins, except in the case of spin-rotational relaxation. Kivelson[15] has given a detailed discussion of electron spin relaxation; he notes that electron spin relaxation due to the anisotropic g-tensor is analogous to nuclear relaxation due to an anisotropic chemical shift tensor, and that electron spin relaxation due to anisotropic hyperfine interactions is analogous to nuclear relaxation due to anisotropic spin-spin interaction. None of these effects will be discussed in detail here; however, we do note that the perturbation Hamiltonians $\mathscr{H}_p$ are (a) for electron spins,

$$
\mathscr{H}_p = \mu_0 \mathbf{S} \cdot \mathbf{G} \cdot \mathbf{H}
\tag{1.34}
$$

where $\mathbf{S}$ = electron angular momentum, μ_0 = Bohr magneton, $\mathbf{H}$ = applied magnetic field, and $\mathbf{G}$ is the traceless g-tensor; and

$$
\mathscr{H}_p = \hbar \mathbf{I} \cdot \mathbf{A}_e \cdot \mathbf{S}
\tag{1.35}
$$

where $\mathbf{A}$ is the anisotropic hyperfine coupling tensor; (b) for nuclear spins,

$$
\mathscr{H}_p = -\gamma \hbar \mathbf{I} \cdot \underline{\sigma} \cdot \mathbf{H}'
\tag{1.36}
$$

where $\underline{\sigma}$ is the anisotropic chemical shift tensor; and

$$\mathcal{H}_p = \hbar \mathbf{I} \cdot \mathbf{A}_n \cdot \mathbf{I}' \tag{1.37}$$

All of these perturbation terms are time-dependent because the coupling tensors are fixed to the molecular frame and thus rotate with the molecules. Although they consequently contribute to spin relaxation, the nuclear terms are relatively unimportant. Since the rotation of any (traceless) second-rank tensor is describable by the $D_{km}{}^j(\Omega)$ with $j = 2$, it can be seen that the correlation times associated with these relaxation mechanisms will be proportional to the zeroth moments of correlation functions involving $D_{km}{}^2(\delta\Omega)$, when the Larmor frequency is small compared to the inverse of the correlation time; consequently, the detailed theory is quite similar to that for the dipolar correlation time given by (1.21). Thus, a model for rotational motion that yields an expression for the time dependence of the functions $\langle D_{km}^2(\delta\Omega) \rangle$ will in principle allow one to compute a variety of magnetic relaxation times.

C. Raman and Depolarized Light Scattering

Of the other observables that depend upon the reorientation of a second-rank molecule-fixed tensor, the Raman effect is undoubtedly the most important. The theoretical expressions for Raman spectra in dense phases[1,16] can be written in a way that is quite similar to the formalism for frequency-shifted light scattering.[17] Indeed, the differences between these two phenomena are minor, having to do primarily with the magnitudes of the experimental frequency shifts, and with the nature of the corrections necessary to extract rotational information from the raw experimental data. In both cases, one can calculate the relative intensity $\hat{I}(\omega)$ from the experimental results by evaluating:

$$\hat{I}(\omega) = \frac{I(\omega)}{\int_{\text{band}} I(\omega)\, d\omega} \tag{1.38}$$

where ω is the frequency shift in radians/sec relative to the frequency of the incident light and $I(\omega)$ is the fraction of the incident radiation which is scattered into unit solid angle in unit interval of ω. The theory of nonresonant vibrational Raman scattering can be written in a form that gives $\hat{I}(\omega)$ as

$$\hat{I}(\omega) = \frac{\int_{-\infty}^{\infty} \langle [\boldsymbol{\varepsilon}_o \cdot \underline{\alpha}^{(v)}(0) \cdot \boldsymbol{\varepsilon}_s][\boldsymbol{\varepsilon}_o \cdot \underline{\alpha}^{(v)}(t) \cdot \boldsymbol{\varepsilon}_s] e^{-i\omega t} \rangle\, dt}{2\pi \langle [\boldsymbol{\varepsilon}_o \cdot \underline{\alpha}(0) \cdot \boldsymbol{\varepsilon}_s]^2 \rangle} \tag{1.39}$$

where $\underline{\alpha}^{(v)}$ is an off-diagonal matrix element of the polarizability tensor between the ground vibrational state and vibrational state v with vibrational energy $\hbar\omega_v$; and $\boldsymbol{\varepsilon}_o$, $\boldsymbol{\varepsilon}_s$ are unit vectors denoting the orientations of the electric vectors of the incident and observed scattered radiation, respectively.

The time dependence of $\underline{\alpha}^{(v)}$ arises from several sources:

1. The polarizability tensor $\underline{\alpha}_v$ is fixed in the molecular frame, and varies with time as the molecule rotates; this can give rise to a rotational band centered on ω_v.

2. Time-dependent distortions of the molecular frame are caused by molecular collisions (in other words, by strong unbalanced intermolecular forces) in dense phases. This gives rise to a broadening of the vibrational line because the collisions interrupt the harmonic oscillation at ω_v; in addition, time-dependent changes in the values of the components of $\underline{\alpha}^{(v)}$ result which give rise to collision-induced wings in the rotational bands.

3. If one can eliminate the intensity in the observed vibrational Raman line profile due to the factors mentioned in (2),[18, 19] the remaining intensity may be affected by a variety of other complicating factors,[20] such as broadening due to spectral slit widths, to isotopic splitting of the vibrational line, and to anharmonic hot bands.

Thus, the extraction of unambiguous information about rotational motion in dense phases from Raman spectra is far from trivial. Assuming that the various corrections have been applied at least approximately, we now wish to examine the rotational contribution in more detail.

We first note that it is possible to compute the entire rotational Raman correlation function $C_R(t)$ (rather than the correlation time) by evaluating

$$C_R(t) = \int_{\text{band}} \hat{I}(\omega) e^{i\omega t} \, d\omega \tag{1.40}$$

Since the Raman correlation time τ_R is defined to be the zeroth moment of $C_R(t)$, it is easy to show that

$$\tau_R = \pi \hat{I}(0) \tag{1.41}$$

(Note that $\hat{I}(\omega)$ has units of inverse frequency.)

As defined in (1.40) and (1.41), both $C_R(t)$ and τ_R depend upon many variables, including (a) the relative orientation of ε_o and ε_s, the polarization vectors of the incident and observed scattered light; (b) the particular normal mode involved in the transition, since this determines the nature of the molecule-fixed $\underline{\alpha}^{(v)}$; (c) the time dependence of the reorientation angles $\delta\Omega$ for the molecule moving under the influence of intermolecular torques. In order to show explicitly how these factors enter and to express the functions of molecular orientation in terms of the $D_{km}{}^2(\Omega)$ which are appropriate for second-rank tensors, we shift to the spherical (rather than the Cartesian) basis. A laboratory coordinate system is defined with z-axis perpendicular to the scattering plane (i.e., the plane defined by the propagation vectors $\mathbf{k}_o$ and $\mathbf{k}_s$)

and with xz plane parallel to the plane defined by $\mathbf{k}_s$, $\boldsymbol{\varepsilon}_o$. In this case, the spherical vectors $\boldsymbol{\varepsilon}_o$ and $\boldsymbol{\varepsilon}_s$ become

$$\boldsymbol{\varepsilon}_o = \begin{vmatrix} -\sqrt{\tfrac{1}{2}} \sin \psi_o e^{i\theta} \\ \cos \psi_o \\ \sqrt{\tfrac{1}{2}} \sin \psi_o e^{-i\theta} \end{vmatrix} \tag{1.42}$$

$$\boldsymbol{\varepsilon}_s = \begin{vmatrix} -\sqrt{\tfrac{1}{2}} \sin \psi_s \\ \cos \psi_s \\ \sqrt{\tfrac{1}{2}} \sin \psi_s \end{vmatrix} \tag{1.43}$$

where the scattering angle θ is the angle between $\mathbf{k}_o$ and $\mathbf{k}_s$, and ψ_o, ψ_s are the angles between the normal to the scattering plane and the polarization vectors of the initial and scattered light, respectively.

An arbitrary molecule-fixed second-rank tensor $\mathbf{A}$ can be expressed in a space-fixed frame and in the spherical basis by computing the matrix product:

$$\underline{\alpha}^{(v)}_{\text{space}} = \mathbf{D}\underline{\alpha}^{(v)}_{\text{molec}} \mathbf{D}^{-1} \tag{1.44}$$

where $\mathbf{D}$ is the 3×3 array of $D_{km}^{\ 1}(\Omega)$ and $\underline{\alpha}_{\text{molec}}$ is a symmetric tensor expressed in the spherical basis:

$$\underline{\alpha}^{(v)}_{\text{molec}} = \bar{\alpha}^{(v)}\mathbf{1} + \underline{\alpha}_t^{(v)} \tag{1.45}$$

where $\bar{\alpha}^{(v)} = (\alpha_{xx}^{(v)} + \alpha_{yy}^{(v)} + \alpha_{zz}^{(v)})/3$ and

$$\underline{\alpha}_t^{(v)} = \begin{vmatrix} -\tfrac{1}{3}\alpha_2^{(v)} & -\alpha_3^{(v)} & \alpha_1^{(v)} - \alpha_4^{(v)} \\ -\alpha_3^{*(v)} & \tfrac{2}{3}\alpha_2^{(v)} & \alpha_3^{(v)} \\ \alpha_1^{(v)} + \alpha_4^{(v)} & \alpha_3^{*(v)} & -\tfrac{1}{3}\alpha_2^{(v)} \end{vmatrix} \tag{1.46}$$

where

$$\alpha_1^{(v)} = \tfrac{1}{2}(\alpha_{yy}^{(v)} - \alpha_{xx}^{(v)}) \tag{1.47}$$

$$\alpha_2^{(v)} = \alpha_{zz}^{(v)} - \tfrac{1}{2}(\alpha_{xx}^{(v)} + \alpha_{yy}^{(v)}) \tag{1.48}$$

$$\alpha_3^{(v)} = \frac{1}{2\sqrt{2}}(\alpha_{xz}^{(v)} + i\alpha_{yz}^{(v)}) \tag{1.49}$$

$$\alpha_4^{(v)} = \frac{i}{2}\alpha_{xy}^{(v)} \tag{1.50}$$

A lengthy expression for $\underline{\alpha}^{(v)}_{\text{space}}$ results when the matrix multiplications of (1.44) are performed with the aid of (1.45) and (1.46). Explicit results are given elsewhere;[21] for present purposes, it is sufficient to note that $\bar{\alpha}^{(v)}$ is invariant to changes in molecular orientation, but the product $\mathbf{D}\underline{\alpha}_t^{(v)}\mathbf{D}^{-1}$

gives a tensor with elements that vary as $D_{k,m}{}^2(\Omega)$. Consequently, the rotational correlation functions and correlation times for Raman line broadening are dependent upon the time dependence of angular functions with $l = 2$; no rotational broadening arises from the $\bar{\alpha}^{(v)}$ part of the problem.

It is not feasible to progress further with the general formalism for Raman line shapes. Ordinarily, considerable simplifications are introduced by appropriate experimental choices of ε_o, ε_s. In addition, $\underline{\alpha}^{(v)}$ for a specific vibrational transition can be selected to have only a few nonzero elements; this further simplifies the mathematics.

We now turn to the theory for the frequency-shifted scattering of light at frequencies lower than those typical for vibrational Raman scattering. Because a small frequency shift is to be associated with a relatively slow molecular motion, we anticipate that translational and rotational displacements will be the source of such scattering rather than vibrational normal modes modulated by rotational motion. Since no change in vibrational state occurs for small frequency shifts, effects of vibrational relaxation, isotopic splitting, and hot bands are no longer present; collision induced scattering still contributes to the total and should be corrected for if one wishes to study rotational motion. The basic equation[17] for the scattering intensity is quite similar to the vibrational Raman equation (1.39); we write it here as

$$2\pi \hat{I}(\omega) = \int_{-\infty}^{\infty} C_{LS}(t) e^{i\omega t}\, dt \qquad (1.51)$$

where $C_{LS}(t)$ is a (normalized) light scattering correlation function equal to

$$C_{LS}(t) = \frac{1}{Q}\left\langle \sum_j (\varepsilon_o \cdot \underline{\alpha}_i(0) \cdot \varepsilon_s)(\varepsilon_o \cdot \underline{\alpha}_j(t) \cdot \varepsilon_s) e^{i\boldsymbol{\kappa}\cdot \mathbf{r}_{ij}(t)} \right\rangle \qquad (1.52)$$

where $\boldsymbol{\kappa} = \mathbf{k}_s - \mathbf{k}_o$, and Q is a normalizing factor given by

$$Q = \left\langle \sum_j (\varepsilon_o \cdot \underline{\alpha}_i(0) \cdot \varepsilon_s)(\varepsilon_o \cdot \underline{\alpha}_j(0) \cdot \varepsilon_s) e^{i\boldsymbol{\kappa}\cdot \mathbf{r}_{ij}(0)} \right\rangle \qquad (1.53)$$

The sums over j in (1.52) and (1.53) are over all molecules in the ensemble, and serve to illustrate the feature that correlated rotational and (especially) translational motions can play an important role in determining low-frequency scattered intensities. The intermolecular vector $\mathbf{r}_{ij}(t) = \mathbf{r}_j(t) - \mathbf{r}_i(0)$, where $\mathbf{r}_i(0)$, $\mathbf{r}_j(t)$ denote the center-of-mass positions of molecules i and j; and $\underline{\alpha}_i(0)$, $\underline{\alpha}_j(t)$ are now the *permanent* polarizability tensors of molecules i and j.

In computing the ensemble averages in (1.52) and (1.53) it is generally assumed that translational and rotational motions are not correlated; the sum is also split into a "self"-scattering term and a "correlated"-scattering term:

$$C_{LS}(t) = \frac{1}{Q} \langle (\boldsymbol{\varepsilon}_o \cdot \underline{\alpha}_i(0) \cdot \boldsymbol{\varepsilon}_s)(\boldsymbol{\varepsilon}_o \cdot \underline{\alpha}_i(t) \cdot \boldsymbol{\varepsilon}_s) \rangle \langle e^{i\boldsymbol{\kappa} \cdot \mathbf{r}(t)} \rangle$$
$$+ \sum_{j \neq i} \langle (\boldsymbol{\varepsilon}_o \cdot \underline{\alpha}_i(0) \cdot \boldsymbol{\varepsilon}_s)(\boldsymbol{\varepsilon}_o \cdot \underline{\alpha}_j(t) \cdot \boldsymbol{\varepsilon}_s) \rangle \langle e^{i\boldsymbol{\kappa} \cdot \mathbf{r}_{ij}(t)} \rangle \qquad (1.54)$$

where $\mathbf{r}(t)$ is the displacement of a molecule in time t. Since (1.45) is applicable to the permanent polarizability as well as to $\underline{\alpha}^{(v)}$, we can write

$$\langle (\boldsymbol{\varepsilon}_o \cdot \underline{\alpha}_i(0) \cdot \boldsymbol{\varepsilon}_s)(\boldsymbol{\varepsilon}_o \cdot \underline{\alpha}_j(t) \cdot \boldsymbol{\varepsilon}_s) \rangle = \bar{\alpha}^2 (\boldsymbol{\varepsilon}_o \cdot \boldsymbol{\varepsilon}_s)^2 + \langle (\boldsymbol{\varepsilon}_o \cdot \underline{\alpha}_{ti}(0) \cdot \boldsymbol{\varepsilon}_s)(\boldsymbol{\varepsilon}_o \cdot \underline{\alpha}_{tj}(t) \cdot \boldsymbol{\varepsilon}_s) \rangle$$
$$(1.55)$$

(Cross terms involving $\bar{\alpha}\, \underline{\alpha}_t$ vanish after angle averaging in an isotropic fluid.) When (1.55) is substituted into both parts of (1.54), it becomes possible to sort out the contributions to the numerator of (1.52). In many cases of interest, the time scale of the translational motion will be long relative to that of the rotation; 10^{-6} sec for translation and 10^{-11} sec for rotation are typical. This leads us to write

$$C_{LS}(t) = \frac{1}{Q}(I + II) \qquad (1.56)$$

where

$$I = \bar{\alpha}^2 \{ \langle e^{i\boldsymbol{\kappa} \cdot \mathbf{r}(t)} \rangle + \rho \langle e^{i\boldsymbol{\kappa} \cdot \mathbf{r}_{12}(t)} \rangle \} (\boldsymbol{\varepsilon}_o \cdot \boldsymbol{\varepsilon}_s)^2 \qquad (1.57)$$

When Fourier transformed, this term gives Rayleigh light scattering. It is largest when $\boldsymbol{\varepsilon}_o$ is parallel to $\boldsymbol{\varepsilon}_s$ (polarized scattering) and is absent when the two vectors are perpendicular. Its frequency dependence is entirely due to translational displacements and thus is of no further interest in this discussion of rotational motion.

Term II can be written as

$$II = \rho k T \kappa_T \Bigg\{ \langle (\boldsymbol{\varepsilon}_o \cdot \underline{\alpha}_t(0) \cdot \boldsymbol{\varepsilon}_s)(\boldsymbol{\varepsilon}_o \cdot \underline{\alpha}_t(t) \cdot \boldsymbol{\varepsilon}_s) \rangle$$
$$+ \Bigg\langle \sum_{j \neq i} (\boldsymbol{\varepsilon}_o \cdot \underline{\alpha}_{ti}(0) \cdot \boldsymbol{\varepsilon}_s)(\boldsymbol{\varepsilon}_o \cdot \underline{\alpha}_{tj}(t) \cdot \boldsymbol{\varepsilon}_s) \Bigg\rangle \Bigg\} \qquad (1.58)$$

where it has been assumed that $\mathbf{r}(t) = 0$ and $\mathbf{r}_{12}(t) = \mathbf{r}_{12}(0)$ for times as short as the rotational decay time; note that

$$\rho k T \kappa_T = 1 + \rho \int (g(r) - 1)4\pi r^2 \, dr \qquad (1.59)$$

where $g(r)$ is the equilibrium center-of-mass correlation function. If the polarization vectors of the incident and scattered light are perpendicular,

$\boldsymbol{\varepsilon}_o \cdot \boldsymbol{\varepsilon}_s = 0$ and one can extract a rotational correlation function $C_{LS}(t)$ from the data which is equal to:

$$C_{LS}(t) = \frac{\langle (\boldsymbol{\varepsilon}_o \cdot \underline{\alpha}_t(0) \cdot \boldsymbol{\varepsilon}_s)(\boldsymbol{\varepsilon}_o \cdot \underline{\alpha}_t(t) \cdot \boldsymbol{\varepsilon}_s)) \rangle + \left\langle \sum_{j \neq i} (\boldsymbol{\varepsilon}_o \cdot \underline{\alpha}_{ti}(0) \cdot \boldsymbol{\varepsilon}_s)(\boldsymbol{\varepsilon}_o \cdot \underline{\alpha}_{tj}(t) \cdot \boldsymbol{\varepsilon}_s) \right\rangle}{\langle (\boldsymbol{\varepsilon}_o \cdot \underline{\alpha}_t(0) \cdot \boldsymbol{\varepsilon}_s)^2 \rangle + \left\langle \sum_{j \neq i} (\boldsymbol{\varepsilon}_o \cdot \underline{\alpha}_{ti}(0) \cdot \boldsymbol{\varepsilon}_s)(\boldsymbol{\varepsilon}_o \cdot \underline{\alpha}_{tj}(0) \cdot \boldsymbol{\varepsilon}_s) \right\rangle}$$

$$(1.60)$$

For the most part, models of rotational motion suggested to date ignore the possibility of correlated reorientations, even in dense phases.[22] To the extent that this is true, the sums over j in the numerator and denominator of (1.60) vanish. The resulting expression can be further simplified if it is realized that the permanent polarizability is often a diagonal tensor in the principal axis system of the molecule. In this case, the product $(\boldsymbol{\varepsilon}_o \cdot \underline{\alpha}_t \cdot \boldsymbol{\varepsilon}_s)$ can be written as

$$(\boldsymbol{\varepsilon}_o \cdot \underline{\alpha}_t(t) \cdot \boldsymbol{\varepsilon}_s) = \sum_{k=-2}^{2} \left[\alpha_2 D_{k,0}^2(\Omega(t)) - \sqrt{\frac{3}{2}} \alpha_1 (D_{k,2}^2(\Omega(t)) + D_{k-2}^2(\Omega(t))) \right] d_k^*$$

$$(1.61)$$

where the spherical components of $\boldsymbol{\varepsilon}_o$ and $\boldsymbol{\varepsilon}_s$ have been combined as:

$$d_0 = -\tfrac{1}{3}\varepsilon_{1o}^* \varepsilon_{1s} + \tfrac{2}{3}\varepsilon_{0o}^* \varepsilon_{0s} - \tfrac{1}{3}\varepsilon_{-1o}^* \varepsilon_{-1s}$$
$$d_{\pm 1} = \tfrac{1}{3}(\varepsilon_{0o}^* \varepsilon_{\pm 1s} - \varepsilon_{\mp 1o}^* \varepsilon_{0s})$$
$$d_{\pm 2} = \tfrac{2}{3}\varepsilon_{\mp 1o}^* \varepsilon_{\pm 1s}$$

$$(1.62)$$

In evaluating the ensemble averages of (1.60), one can use (1.5) and (1.16) to show that

$$\langle (\boldsymbol{\varepsilon}_o \cdot \underline{\alpha}_t(0) \cdot \boldsymbol{\varepsilon}_s)(\boldsymbol{\varepsilon}_o \cdot \underline{\alpha}_t(t) \cdot \boldsymbol{\varepsilon}_s) \rangle =$$

$$\frac{1}{5} \left\langle \alpha_2{}^2 D_{00}{}^2(\delta\Omega) + \frac{3}{2}\alpha_1{}^2 \sum_{sgn} D_{\pm 2, \pm 2}^2(\delta\Omega) - \frac{3}{2}\alpha_1\alpha_2 \sum_{sgn} D_{0, \pm 2}^2(\delta\Omega) + D_{\pm 2, 0}^2(\delta\Omega) \right\rangle$$

$$\times \sum_k d_k d_k^* \qquad (1.63)$$

where sgn denotes a sum over all sign permutations, and

$$\sum_k d_k d_k^* = \tfrac{1}{3} + \tfrac{1}{9}[\cos\psi_0 \cos\psi_f + \sin\psi_0 \sin\psi_f \cos\theta]^2 \qquad (1.64)$$

Equation (1.64) is needed only for experimental arrangements where $\boldsymbol{\varepsilon}_o \cdot \boldsymbol{\varepsilon}_s \neq 0$; in the case under consideration here, this factor cancels in numerator and denominator to give:

$$C_{LS}(t) =$$

$$\frac{\left\langle \alpha_2{}^2 D_{00}{}^2(\delta\Omega) + \frac{3}{2}\alpha_1{}^2 \sum_{sgn} D_{\pm 2 \pm 2}^2(\delta\Omega) - \frac{3}{2}\alpha_1\alpha_2 \sum_{sgn} D_{0 \pm 2}^2(\delta\Omega) + D_{\pm 20}^2(\delta\Omega) \right\rangle}{\alpha_2{}^2 + 3\alpha_1{}^2}$$

(1.65)

It should be noted here that the assumptions made in deriving (1.65) are known to fail in a number of specific cases; in particular, translation-rotation coupling in liquids such as aniline, quinoline, and nitrobenzene[23] gives rise to κ-dependent features in the depolarized scattering which are completely absent when one Fourier transforms (1.65). Theoretical treatments of such systems have been presented which account for the observations.[24] These calculations are based on hydrodyanmical arguments concerning the rotation-translation coupling and have not yet led to quantitative results that would relate these experiments to molecular properties.[25]

D. Quasielastic Neutron Scattering

Frequency-shifted scattering also occurs in neutron diffraction experiments. One of the most striking differences between neutron and light scattering arises from the absence of a selection rule that would restrict the values of j in the $D_{km}{}^j(\delta\Omega)$ that contribute to the rotational scattering. To see this, we start with the standard expression[26,27] for the differential neutron scattering cross-section that gives the scattered flux into unit solid angle $d\Omega$ and unit energy transfer E

$$\frac{d^2\sigma}{d\Omega\, dE} = \frac{1}{h}\frac{k_s}{k_o} e^{-E/2kT - \kappa^2 h^2/8mkT} \int_0^\infty dt\, e^{i\omega t} \sum_{m,n} b_m^* b_n \langle e^{i\kappa\cdot(\mathbf{r}_n(t) - \mathbf{r}_m(0))}\rangle \quad (1.66)$$

where b_n is the bound-atom scattering cross-section of nucleus n and $\mathbf{r}_n(t)$ is the position of the nucleus at time t. As before, $\kappa = \mathbf{k}_s - \mathbf{k}_o$, where $\mathbf{k}_o$, $\mathbf{k}_s$ are the propagation vectors of the incident and scattered beams. The usual procedure now is to write

$$\sum_{m,n} b_m b_n e^{i\kappa\cdot[\mathbf{r}_n(t) - \mathbf{r}_m(0)]} = (\overline{b^2} - \overline{b}^2)F_s(\kappa, t) + \overline{b}^2 F(\kappa, t) \quad (1.67)$$

where

$$\overline{b^2} = \frac{1}{N}\sum_n b_n{}^2 \quad (1.68)$$

$$\overline{b}^2 = \frac{1}{N^2}\sum_{m \neq n} b_m b_n \quad (1.69)$$

with N equal to the number of different nuclei present in the molecule;

$$F_s(\kappa, t) = \frac{1}{b^2} \sum_n b_n^2 \langle e^{i\mathbf{\kappa} \cdot \delta \mathbf{r}_n(t)} \rangle \tag{1.70}$$

where $\delta\mathbf{r}_n(t)$ is the displacement of nucleus n in time t;

$$F(\kappa, t) = \frac{1}{b^2} \sum_{m,n} b_m b_n \langle e^{i\mathbf{\kappa} \cdot (\mathbf{r}_n(t) - \mathbf{r}_m(0))} \rangle \tag{1.71}$$

The first term on the right-hand side of (1.67) is called the incoherent scattering, and the second, the coherent scattering. Clearly, the incoherent scattering term is the simplest to interpret because it involves only single-particle motion, whereas the coherent term involves the decay of time-dependent mutual correlations in nuclear position. Depending upon the substance under investigation, the measured scattering can be dominated by the incoherent part, as is generally the case with proton-containing molecules, or it can be a mixture of coherent and incoherent terms.

For simplicity, we will explicitly treat $F_s(\kappa, t)$; most of the argument can be applied to $F(\kappa, t)$ as well. We are interested only in the low-energy transfer scattering (denoted by "quasielastic") which is caused by rotational and translational motions only. Thus, we can write

$$\delta\mathbf{r}_n(t) = \delta\mathbf{r}_{cm}(t) + \mathbf{R}_n(t) - \mathbf{R}_n(0) \tag{1.72}$$

where $\delta\mathbf{r}_{cm}(t)$ is the center-of-mass displacement of the molecule containing nucleus n, and $\mathbf{R}_n(t)$ is the position of nucleus n relative to the center of mass. We expand[28,29] the exponential as

$$e^{i\mathbf{\kappa} \cdot \mathbf{R}_n(t)} = \sum_{l=0}^{\infty} (2l + 1)i^l j_l(\kappa R_n) \sum_r D_{0r}{}^l(\Omega(t)) D_{r0}{}^l(\eta_n \zeta_n) \tag{1.73}$$

where j_l is a spherical Bessel function, R_n, η_n, ζ_n are the coordinates of nucleus n relative to the molecular axes, and the molecular orientation at time t is denoted by $\Omega(t)$, as usual. Assuming that translation and rotation are uncorrelated and that initial orientations are random, the incoherent scattering function $F_s(\kappa, t)$ can now be written as

$$F_s(\kappa, t) = \langle e^{i\mathbf{\kappa} \cdot \delta\mathbf{r}_{cm}(t)} \rangle C_{inc}(t, \kappa) \tag{1.74}$$

where the incoherent nuclear rotational correlation function is given by

$$C_{inc}(t, \kappa) = \sum_{l,r,r'} \sigma_{l,r,r'} \langle D_{rr'}^l(\delta\Omega) \rangle \tag{1.75}$$

with

$$\sigma_{l,r,r'} = \frac{(2l + 1)}{b^2} \sum_n b_n^2 [j_l(\kappa R_n)]^2 D_{r0}{}^l(\eta_n, \zeta_n) D_{r'0}^{*l}(\eta_n, \zeta_n) \tag{1.76}$$

The coherent scattering can be expressed in a similar fashion,[29] but explicit inclusion of orientational correlations gives rise to relatively complex expressions that will not be reproduced here. Of course, angular correlations can play an important role in elastic neutron scattering,[27,29,30] and thus should be included in analyses of coherent inelastic scattering; however, more sophisticated models than those considered in this paper are needed (together with very precise, detailed experimental data), so we will limit the discussion to incoherent scattering.

Note that (1.75) indicates that a quantitative model for incoherent neutron scattering requires $\langle D_{rr'}^l(\delta\Omega)\rangle$ for all l, r, r' (or at least, for all values of the indices corresponding to nonnegligible $\sigma_{l,r,r'}$).

This is only one of several approaches[31] that have been suggested as solutions to the problem of interpreting quasielastic neutron scattering data; its chief virtue here is the appearance of an expression for the rotational contribution that is consistent with our formulation of the theory for other observable quantities.

E. Infrared Absorption and Dielectric Relaxation

Vibration-rotation bands in the infrared spectra of dense phases can be interpreted in a way that is a close analog to the treatment of Raman spectra; dielectric relaxation (or far infrared absorption) in polar fluids is also analogous to depolarized light scattering. Specifically, the normalized intensity $\hat{I}(\omega)$ in an infrared vibration-rotation band can be expressed as[32]

$$\hat{I}(\omega) = \frac{1}{2\pi} \int_{-\infty}^{\infty} e^{i(\omega - \omega_v)t} C_{IR}(t)\, dt \qquad (1.77)$$

where ω_v is the vibrational frequency and the correlation function is

$$C_{IR}(t) = \frac{\langle(\boldsymbol{\varepsilon}_o \cdot \mathbf{m}^{(v)}(0))(\boldsymbol{\varepsilon}_o \cdot \mathbf{m}^{(v)}(t))\rangle}{\langle(\boldsymbol{\varepsilon}_o \cdot \mathbf{m}^{(v)}(0))^2\rangle} \qquad (1.78)$$

where $\boldsymbol{\varepsilon}_o$ is a unit vector parallel to the electric vector of the incident radiation and $\mathbf{m}^{(v)}(t)$ is the transition dipole moment for the band; if the incident radiation is unpolarized,

$$C_{IR}(t) = \frac{\langle\mathbf{m}^{(v)}(0) \cdot \mathbf{m}^{(v)}(t)\rangle}{\langle\mathbf{m}^{(v)}(0) \cdot \mathbf{m}^{(v)}(0)\rangle} \qquad (1.79)$$

The complicating factors enumerated in our discussion of the Raman problem are present in infrared spectra as well; vibrational broadening, isotope frequency splittings, and collision-induced effects are among the most important. Assuming that corrections for these factors can be made,[33–35] one finds that the normalized orientational correlation function that determines the spectral band shape is

$$C_{IR}(t) = \langle \cos \delta\theta^{(v)}(t) \rangle \qquad (1.80)$$

where $\delta\theta^{(v)}(t)$ is the reorientation angle of the transition dipole in time t. Since $\cos \delta\theta^{(v)}(t) = D_{00}{}^{1}(\delta\Omega^{(v)})$, (1.17) indicates that

$$C_{IR}(t) = C_{1,0}^{(v)}(t) \qquad (1.81)$$

The previous analysis then gives

$$C_{IR}(t) = \sum_{r,r'} D_{r0}{}^{1}(\eta, \zeta) D_{r'0}^{*1}(\eta, \zeta) \langle D_{rr'}^{*1}(\delta\Omega) \rangle \qquad (1.82)$$

where η, ζ now specify the orientation of the transition moment relative to the molecular principal axis system, and $\delta\Omega$ denotes the angles of reorientation of the molecular axes in time t.

As it is usually understood, dielectric relaxation experiments consist of measurements of the frequency dependence of the dielectric constant at frequencies extending from zero into the microwave region. The imaginary part of the complex dielectric constant is a measure of the energy absorption in the system. If the ensemble being studied is made up of polar molecules, the energy absorption arises from finite reorientation times of the molecules under the influence of the applied field, as was pointed out many years ago.[36] In fact, far infrared absorption in polar fluids merely consists of an extension of the frequency range of dielectric relaxation experiments.[32] Essentially, one is measuring $I(\omega)$ in both cases, where the intensity (weighted by $\omega(1 - \exp(-h\omega/kT)^{-1})$ is, as usual, the Fourier transform of a correlation function $C_D(t)$. The formal expression for this dielectric correlation function is well known:

$$C_D(t) = \frac{\langle \mathbf{M}(0) \cdot \mathbf{M}(t) \rangle}{\langle \mathbf{M}(0) \cdot \mathbf{M}(0) \rangle} \qquad (1.83)$$

where $\mathbf{M}(t)$ is the macroscopic moment of the fluid at time t. Of course, one wishes to relate this expression to molecular properties, and particularly, to the time-dependent behavior of $\mathbf{\mu}_i(t)$, the permanent dipole moment of molecule i at time t in the fluid. In fact, the correlation function of primary interest is

$$C_\mu(t) = \frac{\left\langle \sum_j \mathbf{\mu}_i(0) \cdot \mathbf{\mu}_j(t) \right\rangle}{\left\langle \sum_j \mathbf{\mu}_i(0) \cdot \mathbf{\mu}_j(0) \right\rangle} \qquad (1.84)$$

The exact nature of the relationship between $C_D(t)$ and $C_\mu(t)$ is a subject of considerable controversy at present.[37,38] It emerges that both the microscopic correlation time and $C_\mu(t)$ itself will be different from the macroscopic quantities;[37-39] for instance, the two relaxation times can differ by as much as a factor of two.

Inasmuch as the models to be discussed are representations of molecular rotation, we focus here on the molecular correlation function $C_\mu(t)$. In contrast to the infrared correlation function (but in resemblance to the depolarized light scattering correlation function), (1.84) contains contributions from correlated mutual dipole moments. We show this explicitly by writing

$$C_\mu(t) = \frac{1}{Q_\mu}\left\{\langle\cos\delta\Omega(t)\rangle + \left\langle\sum_j\cos\delta\theta_{ij}(t)\right\rangle\right\} \qquad (1.85)$$

where

$$Q_\mu = 1 + \left\langle\sum_j\cos\delta\theta_{ij}(0)\right\rangle \qquad (1.86)$$

and $\delta\theta_{ij}(t)$ is the angle between the permanent dipole on molecule i at time t and that on molecule j at time zero. The "self"-correlation term can be written in the form of (1.17); when the mutual correlations are written in terms of the relative reorientation angles $\delta\Omega_{ij}$, one finds

$$C_\mu(t) = \frac{1}{Q_\mu}\left\{\sum_{rr'}D_{r0}{}^1(\eta,\zeta)D_{r'0}^{*1}(\eta,\zeta)\langle D_{rr'}^1(\delta\Omega)\rangle\right.$$

$$\left. + \left\langle\sum_j\sum_{k,m}D_{kr}{}^1(\Omega_i(0))D_{km}^{*1}(\Omega_j(0))D_{mr'}^{*1}(\delta\Omega_j)\right\rangle\right\} \qquad (1.87)$$

where η, ζ are now the orientation angles of the permanent dipole in the molecule frame. This expression simplifies greatly if the two terms in brackets have the same time dependence, regardless of correlations between the orientations $\Omega_i(0)$ and $\Omega_j(0)$. In that case,

$$C_\mu(t) = \sum_{rr'}D_{r0}{}^1(\eta,\zeta)D_{r'0}^{*1}(\eta,\zeta)\langle D_{rr'}^1(\delta\Omega)\rangle \qquad (1.88)$$

Equation (1.88) is also obtained if it is assumed that the orientations of molecules i and j are uncorrelated at $t = 0$.

F. Summary

Although this survey of orientation-dependent phenomena is far from complete, we have included a number of the experiments of current interest; more importantly, we have attempted to phrase the theory in a way that clearly shows how rotational motion is related to the observable quantities.

In all cases, single molecule motion plays an important role. Even though correlated rotations also appear, the models to be discussed are primarily designed to deal with the rotation of a single molecule in the potential field of its neighbors. (Limited progress has been attained in the theory of correlated rotations, and it seems that increasing efforts will be made along these lines in the near future.)

It is appropriate to make one other general point here: except for the case where the rates of molecular reorientation are isotropic, the various experimental correlation times and correlation functions will be sensitive to the orientation of the molecule-fixed vector (or tensor, in some cases) as well as to the value of l that applies. In many early studies, conclusions drawn from comparisons of different kinds of correlation time for a molecule in the liquid at a particular temperature and pressure have neglected this sensitivity and consequently are erroneous in some cases.

II. INERTIAL MODELS FOR REORIENTATION

A. Equations of Motion

Consider a rigid molecule whose rotations obey the laws of classical mechanics. In a dense phase, it moves under the influence of a fluctuating torque $N(t)$ due to the angle dependence of its interactions with other molecules. The equations of motion of a rigid body are well-known,[40] and are written here as

$$\dot{L}_x = \left(\frac{1}{I_{zz}} - \frac{1}{I_{yy}}\right)L_y L_z + N_x$$

$$\dot{L}_y = \left(\frac{1}{I_{xx}} - \frac{1}{I_{zz}}\right)L_z L_x + N_y \qquad (2.1)$$

$$\dot{L}_z = \left(\frac{1}{I_{yy}} - \frac{1}{I_{xx}}\right)L_x L_y + N_z$$

where $\mathbf{L}$ is the angular momentum in the body-fixed frame (as we will see, this is convenient for ensemble averaging), $\mathbf{N}$ is the torque on the molecule, and $\mathbf{I}$ is the moment of inertia tensor (diagonal in the principal axis system). Note that

$$N_i = -\left(\frac{\partial V}{\partial \psi_i}\right) \qquad (2.2)$$

where V is the potential energy due to the interactions of a molecule with its neighbors and $d\psi_i$ is the angular displacement of the molecule about its ith principal axis.

We will discuss approximations to the torque $\mathbf{N}(t)$ later; at this point, we point out that it can take on an extremely wide range of values. Even in dense phases, torques will approach zero as the intermolecular interaction potential approaches spherical symmetry; when the molecule is nonspherical, the torques about a particular axis can be extremely small if the axis is one of high symmetry. Thus, the free or nearly free rotation that occurs when the torques are small is not necessarily an unrealistic model for molecular motion in dense phases. Furthermore, it is quite possible that reorientation will be anisotropic, either because of differences in the elements of the moment of inertia tensor when the rotation is unhindered, or because of differences in the components of the torque vector when it is the dominant factor in the equations of motion.

B. Free Rotors

We first attempt to solve the problem of torque-free rotation of a classical ensemble of rigid rotors. Inasmuch as the equations in (2.1) have never been solved in closed form for an asymmetric top, we simplify by assuming that $I_{xx} = I_{yy} = I$. For the symmetric top, the equations in (2.1) now become

$$\dot{L}_x = \left(\frac{1}{I_{zz}} - \frac{1}{I}\right) L_y L_z$$

$$\dot{L}_y = \left(\frac{1}{I_{zz}} - \frac{1}{I}\right) L_z L_x \qquad (2.3)$$

$$\dot{L}_z = 0$$

The solutions to these equations are well-known; since the angular momentum about the body z-axis is constant, one has

$$\gamma(t) = \gamma(0) + \frac{L_z}{I_{zz}} t \qquad (2.4)$$

The time dependence of the Euler angles α, β is somewhat more complicated because of the precessional motion about a space-fixed axis that results. In any case, this is not the most convenient approach for an ensemble of symmetric tops.

In fact, the dynamics of an ensemble are most conveniently handled using the formalism of the Liouville equation. Thus, we define $W(\Gamma(t), t, \Gamma(0))$ as the conditional probability density that a molecule will have rotational phase space variables $\Gamma(t)$ at time t if it was known to have variables $\Gamma(0)$ at time zero. This function obeys the equation of motion,

$$\frac{\partial W}{\partial t} = -i\mathscr{L}W \qquad (2.5)$$

where $\mathscr{L}$ is the Liouville operator. For the symmetric top, it is convenient to take the phase-space variables to be the Euler angles and the body-fixed angular momentum. In writing $\mathscr{L}$ in terms of these quantities, it is convenient to work in a frame that precesses with respect to the body-fixed frame with angular velocity $\omega_p = (I_{zz}^{-1} - I^{-1})L_z$. One then finds[41]

$$\mathscr{L} = \frac{1}{I} \sum_{\mu=-1}^{1} L_\mu l_\mu \tag{2.6}$$

where L_μ is the spherical component of L in the precessing frame and l_μ is the body-fixed component of the quantum mechanical angular momentum operator (apart from a factor of $\hbar$).

After some manipulation, (2.5) can be solved for $W(\Gamma(t), t, \Gamma(0))$; the result is

$$W(\Gamma(t), t, \Gamma(0)) = \sum_{j=0}^{\infty} \sum_{r,r'=-j}^{j} \frac{2j+1}{8\pi^2} D_{rr'}^j(\delta\Omega) f_{rr'}^j(\mathbf{L}(t), t; \mathbf{L}(0)) \tag{2.7}$$

where

$$f_{rr'}^j(\mathbf{L}(t), t; \mathbf{L}(0)) = \exp\left(-ir'b \times L(0)t/I\right) \exp\left(i(r - r')\phi_L(0)\right)$$

$$\times \sum_{m=-j}^{j} \exp\left(imL(0)t/I\right) d_{rm}^j(\theta_L(0)) d_{r'm}^j(\theta_L(0))$$

$$\times \prod_{\mu=-1}^{1} \delta(L_\mu(t) - L_\mu(0)) \exp\left(i\mu b \times L(0)t/I\right) \tag{2.8}$$

where $b = (I - I_{zz})/I$, and $\phi_L(0)$, $\theta_L(0)$, $L(0)$ are the spherical polar coordinates of the body-fixed $\mathbf{L}$ at $t = 0$.

The product of Dirac delta functions in (2.8) expresses the conservation of the precessing body-fixed angular momentum for a free rotor; the appearance of the other factors in that expression is dictated by the requirement that the distribution function reproduce the known orientational motion of a free rotor with given initial angular momentum.

The expansion of (2.7) in terms of the $D_{rr'}^j(\delta\Omega)$ is quite general and, in view of the formalism used for the rotational correlation functions listed in the Introduction, is also very convenient. It expresses the facts that (1) the $D_{rr'}^j(\delta\Omega)$ form a complete set of orthogonal functions, and (2) the conditional distribution function for molecules in an isotropic liquid depends only upon the reorientations and not upon initial set of Eulerian angles. (Of course, the form of the coefficients $f_{rr'}^j$ depends very much upon the model; as we will see, the $f_{rr'}^j$ can even be independent of angular momentum, especially in the case of "jump" or rotational diffusion models.)

A derivation of the conditional distribution function $W(\Gamma(t), t; \Gamma(0))$ for a freely rotating linear molecule is analogous to the symmetric top problem;[41] the chief difference is that it is preferable to work with space-fixed angular velocity in the linear molecule problem rather than body-fixed. Thus, the phase space variables are $\omega(t)$ and $\phi_\omega(t)$, the magnitude and azimuthal orientation angle of the angular velocity $\omega(t)$ at time t, and $\delta\alpha$, $\delta\beta$, the azimuthal and polar angular displacements of the molecule in time t. The distribution function can be written in two equivalent ways. The simplest is

$$W(\Gamma(t), t, \Gamma(0)) = \delta\left(\delta\alpha - \phi_\omega(0) + \frac{\pi}{2}\right)\delta(\cos\delta\beta - \cos(\omega(0)t))\delta(\omega(t) - \omega(0))$$

(2.9)

The delta functions express the conservation of space-fixed angular velocity ω and also, the fact that ω is perpendicular to the plane of rotation with fixed orientation $\delta\alpha$, and that the polar angle $\delta\beta$ changes at a constant rate equal to $\omega(0)$. For some purposes, it is convenient to expand the distribution function in the complete set of functions $D_{r,0}^j(\delta\alpha, \delta\beta, 0)$; in this case, one obtains the second formulation:

$$W(\Gamma(t), t; \Gamma(0)) = \sum_{j,r} \frac{2j+1}{4\pi} D_{r,0}^j(\delta\alpha, \delta\beta, 0) f_r^j(\omega(t), t; \omega(0))$$

(2.10)

where

$$f_r^j(\omega(t), t; \omega(0)) = \sum_m \exp(im\omega(0)t) d_{0m}^j\left(\frac{\pi}{2}\right) d_{rm}^j\left(\frac{\pi}{2}\right)$$
$$\times \exp[-ir\phi_\omega(0)]\delta(\omega(t) - \omega(0))$$

(2.11)

Equations (2.7) and (2.10) can now be utilized to compute correlation functions for ensembles of freely rotating molecules; in real systems, these are useful not only in the limit of vanishing torques but also in certain experiments (especially inelastic neutron scattering) where the time scale of the measurements is so short that one learns only about the rotational motion in time intervals where the torques have not yet had an appreciable influence upon the dynamics. For the symmetric top, one has

$$\langle D_{rr}^{*\,l}(\delta\Omega)\rangle = \int D_{rr}^{*j}(\delta\Omega)W(\Gamma(t), t; \Gamma(0)) \, d\Gamma(t)W^0(\Gamma(0)) \, d\Gamma(0)$$

(2.12)

where $W^0(\Gamma(0))$ is the probability density for $\Gamma(0)$ at $t = 0$; in fact, our use of reorientation angles and isotropy for the system at initial time means that this probability density is just $f^{(0)}(L(0))$, the Maxwellian distribution for the angular momentum of a classical symmetric top, or $f^{(0)}(\omega(0))$, the

Maxwellian distribution for the angular velocity of a linear molecule. The explicit expressions are

$$f^{(0)}(\mathbf{L}(0))\, d\mathbf{L}(0) = \frac{1}{(2\pi kT)^{3/2} II_{zz}^{1/2}} \exp\left[-\frac{L^2(0)}{2IkT}(1 + b\cos^2\theta_L(0))\right]$$

$$\times\, d\phi_L(0)\sin\theta_L(0)\, d\theta_L(0) L^2(0)\, dL(0) \qquad (2.13)$$

and

$$f^{(0)}(\boldsymbol{\omega}(0))\, d\boldsymbol{\omega}(0) = \left(\frac{I}{2\pi kT}\right)\exp\left[-\frac{I\omega^2(0)}{2kT}\right] d\phi_\omega(0)\omega(0)\, d\omega(0) \qquad (2.14)$$

Thus, an integration over $\Gamma(t)$ in (2.12) gives

$$\langle D_{rr'}^{*j}(\delta\Omega)\rangle = \int \langle f_{rr'}^j(\mathbf{L}(t); t; \mathbf{L}(0))\rangle_{\mathbf{L}(t)} f^{(0)}(\mathbf{L}(0))\, d\mathbf{L}(0) \qquad (2.15)$$

for the symmetric top. In the linear molecule case, one can show that

$$\langle D_{rr'}^{*j}(\delta\Omega)\rangle = \delta_{r',0}\int \exp\left[ir\phi_\omega(0)\right] d_{r0}{}^j(\omega(0)t) f^{(0)}(\omega(0))\, d\omega(0) \qquad (2.16)$$

$$= \frac{\delta_{r',0}\,\delta_{r,0}\,I}{kT}\int d_{00}{}^j(\omega(0)t) \exp\left[-\frac{I\omega^2(0)}{2kT}\right]\omega(0)\, d\omega(0) \qquad (2.17)$$

Since $d_{00}{}^l(x)$ is the lth Legendre polynomial with argument $\cos x$, the integral in (2.17) can be expressed in terms of Kummer's confluent hypergeometric function.[41,42] The final evaluations of both (2.15) and (2.17) must be done numerically; only in the special case of the freely rotating spherical top is it possible to derive an analytic expression for the expectation value. The explicit expression in this case is

$$\langle D_{rr'}^{*j}(\delta\Omega)\rangle = \frac{\delta_{rr'}}{2j+1}\sum_{m=-j}^{j}(1 - m^2 t^{*2})e^{-m^2 t^{*2}/2} \qquad (2.18)$$

where t^*, the time in reduced units, is defined by

$$t^* = t(kT/I)^{1/2} \qquad (2.19)$$

Note that (2.18) indicates that $\langle D_{rr'}^{*j}(\delta\Omega)\rangle$ is independent of the index r; this arises from the fact that the choice of principal axes for a spherical top is arbitrary; the time dependence of $D_{00}{}^j(\delta\Omega^{(v)})$ must be the same regardless of the values assigned to η, ζ (e.g., see (1.17)).

At the short times where one might expect the free rotor model to be applicable in real systems, (2.18) can be approximated by a Gaussian function of time:

$$\langle D_{rr'}^{*j}(\delta\Omega)\rangle = \exp\left[-\tfrac{1}{2}j(j+1)t^{*2}\right] \qquad (2.20)$$

In fact, it is possible to derive short-time Gaussian approximations to these functions for the linear molecule and symmetric top as well. The results can be written as

$$\langle D_{rr}^{*j}(\delta\Omega)\rangle = \exp\left(-\alpha_{j,r} t^{*2}\right) \tag{2.21}$$

where

$$\alpha_{j,0} = \tfrac{1}{2}j(j+1) \tag{2.22}$$

for the symmetric top and the linear molecule;

$$\alpha_{j,r} = 0 \text{ for } r \neq 0, \text{ linear molecule} \tag{2.23}$$

For $j = 1$ and 2 only, the symmetric top results are

$$\begin{aligned}
\alpha_{1,1} &= (2 + 3b + b^2)/2(1 + b) \\
\alpha_{2,1} &= (6 + 7b + b^2)/2(1 + b) \\
\alpha_{2,2} &= (3 + 5b + 2b^2)/2(1 + b)
\end{aligned} \tag{2.24}$$

Figure 1 shows the exact and the Gaussian approximation to the spherical top correlation functions $C_{l,r}(t^*)$ for several values of l; it can be seen that the approximate form actually is rather accurate for $t^* < 2$, and becomes more accurate as l increases. Inasmuch as the root mean square rotational frequency of a free molecule is on the order of $(kT/I)^{1/2}$, it can be concluded that the Gaussian will not fail noticeably unless one needs to know the correlation function after the molecule has gone through roughly two complete rotations. Of course, the long-time behavior strongly affects the calculation of the correlation time τ_l; in fact, this quantity is always infinite for an ensemble of classical free spherical tops because it is the integral of the correlation function over all times, and

$$\lim_{t^* \to \infty} \langle D_{rr}^{*j}(\delta\Omega)\rangle = \frac{1}{2j+1} \tag{2.25}$$

On the other hand, if there is a gradual decay in correlations in a dense medium, it is not totally unreasonable to assume that the correlation times obtainable from the Gaussian approximation may be of the correct order of magnitude when the intermolecular torques are small. In essence, one is assuming that the long-time rises shown in Fig. 1 for the exact free-rotor correlation functions are absent. In that case, one finds

$$\tau_{l,r} = \frac{1}{2}\left[\frac{\pi I}{\alpha_{j,r} kT}\right]^{1/2} \tag{2.26}$$

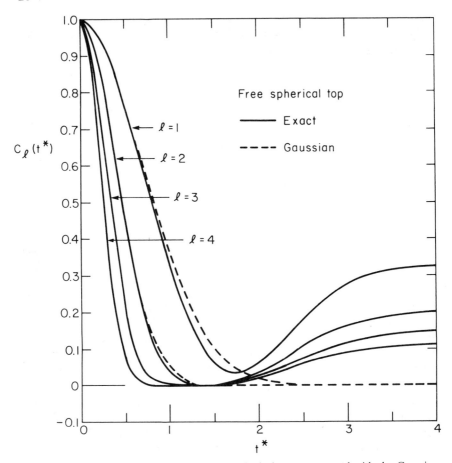

Fig. 1. Exact correlation functions for a free spherical rotor compared with the Gaussian approximation given by (2.20). Reduced time $t^* = t\sqrt{kT/I}$.

C. Extended Diffusion Models

The "Gaussian" calculation of rotational correlation times suffers from several obvious defects: it is far from clear when, if ever, (2.26) will be applicable; furthermore, an adequate theory would be capable of representing the smooth transition from free rotation to highly hindered rotation. Such a model was first proposed by Gordon[43] and has since been extended by several authors.[44-46] In these calculations the following assumptions are made:

1. The molecules undergo a succession of "collisions" of negligible duration compared to the time between "collisions."

2. Molecules undergo free rotation between "collisions" which occur with a frequency β.

3. During a "collision," molecular orientations are unchanged, but the direction of the angular momentum vector is randomized after averaging over collisional variables (such as the impact parameter).

4. In addition, the magnitude of $\mathbf{L}$ can be randomized (denoted as J-diffusion) or it can remain unchanged (denoted by M-diffusion).

These models can thus be described as random walks of the angular momentum vector, with a step frequency β. Although we will have more to say later about stochastic processes for rotational degrees of freedom, it is important to note at the outset that the motion in any collisional step is independent of the previous values of the angular momenta (excepting, of course, the magnitude of $\mathbf{L}$ in M-diffusion). This leads to significant simplifications in the mathematics; unfortunately, molecular dynamic computations of molecular rotation in dense fluids indicate that persistence of angular momenta is appreciable in the sense that reversals of the direction of $\mathbf{L}$ in a "collision" are more probable than slight deflections in the orientation of this vector.[47] A final preliminary remark is that "randomization" here means that the distribution of $\mathbf{L}$ becomes Maxwellian, that is, this is the most random distribution possible in an isothermal ensemble.

Although the full calculation of the correlation functions for the extended diffusion models is too complicated to reproduce here, a discussion of some of the principal features of the argument is helpful in understanding the approximations made as well as the results obtained. These points can be adequately illustrated by a calculation of $C_l(t) = \langle D_{00}{}^l(\delta\Omega) \rangle$. In order to do this, we make a "collisional" expansion:

$$C_l(t)]_M = \sum_{N=0}^{\infty} P_N(t)C_l(t; N)]_M \tag{2.27}$$

where $P_N(t)$ is the fraction of molecules that have undergone N and only N "collisions" in time interval t, and $C_l(t; N)]_M$ is the M-diffusion correlation function for such molecules. (Clearly, an analogous equation can be written for the J-diffusion model.) If successive "collisions" are uncorrelated, $P_N(t)$ is given by the Poisson distribution:[48]

$$P_N(t) = \frac{(\beta t)^N}{N!} e^{-\beta t} \tag{2.28}$$

Thus,

$$C_l(t)]_M = e^{-\beta t} C_l(t; 0) + \beta t \, e^{-\beta t} C_l(t; 1) + \cdots \tag{2.29}$$

where $C_l(t, 0)$ is the correlation function for the uncollided molecules, and is thus one of the free-rotor correlation functions derived previously. For simplicity, we will limit the treatment to linear molecules. An explicit expression for $C_l(t; 0)$ can be obtained from (2.17); we write this as

$$C_l(t; 0) = \langle d_{00}{}^l(\omega^{(0)}t)\rangle_{\omega^{(0)}} \tag{2.30}$$

where $\omega^{(0)}$ is the angular velocity of a molecule that has undergone zero collisions in time t and is thus equal to $\omega(0)$, the initial angular velocity.

Consider now the molecules that have collided once and only once at a time t_1 within the interval t. In treating the one-collision term in $C_l(t)$, it is most convenient to phrase the problem in a way that will allow us to use the simple form of $W(\Gamma(t), t; \Gamma(0))$ shown in (2.9). Thus, the function $D_{00}{}^l(\delta\Omega)$ after the collision at t_1 has occurred will be written in a set of space-fixed axes that are coincident at time t_1 with the molecular orientation $\Omega(t_1)$. We use the spherical harmonic addition theorem to write

$$D_{00}{}^l(\delta\Omega) = \sum_r D_{r0}{}^l(\delta\Omega_+)D_{r0}^{*l}(\Omega(t_1)) \tag{2.31}$$

where $\delta\Omega_+$ denotes the reorientation occurring in the time interval $t - t_1$ after the collision, relative to space-fixed axes with orientation $\Omega(t_1)$. Since the angles $\Omega(t_1)$ are actually $\delta\Omega_-$, the reorientation of the molecule in time t_1 relative to the original set of space-fixed axes, we can now average over $\delta\Omega_+$ and over $\delta\alpha_-$, $\delta\beta_-$ with the aid of (2.9) to find

$$C_l(t; 1) = \left\langle \sum_r \exp\left[-ir(\phi_\omega^{(1)} - \phi_\omega^{(0)})\right] d_{r0}{}^l(\omega^{(1)}(t - t_1)) d_{r0}^{*l}(\omega^{(0)}t_1) \right\rangle_{t_1, \omega^{(1)}, \omega^{(0)}} \tag{2.32}$$

where $\omega^{(0)}$, $\omega^{(1)}$ denote the original angular velocity and the angular velocity just after the collision at t_1, respectively. In both the M- and J-diffusion models, it is assumed that $\phi_\omega^{(1)}$ and $\phi_\omega^{(0)}$ are distributed randomly; an average over these variables reduces (2.32) to

$$C_l(t; 1) = \langle d_{00}{}^l(\omega^{(1)}(t - t_1)) d_{00}{}^l(\omega^{(0)}(t_1))\rangle_{t_1, \omega^{(1)}, \omega^{(0)}} \tag{2.33}$$

An average over t_1 is required because the collision can occur at any time between 0 and t; thus,

$$C_l(t; 1) = \frac{1}{t} \int_0^t dt_1 \langle d_{00}{}^l(\omega^{(1)}(t - t_1)) d_{00}{}^l(\omega^{(0)}(t_1))\rangle_{\omega^{(0)}, \omega^{(1)}} \tag{2.34}$$

The difference between the M- and J-diffusion models appears when one averages over angular speeds; in the M-diffusion case,

$$\langle F\rangle_{\omega^{(0)}, \omega^{(1)}} = \int F\delta(\omega^{(0)} - \omega^{(1)})f^{(0)}(\omega^{(0)}) d\omega^{(1)} d\omega^{(0)} \tag{2.35}$$

whereas J-diffusion corresponds to

$$\langle F \rangle_{\boldsymbol{\omega}^{(0)}, \boldsymbol{\omega}^{(1)}} = \int F f^{(0)}(\boldsymbol{\omega}^{(1)}) f^{(0)}(\boldsymbol{\omega}^{(0)}) \, d\boldsymbol{\omega}^{(0)} \, d\boldsymbol{\omega}^{(1)} \qquad (2.36)$$

If we drop the superscripts for clarity, we can now write

$$C_l(t; 1)]_M = \frac{1}{t} \int_0^t dt_1 \langle d_{00}{}^l(\omega(t - t_1)) \, d_{00}{}^l(\omega t_1) \rangle_\omega \qquad (2.37)$$

$$C_l(t; 1)]_J = \frac{1}{t} \int_0^t dt_1 \langle d_{00}{}^l(\omega(t - t_1)) \rangle_\omega \langle d_{00}{}^l(\omega t_1) \rangle_\omega \qquad (2.38)$$

Although the illustrative calculation given here is limited to the linear molecule, it can be modified to deal with the symmetric top case as well. The key to the calculation of $C_l(t; 1)$ is the use of (2.31) for the linear molecule or (1.5) for the symmetric top to express functions of the total angular change in terms of the angular displacements in the two diffusional steps before and after the "collision" at t_1. This allows one to use the free-rotor phase-space distribution functions given in (2.7) and (2.9) together with the distributions of angular momentum after collision that are specified by the model to calculate the one-collision terms.

Furthermore, a natural extension of this idea will yield a general approach to the calculation of the $3, 4, \ldots, N$ collision terms. In particular, repeated use of the addition theorems for $D_{k,m}^j$ or spherical harmonics at each collision gives rise to an equation giving $D_{km}{}^j(\delta\Omega)$ in terms of a sum of products of the functions for each diffusional step. Ensemble averaging can then be performed for each step, to give an expression for $D_{km}{}^j(\delta\Omega)$ at specific values of the collision times $t_1, t_2, \ldots, t_N$.

In order to evaluate the multiple time integrals involved in averaging over all collision times within the interval t, it is necessary to first calculate the Fourier transform of the correlation function. To see this most clearly, we will evaluate the spectral intensity $I(\omega)$ for the truncated collisional expansion of the linear molecule $C_l(t)$ considered above:

$$\hat{I}(\omega) = \frac{1}{2\pi} \int_{-\infty}^{\infty} e^{i\omega t - \beta|t|} [C_l(|t|; 0) + \beta|t| C_l(|t|; 1)] \, dt \qquad (2.39)$$

where we have explicitly indicated that $C_l(t)$ is an even function of t by inserting absolute value signs. Let us define the complex Laplace transform $\mathscr{I}_l(\omega)$, the cosine transform $\mathscr{C}_l(\omega)$, and the sine transform $\mathscr{S}_l(\omega)$ by:

$$\langle \mathscr{I}_l(\omega) \rangle = \langle \mathscr{C}_l(\omega) \rangle + i \langle \mathscr{S}_l(\omega) \rangle = \int_0^\infty e^{(i\omega - \beta)t} \mathscr{C}_l(t; 0) \, dt \qquad (2.40)$$

where the brackets denote an average over $\omega^{(0)}$. Evidently, the zero-collision contribution to $\hat{I}(\omega)$ is just $\langle \mathscr{C}_l(\omega) \rangle / \pi$. The one-collision term $\hat{I}(\omega; 1)$ can be written as

$$\hat{I}(\omega; 1) = \frac{\beta}{\pi} \text{ Real part}$$

$$\left\langle \int_0^\infty e^{(i\omega - \beta)(t - t_1)} d_{00}{}^l(\omega^{(1)}(t - t_1)) \, dt \int_0^t e^{(i\omega - \beta)t_1} d_{00}{}^l(\omega^{(0)}t_1) \, dt_1 \right\rangle_{\omega^{(0)}, \omega^{(1)}} \quad (2.41)$$

The convolution theorem can be applied to this expression to show that

$$\hat{I}(\omega; 1) = \frac{\beta}{\pi} \text{ Real part } \langle \mathscr{I}_l(\omega) \rangle^2 \quad (2.42)$$

$$= \frac{\beta}{\pi} [\langle \mathscr{C}_l(\omega) \rangle^2 - \langle \mathscr{S}_l(\omega) \rangle^2] \quad (2.43)$$

for the J-diffusion model. Thus

$$\hat{I}(\omega)]_J = \frac{1}{\pi} [\langle \mathscr{C}_l(\omega) \rangle + \beta(\langle \mathscr{C}_l(\omega) \rangle^2 - \langle \mathscr{S}_l(\omega) \rangle^2)] \quad (2.44)$$

The spectral intensity calculated from the M-diffusion model differs only in the averaging over angular velocity $\omega^{(0)}$, $\omega^{(1)}$; one finds

$$\hat{I}(\omega)]_M = \frac{1}{\pi} \langle \mathscr{C}_l(\omega) + \beta[\mathscr{C}_l{}^2(\omega) - \mathscr{S}_l{}^2(\omega)] \rangle \quad (2.45)$$

This technique can be extended to compute $\hat{I}(\omega, N)$ for N arbitrarily large; the explicit results are

$$\hat{I}(\omega, N)]_J = \frac{\beta^{N-1}}{\pi} \text{ Real part } \langle \mathscr{I}_l(\omega) \rangle^N \quad (2.46)$$

$$\hat{I}(\omega, N)]_M = \frac{\beta^{N-1}}{\pi} \text{ Real part } \langle \mathscr{I}_l(\omega)^N \rangle \quad (2.47)$$

The sums over N can now be carried out to give

$$\hat{I}(\omega)]_J = \frac{1}{\pi} \frac{\langle \mathscr{C}_l(\omega) \rangle - \beta(\langle \mathscr{C}_l(\omega) \rangle^2 + \langle \mathscr{S}_l(\omega) \rangle^2)}{1 - 2\beta \langle \mathscr{C}_l(\omega) \rangle + \beta^2(\langle \mathscr{C}_l(\omega) \rangle^2 + \langle \mathscr{S}_l(\omega) \rangle^2)} \quad (2.48)$$

$$\hat{I}(\omega)]_M = \frac{1}{\pi} \left\langle \frac{\mathscr{C}_l(\omega) - \beta(\mathscr{C}_l{}^2(\omega) + \mathscr{S}_l{}^2(\omega))}{1 - 2\beta \mathscr{C}_l(\omega) + \beta^2(\mathscr{C}_l{}^2(\omega) + \mathscr{S}_l{}^2(\omega))} \right\rangle \quad (2.49)$$

These expressions are valid for all values of β; the correlation functions obtained by Fourier transforming equations (2.48) and (2.49) vary smoothly from the free-rotor curves at $\beta = 0$ up to a form at large β which is characteristic of the random reorientational walk that will be discussed in Section III. Correlation functions for a range of β values are shown in Figs. 2 and 3.

Expressions for rotational correlation times for the extended diffusion models are readily obtained from the normalized spectral intensity at zero frequency; one finds

$$\tau_l = \frac{\mathscr{C}_l(0)}{1 - \beta \mathscr{C}_l(0)} \qquad (M\text{-diffusion}) \qquad (2.50)$$

$$= \frac{\langle \mathscr{C}_l(0) \rangle}{1 - \beta \langle \mathscr{C}_l(0) \rangle} \qquad (J\text{-diffusion}) \qquad (2.51)$$

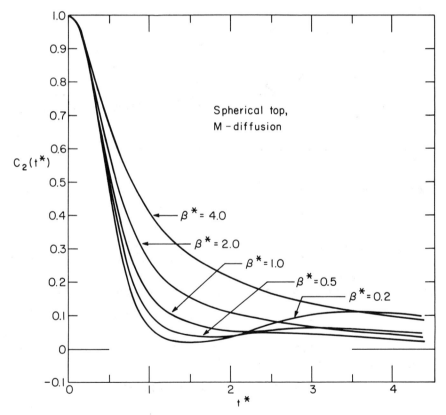

Fig. 2. Correlation functions for the $l = 2$ case of the M-diffusion model of a spherical top. Curves shown are for various values of the reduced collision frequency $\beta^* = \beta\sqrt{I/kT}$.

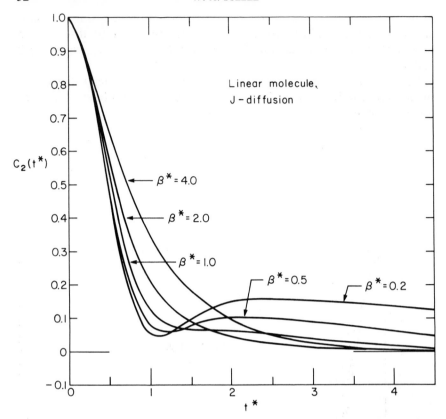

Fig. 3. Correlation functions for the $l = 2$ case of J-diffusion model of a linear molecule.

Plots of some reduced correlation times $\tau_l^* = \tau_l(I/kT)^{1/2}$ versus the reduced collision frequency $\beta^* = \beta(I/kT)^{-1/2}$ are shown in Fig. 4. In general, τ_l^* approaches infinity as β^* approaches zero, due to the fact that free rotor correlation functions approach constant values at long times, as is illustrated in Fig. 2 for spherical tops. The exception to this generalization is that

$$\lim_{\beta \to 0} \tau_l^* = 0 \qquad \text{for } l \text{ odd, linear molecules} \tag{2.52}$$

Figure 5 shows how these limits are approached for several cases; Fig. 4 indicates that (2.52) requires not only that

$$\lim_{t^* = \infty} C_l(t^*) = 0 \qquad \text{for } l \text{ odd, linear molecules} \tag{2.53}$$

but also that the correlation function becomes negative for large t^*; this negative region is absent from the spherical top $C_1(t^*)$ (not shown here).

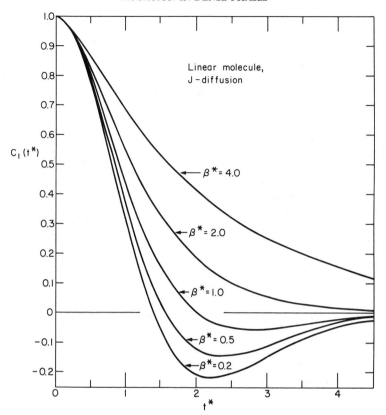

Fig. 4. Correlation functions for the $l = 1$ case of J-diffusion model of a linear molecule.

Correlation times for symmetric top molecules approach infinity as β approaches zero; nevertheless, their correlation functions can exhibit negative regions, depending upon the relative moments of inertia.

The correlation times plotted in Fig. 5 suggest that J-diffusion correlation functions at $\beta^* > 2$ may not be particularly sensitive to the inertial ratio (I/I_{zz}), in contrast to the M-diffusion model. Detailed comparisons show that show that this is indeed the case.

Correlation times obtained from the Gaussian approximation to τ_l^* (2.26) are

$$\tau_1^* = 0.889$$
$$\tau_2^* = 0.512 \tag{2.54}$$

If these are to be equal to the minimum correlation time in the plots of Fig. 5, we see that this approximation fails completely for the $l = 1$, linear

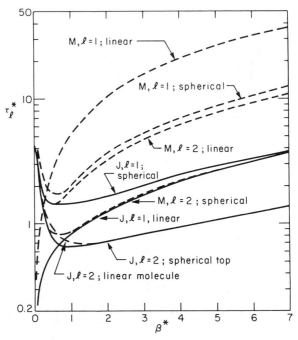

Fig. 5. Reduced orientational correlation times for $l = 1$ and $l = 2$ as a function of the reduced collision frequency β^*. Both versions of the extended diffusion model are shown, for linear as spherical top molecules.

molecule case; furthermore, the predicted values are lower than the minima for the other inertial models, but are at least of the correct order of magnitude.

If it is assumed that other theories will yield minimum correlation times similar to those for the extended diffusion models it can now be concluded that the difference between the Gaussian τ_l^* and an experimental value can be as large as a factor of two without invalidating a description of the motion as "nearly free rotation," if one is willing to accept $\beta^* \simeq 1$ as "nearly free rotation"; our definition of reduced collision frequency implies that an average molecule undergoes roughly a complete rotation between collisions when $\beta^* = 1$.

The curves shown in Figs. 2 to 4 all start out as the Gaussian decays predicted in (2.21) for free rotors; as t^* increases, the curves decay less rapidly than the Gaussians due to the partial and complete reversals in direction assumed in the extended diffusion models; it is evident that this tendency is noticeable at shorter t^* for the larger β^* values. This is one of the most physically appealing features of these models. Although it is not evident from the curves in Figs. 3 to 5 (but is illustrated in Fig. 9), another property of the results obtained (or at least, those for the J-diffusion model) is that the curves for large β^* are close to the exponential functions of t^* anticipated for the isotropic random walk model in angle space that will be discussed in Section III.

Although no method for directly observing them exists at present, angular momentum correlation functions can also be calculated for the extended diffusion models. For instance, we can define

$$A(t) = \langle \mathbf{L}(0) \cdot \mathbf{L}(t) \rangle \tag{2.55}$$

Since the models predict that $\mathbf{L}$ is randomized after a single collision, we can immediately write

$$A(t) = \langle \mathbf{L}(0) \cdot \mathbf{L}(t) \rangle_{\text{free}}\, e^{-\beta t} \tag{2.56}$$

where $\exp(-\beta t)$ is, as before, the fraction of uncollided molecules after time t. For the linear molecule and the spherical top, $\mathbf{L}(t)$ is conserved for a free rotor; precessional motion should be included in evaluating (2.56) for a symmetric top molecule.

It is evident from (2.56) that the collision frequency β is equal to the inverse of the angular momentum correlation time τ_L; furthermore, models that give angular momentum correlation functions of the form of (2.56) will yield the M-diffusion model; one must also assume that $\langle L(0)L(t) \rangle$ decays as $\exp(-\beta t)$ in order to derive the J-diffusion model. In principle, these ideas may enable one to by-pass the "collision" picture to some extent. Experimental determinations of $A(t)$ are possible,[49] but only at very short times. On the other hand, this correlation function is readily derived from molecular dynamics computations for ensembles of rotating molecules.[47,50] Although negative values of $A(t)$ have been found in these studies, it seems that modifications of the extended diffusion models to include such behavior may be quite difficult to derive. Primarily, this is due to the fact that the convolution theorem that enables one to evaluate the N-collision terms is no longer applicable when persistence of velocity effects result in the coupling of one diffusional step to the previous one.

D. Collision Frequencies in Dense Systems

To complete the theoretical picture described by the extended diffusion models, one would like to be able to relate the reduced collision frequency β^* to the properties of the system. Although the current method of estimating this quantity is quite crude, it does provide some indication of a pathway to a more precise calculation. We proceed by assuming that the angular momentum will be randomized only if the collision is violent enough for the molecules to come within a distance σ of one another; we thus represent the real nonspherical molecules by a spherically symmetric approximation. The rate of arrival at a "contact" distance σ in a dense fluid is well known. The requisite expression was first obtained by Enskog[51] for hard spheres, assuming that successive collisions are uncorrelated, and was recently generalized to arbitrary (spherical) interaction laws.[52] The results of these calculations can be written as

$$\beta = 4\rho\sigma^2 \left(\frac{\pi k T}{m}\right)^{1/2} g(\sigma) \tag{2.57}$$

Consequently,

$$\beta^* = 4\rho\sigma^2 \left(\frac{\pi I}{m}\right)^{1/2} g(\sigma) \tag{2.58}$$

where $g(\sigma)$ is the value of the pair correlation function at contact. It is evident that the barrier to numerical evaluation of (2.58) is primarily lack of knowledge of $g(\sigma)$ at the contact distance σ. Nevertheless, it is possible to make a rough estimate of β^* for a molecule such as N_2 in the liquid. If we take $\sigma = 3.7$ Å, $\rho = 1.7 \cdot 10^{22}$ molecules/cm^3, and $g(\sigma) = 5$ (a value appropriate for a hard-sphere fluid at $\rho\sigma^3 = 0.9$), we find $\beta^* = 5$. This is the correct order of magnitude; however, the fit of experimental data to J-diffusion theory[53] indicates that $\beta^* \simeq 1.5$ best represents rotational motion in liquid N_2 at $80°K$.[50] The reason for the discrepancy lies in approximation of the true molecular shape by a sphere and is inherent in many of the current calculations of reorientation times. In fact, spherical molecules do not exert torques upon one another, and consequently, no collision between such molecules will ever bring about a change in angular momentum. It is anticipated that molecular shape (or the angle-dependence of molecular interactions) will play an important role in determining the frequency of collisions that are effective in altering $\mathbf{L}$; for example, formal equations for the relaxation times due to collisions between square-well ellipsoidal molecules have recently been presented[54] which should be relevant to the present problem if one can estimate the pair correlation function at contact for a dense fluid of ellipsoidal molecules.

Several other methods of calculating collision frequencies in dense fluids of nonspherical molecules have been suggested;[55,56] these amount to estimates of $g(\sigma)$ based on quasilattice or cell model representations of the structure of the liquid. Although these arguments give collision frequencies that are undoubtedly as satisfactory as those obtained in the spherical molecule calculation, more accurate calculations of β should be feasible; they are certainly desirable.

We defer discussion of some recent comparisons of experimental data with the extended diffusion models to Section IV.

III. STOCHASTIC MODELS FOR REORIENTATION

A. General

Historically, the earliest descriptions of molecular reorientation in dense phases were based on the idea that the motions were highly hindered. It would then be reasonable to assume that reorientation occurred as a series of thermally activated jumps over a potential barrier or barriers. These ideas also implied that the time between jumps could be taken to be long compared to the time to make a jump, which is essentially the converse of the assumption made in the extended diffusion models discussed previously. These models are limited to computations of the time dependence of orientations rather than angular momentum or combinations of angular momentum and orientation such as the functions occurring in the spin-rotation relaxation calculation. Thus, we need not consider phase-space distribution functions; the partially averaged conditional probability density for reorientation $W(\delta\Omega)$ will do, or even a direct calculation of $\langle D_{km}^{l}(\delta\Omega)\rangle$ is satisfactory.

In general, the models are special cases of cyclic Markov chains; one can therefore use some of the ideas of classical probability theory to solve a particular problem. We will follow this line of approach here, and will show solutions of several specific models, in part to illustrate the method and in part because they have been used to analyze experimental results. In order to do this, some mathematical preliminaries are necessary.

Consider the case where a vector **v** rotates about a rotation axis by an angle ε, as illustrated in Fig. 6. (Before proceeding further, we emphasize that the inertial principal axis system is completely irrelevant to a molecule that conforms to a jump model. When motion is entirely determined by the potential energy, principal axes are chosen to reflect the symmetry of the

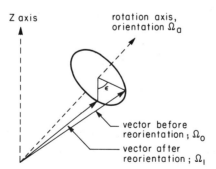

Fig. 6. Coordinate system for reorientation of a molecule-fixed vector about a space-fixed rotation axis with arbitrary orientation relative to the vector; the angle of rotation about this axis is denoted by ε, and the Eulerian orientation angles, by Ω.

potential. Although the two coordinate systems may be coincident in a molecule with an axis of high symmetry, this is seldom the case when the molecule lacks obvious elements of symmetry.)

Referring to Fig. 6, we denote the orientations of the vector of interest by Ω_0, Ω_1 before and after reorientation, respectively, in a space-fixed frame, and by $\Omega_0^{(a)}$, $\Omega_1^{(a)}$ relative to the rotation axis. Thus,

$$D_{k,0}{}^l(\Omega_1^{(a)}) = e^{-ik\varepsilon}D_{k0}{}^l(\Omega_0^{(a)}) \tag{3.1}$$

If Ω_a is the space-fixed orientation of the rotation axis,

$$D_{k0}{}^l(\Omega_1) = \sum_r D_{kr}{}^l(\Omega_a)D_{r0}{}^l(\Omega_1^{(a)}) \tag{3.2}$$

and

$$D_{k0}{}^l(\Omega_0^{(a)}) = \sum_{r'} D_{r'0}^l(\Omega_0)D_{r'r}^{*l}(\Omega_a) \tag{3.3}$$

Equation (3.3) is substituted into (3.1); the result is substituted into (3.2) to give

$$D_{k0}{}^l(\Omega_1) = \sum_{r,r'} e^{-ir\varepsilon}D_{kr}{}^l(\Omega_a)D_{r'r}^{*l}(\Omega_a)D_{r'0}{}^l(\Omega_0) \tag{3.4}$$

Suppose now the vector undergoes a sequence of jumps by angles $\varepsilon^{(1)}$, $\varepsilon^{(2)}$ $\cdots$ about reorientation axes with orientations $\Omega_a^{(1)}$, $\Omega_a^{(2)}$ $\cdots$. Repeated application of these equations will give functions of Ω_n, the orientation of the vector after n jumps. For example,

$$D_{k0}{}^l(\Omega_2) = \sum_{s,s'} \sum_{r,r'} e^{-i(r\varepsilon_1 + s\varepsilon_2)}D_{ks}{}^l(\Omega_a^{(2)})D_{s's}^{*l}(\Omega_a^{(2)})D_{s'r}^l(\Omega_a^{(1)})D_{r'r}^{*l}(\Omega_a^{(1)})D_{r'0}^l(\Omega_0) \tag{3.5}$$

We are now ready to compute angular correlation functions; we will concentrate on functions such as

$$C_l(t) = \langle D_{00}{}^l(\delta\Omega^{(v)})\rangle \tag{3.6}$$

$$= \sum_{N=0}^{\infty} P_N(t)\Big\langle \sum_r D_{r0}^{*l}(\Omega_0)D_{r0}{}^l(\Omega_N)\Big\rangle \tag{3.7}$$

where $P_N(t)$ is the probability that a molecule has undergone N and only N jumps and the brackets denote an average over jump angles and jump axis orientations.

In fact, there are two cases which yield a simple form for (3.5) and its generalization to N jumps: either a random distribution of orientations for the jump axis relative to the vector;[57] or a fixed angle between the two. In the first case, an average over $\Omega_a{}^{(N)}$ for each N causes the generalized form of (3.5) to reduce to

$$
\begin{aligned}
C_l(t) &= \sum_{N=0}^{\infty} P_N(t) \sum_{r=-l}^{l} \frac{\langle e^{-ir\varepsilon}\rangle^N}{2l+1} \\
&= \sum_{N=0}^{\infty} P_N(t) \left\langle\left(\frac{\sin (2l+1)\dfrac{\varepsilon}{2}}{(2l+1)\sin\dfrac{\varepsilon}{2}}\right)^N\right\rangle
\end{aligned} \tag{3.8}
$$

If successive jumps are uncorrelated, $P_N(t)$ is given by the same Poisson distribution that describes the collision probabilities in an ideal system; thus (2.28) is applicable in the present calculation; furthermore, if one defines

$$
A_l = \left\langle \frac{\sin (2l+1)\dfrac{\varepsilon}{2}}{(2l+1)\sin\dfrac{\varepsilon}{2}} \right\rangle \tag{3.9}
$$

then (3.8) becomes

$$
C_l(t) = \exp\{-\beta t[1 - A_l]\} \tag{3.10}
$$

The final evaluation of $C_l(t)$ depends upon the assumed distribution of jump angle ε. We will return to this problem after dealing with the second simple case, which will be called "single axis reorientation" for obvious reasons. Equation (3.5) and its generalizations can be simplified by use of the closure properties of the $D_{km}{}^l$ ((1.7) and (1.8)) if $\Omega_a^{(N)}$ is the same for all N. One finds

$$
D_{k0}{}^l(\Omega_N) = e^{ik(\varepsilon_1+\varepsilon_2+\cdots+\varepsilon_N)}D_{k0}{}^l(\Omega_0) \tag{3.11}
$$

If $\eta^{(v)}$ is the fixed angle between the vector and the axis of rotation, one can now show that

$$
C_l^{(v)}(t) = \sum_{N=0}^{\infty} P_N(t) \sum_{k=-l}^{l} [d_{k0}{}^l(\eta^{(v)})]^2 \left\langle \exp\left(ik\sum_{i=1}^{N}\varepsilon_i\right)\right\rangle \tag{3.12}
$$

If it is assumed that the distribution of jump angles ε_i in the ith step is independent of all other steps, one has

$$
\left\langle \exp\left\{ik\sum_{i=1}^{N}\varepsilon_i\right\}\right\rangle = \langle e^{ik\varepsilon}\rangle^N \tag{3.13}
$$

Since positive and negative jump angles are equally probable, (3.12) now becomes

$$C_1^{(v)}(t) = \sum_{N=0}^{\infty} P_N(t) \left\{ [d_{00}{}^l(\eta^{(v)})]^2 + 2 \sum_{m=1}^{l} \langle \cos m\varepsilon \rangle^N \cdot [d_{m0}{}^l(\eta^{(v)})]^2 \right\} \qquad (3.14)$$

$$= [d_{00}{}^l(\eta^{(v)})]^2 + 2 \sum_{m=1}^{l} [d_{m0}{}^l(\eta^{(v)})]^2 \cdot \exp\{-\beta t[1 - \langle \cos m\varepsilon \rangle]\} \qquad (3.15)$$

It is important to note that we have placed a significant limitation on the model by assuming that the allowed jump angles are independent of the other steps, because this means that the average value of cos $m\varepsilon$ in the ith jump is independent of the vector orientation just before the ith jump.

B. Single Axis Reorientations

A sketch of a possible potential energy variation for single-axis rotation is shown in Fig. 7 as a counter-example to the case just discussed. If the molecule is jumping between three nonequivalent potential wells, it is evident that the probability distribution for ε_{ij}, the angle for a jump from well i to well j, will depend upon the starting point.

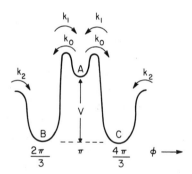

Fig. 7. Schematic sketch of the potential energy as a function of orientation ϕ for rotation against a three-well barrier. The rate constants for changes in orientation are k_0, k_1, and k_2 and the well-depth difference is V.

In order to deal with a situation such as that depicted in Fig. 7, one starts by writing the probabilities for a discrete set of jumps as a matrix. The elements of the matrix $\mathbf{1} - h\mathbf{P}$ are P_{ij}, the probability that a jump occurs from well i to well j in an infinitesimal time interval h. We also define a vector $\mathbf{f}(t)$ with components $f_i(t)$, the fraction of molecules in well i at time t. One then has

$$\mathbf{f}(t + h) = (\mathbf{1} - h\mathbf{P}) \cdot \mathbf{f}(t) \qquad (3.16)$$

The problem is then to calculate $\mathbf{f}(t)$ from $\mathbf{f}(0)$ when t is a finite time interval.

Two mathematically equivalent methods of solution can be followed; the first is to rewrite (3.16) as a set of linear differential equations:

$$\frac{d\mathbf{f}}{dt} = -\mathbf{P}\cdot\mathbf{f} \tag{3.17}$$

The solution of this system of equations is standard,[58] once $\mathbf{P}$ has been specified. For the sake of consistency with other derivations to be presented here, we choose to describe the second of the two approaches, which is based on the idea that one can split the total time into a very large number of infinitesimal intervals. In a Markoffian process, the transition probability matrix is the same in every interval, and is independent of the previous history of the system. Thus, one can write

$$\mathbf{f}(t) = (\mathbf{1} - h\mathbf{P})^s \cdot \mathbf{f}(0) \tag{3.18}$$

where s is the number of intervals in t so that $sh = t$. The matrix $(\mathbf{1} - h\mathbf{P})^s$ is evaluated by first diagonalizing $\mathbf{P}$; if the diagonal form is denoted by $\mathbf{D}$, and the similarity transformation matrix by $\mathbf{V}$, one has

$$\mathbf{V}^{-1}\mathbf{P}\mathbf{V} = \mathbf{D} \tag{3.19}$$

and

$$(\mathbf{1} - h\mathbf{P})^s = \mathbf{V}(\mathbf{1} - h\mathbf{D})^s\mathbf{V}^{-1} \tag{3.20}$$

Furthermore, if one writes the elements of D as

$$D_{ii} = \lambda_i \tag{3.21}$$

the elements of $(\mathbf{1} - h\mathbf{D})^s$ are $(1 - h\lambda_i)^s$ and one can use the Poisson approximation to write

$$\lim_{s\to\infty} (1 - h\lambda_i)^s = e^{-sh\lambda_i} \tag{3.22}$$

$$= e^{-t\lambda_i} \tag{3.23}$$

We note that $\mathbf{P}$ must reduce to the null matrix if all jump probabilities are zero. If the probabilities for jumps to adjacent wells are the only important ones, $\mathbf{P}$ will have the form of a cyclic matrix:[59]

$$\mathbf{1} - h\mathbf{P} = \begin{vmatrix} p_{11} & p_{12} & 0 & 0 & \cdots & & p_{1n} \\ p_{21} & p_{22} & p_{23} & 0 & \cdots & & 0 \\ 0 & p_{32} & p_{33} & p_{34} & \cdots & & 0 \\ \vdots & & & & & & \vdots \\ & & & & & & p_{n-1\,n} \\ p_{n1} & 0 & 0 & 0 & \cdots & p_{nn-1} & p_{nn} \end{vmatrix} \tag{3.24}$$

(Of course, more nonzero elements appear if jumps to nonadjacent wells are included among the possibilities.) The off-diagonal elements of $1 - h\mathbf{P}$ are small numbers equal to hk_{ij}, where k_{ij} is the jump probability per unit time; the diagonal elements are

$$1 - hp_{ii} = 1 - h\sum_j k_{ij} \tag{3.25}$$

(This is due to the fact that (3.25) is the probability that the molecule stays in the ith well and thus must be equal to unity minus the sum of all the jump probabilities.)

As an illustration, we now use the formalism to outline the solution of the three-site model shown schematically in Fig. 7.[60] Only the rate constants k_0, k_1, and k_2 need be specified for a general solution, and indeed, one also has

$$\frac{k_0}{k_1} = \exp\left(-\frac{V}{kT}\right) = x \tag{3.26}$$

where V is the well-depth difference. We set

$$\mathbf{f} = \begin{vmatrix} f_A \\ f_B \\ f_C \end{vmatrix} \tag{3.27}$$

where f_A, f_B, f_C are the fractions of molecules with vectors in sites A, B, and C, respectively. Then

$$1 - h\mathbf{P} = \begin{vmatrix} 1 - 2q_0 & q_0 & q_0 \\ q_1 & 1 - (q_1 + q_2) & q_2 \\ q_1 & q_2 & 1 - (q_1 + q_2) \end{vmatrix} \tag{3.28}$$

where

$$\begin{aligned} hk &= q_0 = hk_0 \\ hk' &= q_1 + 2q_0 = h(k_1 + 2k_0) \\ hk'' &= q_1 + 2q_2 = h(k_1 + 2k_2) \end{aligned} \tag{3.29}$$

Standard matrix methods then give

$$1 - h\mathbf{D} = \begin{vmatrix} 1 & 0 & 0 \\ 0 & 1 - (q_1 + 2q_0) & 0 \\ 0 & 0 & 1 - (q_1 + 2q_2) \end{vmatrix} \tag{3.30}$$

$$\mathbf{V} = \begin{vmatrix} 1 & -2x & 0 \\ 1 & 1 & -1 \\ 1 & 1 & 1 \end{vmatrix} \tag{3.31}$$

Thus, one finds

$(1 - h\mathbf{P})^s =$

$$\frac{1}{1 + 2x} \begin{vmatrix} 1 + 2xe^{-kt} & x(1 - e^{-k't}) & x(1 - e^{-k't}) \\ 1 - e^{-k't} & x + \frac{1}{2}e^{-k't} + (x + \frac{1}{2})e^{-k''t} & x + \frac{1}{2}e^{-k't} - (x + \frac{1}{2})e^{-k''t} \\ 1 - e^{-k't} & x + \frac{1}{2}e^{-k't} - (x + \frac{1}{2})e^{-k''t} & x + \frac{1}{2}e^{-k't} + (x + \frac{1}{2})e^{-k''t} \end{vmatrix}.$$

$$(3.32)$$

Equation (3.32) is the desired result; it has also been obtained by solving the linear differential equations.[60] One can now use these results to compute correlation functions $C_i^{(v)}(t)$; if the $i \to j$ jump means that the vector of interest changes its orientation by θ_{ij}, the correlation function is

$$C_i^{(v)}(t) = \sum_{i,j} p_{ij}(t) f_i(0) d_{00}^{l}(\theta_{ij}) \tag{3.33}$$

In the specific model under consideration here

$$f_B(0) = f_C(0) = \frac{x}{1 + 2x}$$

$$f_A(0) = \frac{1}{1 + 2x} \tag{3.34}$$

In the case of single-axis rotation and three sites equally spaced with angular separations of $2\pi/3$,

$$d_{00}^{l}(\theta_{ij}) = 1, \ i = j$$

$$= [d_{00}^{l}(\eta^{(v)})]^2 + 2 \sum_{r=1}^{l} [d_{r0}^{l}(\eta^{(v)})]^2 \cos\left(\frac{2\pi r}{3}\right), \ i \neq j \tag{3.35}$$

This method of solution can be also used when the reorientation is assumed to consist of single axis rotation to one of ρ equivalent sites. After s intervals of time, matrix diagonalization gives[61,62]

$$p_{ij}(s) = \frac{1}{\rho} \sum_{r=1}^{\rho} \Theta^{r(i-j)} \left[\sum_{v=0}^{\rho-1} q_v \Theta^{rv} \right]^s \tag{3.36}$$

where $q_0, q_1, q_2, \ldots$ denote the probabilities for jumps in a time interval h to a site which is $0, 1, 2, \ldots$ sites away from the starting point, and $\Theta = e^{2\pi i/\rho}$. If we let s approach infinity and remember that $q_0 = 1 - \sum_{v=1}^{\rho-1} q_v$, the Poisson approximation gives (3.36) as

$$p_{ij} = \frac{1}{\rho} \sum_{r=1}^{\rho} \Theta^{r(i-j)} \exp\left[-t \sum_{v=1}^{\rho-1} k_v (1 - \Theta^{rv}) \right] \tag{3.37}$$

If we now limit the model to one with nearest-neighbor jumps only and take the jump angles to be $2\pi/\rho$, this expression yields a correlation function which can be summed over v to give

$$C_l^{(v)}(t) = \sum_{m=-l}^{l} \exp\left[-2kt\left(1 - \cos\frac{2\pi m}{\rho}\right)\right][d_{m0}^l(\eta^{(v)})]^2 \qquad (3.38)$$

Equation (3.38) is equivalent to (3.15) if one sets the jump frequency $\beta = 2k$, where k is the jump frequency to *one* of the two equally accessible sites.

Single-axis models are often used in studies of reorientation in solids of long-chain molecules;[62,63] they have also been invoked to interpret nmr relaxation data for solid H_2S.[64]

The model for a rotor undergoing nearest-neighbor jumps in an eight-well model of cubic symmetry has been solved by Brot and Darmon and compared with experiment;[65] Cukier and Lakatos-Lindenberg[67] have discussed Markoffian reorientations through arbitrary angles, and have also given the solution for the eight-well model of Brot and Darmon; group theoretical aspects of reorientation in a field of arbitrary symmetry have been discussed;[66,67] single-axis jumps have been treated for three- and for four-site models with different well depths and rate constants (i.e., with Fig. 7 modified to show different barrier heights);[60,68] and several models for the reorientation of a pear-shaped molecule in both the body-centered orthorhombic and the tetragonal lattice have been solved.[69] This listing is intended to be indicative rather than exhaustive; since such models are clearly most appropriate for reorientation in molecular crystals, an enormous range of possibilities presents itself. In addition, one can refine the basic picture by allowing for angular librations between jumps.[65]

However, we are primarily concerned here with reorientations in fluids. Although numerous stochastic models have been proposed as representations of this process, it is reasonably clear that the description of whole-molecule reorientations in terms of a single-axis (or site-wise) model is inappropriate, except possibly in the case of a strongly hydrogen-bonded fluid. On the other hand, internal rotations in nonrigid molecules almost certainly conform to these models; the three-site model shown in Fig. 7 is a well-known representation of the rotation about C—C single bonds in aliphatic molecules. Thus, one anticipates using these results to compute correlations of the form $\langle D_{ro}^l(\eta(0), \zeta(0))D_{ro}^{*l}(\eta(t), \zeta(t))\rangle$ that appear when the vector of interest is (a) not parallel to the molecular z-axis and (b) time dependent with orientation angles $\eta(t)$, $\zeta(t)$ in the molecular frame.

C. Random Axis Reorientations

Perhaps the simplest Markoffian alternative to fixed-axis rotation is the random-axis reorientation briefly described above; the correlation function that results is given by (3.10), and it is easy to show that

$$C_{l,m}(t) = C_l(t) \qquad \text{for all } m \tag{3.39}$$

In order to complete the calculation, one needs only to assume some distribution law for the jump angle ε. In the paper where Ivanov suggested the random-axis model, a square distribution law was taken for the probability density $p(\varepsilon)$:

$$p(|\varepsilon|) = \frac{1}{2\sigma}, \bar{\varepsilon} - \frac{\sigma}{2} \leq |\varepsilon| \leq \bar{\varepsilon} + \frac{\sigma}{2} \tag{3.40}$$

$$= 0, \text{ otherwise}$$

with $\bar{\varepsilon} + \sigma/2$ restricted to be less than π because a jump angle greater than π is indistinguishable from the corresponding negative jump. As the mean rotation angle $\bar{\varepsilon}$ increases, A_l for an infinitely narrow distribution ($\sigma \to 0$) decays from unity to an oscillating function that is small compared to unity; as the width becomes broader for fixed $\bar{\varepsilon}$, the oscillations in the functions become less marked. Thus, Ivanov's conclusion is that the correlation time τ_l, which is generally

$$\tau_l = \frac{1}{\beta(1 - A_l)} \tag{3.41}$$

becomes independent of l for sufficiently large $\bar{\varepsilon}$ and small σ. Of course, a number of other plausible distributions are possible; for example, a Gaussian distribution of jump angles:

$$p(\varepsilon) = \{2\pi\langle\varepsilon^2\rangle\}^{-1/2} e^{-\varepsilon^2/2\langle\varepsilon^2\rangle} \tag{3.42}$$

where $\langle\varepsilon^2\rangle^{1/2}$ is now the root mean square jump angle. In this case,

$$A_l = \frac{1}{2l+1}\left[1 + 2\sum_{k=1}^{l} e^{-k^2\langle\varepsilon^2\rangle/2}\right] \tag{3.43}$$

Plots of A_l for this model are shown in Fig. 8; for small rms jump angles, the curves shown there are no different from those for Ivanov's square distribution; as the rms angle approaches π, the curves approach limiting values of $(2l + 1)^{-1}$, and one finds[70]

$$\tau_l = \frac{1}{\beta}\frac{2l+1}{2l} \tag{3.44}$$

The other important limit of the random axis model is at very small values of the average jump angle; one finds that

$$\lim_{\langle\varepsilon^2\rangle=0} A_l = \frac{l(l+1)}{6}\langle\varepsilon^2\rangle + 1 \tag{3.45}$$

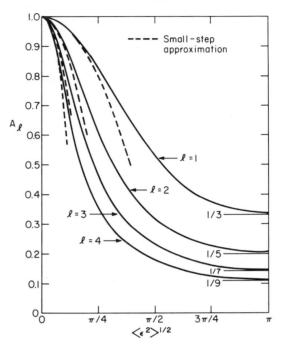

Fig. 8. The constant A_l defined in (3.9) is shown as a function of the rms jump angle for a molecule with a Gaussian distribution of jump angles about a random distribution of orientation axes. Results for the isotropic rotational diffusion model are denoted "small-step approximation." Asymptotic values of A_l in the limit of large-jump angles are also shown.

If one defines a rotational diffusion constant $\mathscr{R}$ by analogy with the translational case,[71]

$$\mathscr{R} = \beta \langle \varepsilon^2 \rangle / 6 \tag{3.46}$$

the correlation function and correlation time become

$$C_l(t) = \exp\left[-l(l+1)\mathscr{R}t\right] \tag{3.47}$$

$$\tau_l = \frac{1}{\mathscr{R}l(l+1)} \tag{3.48}$$

The small-step approximation to A_l is compared with the Gaussian distribution results in Fig. 8; it is interesting to note that the ratio $(1 - A_2)/(1 - A_1)$ (which is equal to 3 for the small-step model) is equal to 2.7 for $\langle \varepsilon^2 \rangle^{1/2}$ as large as 20°; furthermore, $(1 - A_3)/(1 - A_1) = 4.8$ for this rms angle, which hardly differs from the small-step value of 5, at least by present standards of accuracy in the experimental determinations.

One of the basic conceptual difficulties in the large-step models proposed by Ivanov and by a number of other workers[72–75] arises from the assumption that the time required for the molecule to execute a large jump is negligible compared to the time between jumps. There is thus an upper limit on the values of β that can be used to fit experiment without violating this constraint. A rough estimate of t_j, the average time spent in making a jump, gives

$$t_j \simeq \left\{ \langle \varepsilon^2 \rangle \frac{I}{kT} \right\}^{1/2} \tag{3.49}$$

In reduced units, we now see that the allowable jump frequencies are limited by

$$\beta^* \ll \langle \varepsilon^2 \rangle^{1/2} \tag{3.50}$$

On the other hand, we have already discussed the extended diffusion models in Section II; it should be evident that these constraints are not applicable to those models because the motion of the molecules between collisional events is explicitly computed. One can therefore ask whether or not the extended diffusion models become equivalent to the small-step jump model for some range of β^*. (Note that this parameter is a measure of the rate of change of the angular momentum in the extended diffusion models, whereas it measures the rate of change of orientation in the jump models.) One quickly discovers that the degree of correspondence between the two calculations depends both upon the particular version of the model under consideration and upon the criteria used. For example, one can inquire whether or not the ratio of the correlation times for two different l values approaches the value predicted by (3.48). The values of the zeroth moment τ_l^* given in Fig. 5 can be used for this purpose; one finds that

$$\frac{\tau_2}{\tau_1} \simeq 3 \qquad \text{for } \beta^* \geq 4 \tag{3.51}$$

in both the M- and J-diffusion models; additional calculations (not shown) indicate that this result is rather insensitive to the choice of inertial parameters for a symmetric top (i.e., to the value of I_{zz}/I).

Alternatively, a more restrictive criterion for equivalence can be imposed by requiring that the correlation functions for the extended diffusion models approach the exponentially decaying curve that results from the jump model [either (3.47) or (3.10)]. A few representative correlation functions are plotted on a logarithmic scale in Fig. 9 for the extended diffusion models. It is evident there that the curves for the J-diffusion model are reasonably exponential for $\beta^* \geq 2$, if one is willing to overlook the initial Gaussian decays. (The curves in Figs. 3 and 4 indicate that this conclusion is valid *only* for $\beta^* \geq 2$, and detailed calculations confirm this.) On the other hand, the

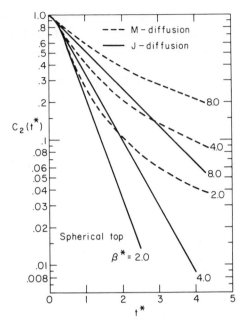

Fig. 9. Logarithmic plots of correlation functions with $l = 2$ for a spherical top obeying the extended diffusion models. Values of the reduced collision frequency are indicated next to the appropriate curves.

M-diffusion correlation functions in Fig. 9 are nonexponential for all β^* values considered, even though the ratios of the correlation times do conform to (3.51).

D. Anisotropic Small-Step Reorientation

The small-step or diffusion model for molecular reorientation in dense systems is by far the best known of those proposed to date. Thus, we will discuss it in terms of more flexible and more sophisticated models than the Ivanov approach. One particularly glaring defect of (3.47) and (3.48) is that they imply isotropic rotational motion, that is, the rotation diffusion constant tensor is equal to $\mathscr{R}\mathbf{1}$. This is seldom the case in real systems. We will see that the rule that is so often used as a test of this model, which is

$$\frac{\tau_l}{\tau_{l'}} = \frac{l'(l' + 1)}{l(l + 1)} \tag{3.52}$$

is not generally applicable when rotational diffusion is anisotropic. Indeed, it is possible to modify the Ivanov model by assuming that $\langle \varepsilon^2 \rangle^{1/2}$ depends upon the orientation of the jump axis relative to the molecular frame. Although this modification of the random-axis model will yield anisotropic rotational diffusion in the small-step limit, we will not pursue this approach further.

A very simple limiting case of anisotropic rotational diffusion can be obtained from the single-axis result given in (3.38); one merely lets the number of sites $\rho \to \infty$ and writes

$$\lim_{\rho \to \infty}\left(1 - \cos\frac{2\pi m}{\rho}\right) = \left(\frac{2\pi}{\rho}\right)^2 \frac{m^2}{2} \tag{3.52}$$

Since $(2\pi/\rho)^2 = \langle \varepsilon^2 \rangle$ for a model limited to jumps between adjacent sites, one can define a one-dimensional rotational diffusion constant $\mathscr{R}_{1D}$ by

$$\mathscr{R}_{1D} = \frac{\beta}{2}\left(\frac{2\pi}{\rho}\right)^2 \tag{3.53}$$

Consequently,

$$C_l^{(\nu)}(t) = \sum_{m=-l}^{l} \exp\left[-\mathscr{R}_{1D}m^2 t\right]\left[d_{m0}{}^l(\eta^{(\nu)})\right]^2 \tag{3.54}$$

It is evident that the l-dependence of the correlation time for this model will be a complicated function of $\eta^{(\nu)}$, the angle between the vector of interest and the axis of rotation. Even in the simplest nontrivial case of $\eta^{(\nu)} = \pi/2$, the result is[76],

$$C_l^{(\nu)}(t) = \frac{1}{2^{2l}} \sum_{u=0}^{l} \exp\left[-\mathscr{R}_{1D}(l - 2u)^2 t\right] \frac{2u![2(l - u)]!}{[u!(l - u)!]^2} \tag{3.55}$$

It is helpful to treat the general problem of anisotropic rotational diffusion in some detail. Since this is a Markoffian process, one can begin by generalizing (3.17) for a discrete set of sites to a quasicontinuous set. Suppose that $W(\delta\Omega)$ is the probability density for a reorientation $\delta\Omega$ in time t; we imagine that the set of infinitesimal increments in angle that make up $\delta\Omega$ form a Hilbert space. The matrix $\mathbf{P}$ in (3.17) then is replaced by a transition density $\Psi(\delta\Omega; \delta\Omega')$ that gives the probability of a change from $\delta\Omega'$ to $\delta\Omega$ in unit time. For continuous distributions, (3.17) becomes

$$\frac{\partial W(\delta\Omega)}{\partial t} = \int \Psi(\delta\Omega; \delta\Omega')W(\delta\Omega')\,d\delta\Omega' \tag{3.56}$$

Rather than following the arguments used in the discrete jump models, we will solve this differential equation for assumed transition probabilities. Since the transition probability for the small-step rotational model must cause the integrand in (3.56) to vanish except for $\delta\Omega'$ in the vicinity of $\delta\Omega$ (i.e., for

$\delta\Omega' - \delta\Omega$ small), we can expand $W(\delta\Omega')$ in a Taylor series around $W(\delta\Omega)$:

$$W(\delta\Omega') = W(\delta\Omega) + \sum_{i=x,y,z} \Delta\psi_i \left(\frac{\partial W}{\partial\psi_i}\right)_{\delta\Omega} + \frac{1}{2} \sum_{i,j=x,y,z} \Delta\psi_i \Delta\psi_j \left(\frac{\partial^2 W}{\partial\psi_i \partial\psi_j}\right)_{\delta\Omega} + \cdots$$

$$(3.57)$$

where $\Delta\psi_i$ denotes a small rotation about the ith molecule-fixed axis. Upon substitution of (3.57) into (3.56), the explicit assumptions of the anisotropic rotational diffusion model are introduced by writing

$$\int \Delta\psi_i \Psi(\delta\Omega, \delta\Omega') \, d\delta\Omega' = 0 \tag{3.58}$$

$$\frac{1}{2} \int \Delta\psi_i \Delta\psi_j \Psi(\delta\Omega, \delta\Omega') \, d\delta\Omega' = \mathscr{R}_{ij} \tag{3.59}$$

All higher moments of the transition probability are assumed to be zero. In (3.59), $\mathscr{R}_{ij}$ is an element of the rotational diffusion constant tensor.

Since no constraint has yet been placed upon the orientation of the molecule-fixed axes in (3.57)–(3.59), we can quite generally take them so as to diagonalize **R**. Thus, (3.56) becomes:

$$\frac{\partial W}{\partial t} = \sum_{i=x,y,z} \mathscr{R}_{ii} \frac{\partial^2 W}{\partial\psi_i^2} \tag{3.60}$$

The solutions of this equation have been derived and listed by many authors,[77-83] starting with Perrin in 1934. It is convenient to expand the probability distribution $W(\delta\Omega)$ is terms of the complete set of functions (for an isotropic distribution at $t = 0$):

$$W(\delta\Omega) = \sum_j \sum_{r,r'=-j}^{j} f_{rr'}^j(t) D_{rr'}^j(\delta\Omega) \tag{3.61}$$

The boundary condition is

$$\lim_{t\to 0} W(\delta\Omega) = \delta(\delta\Omega) \tag{3.62}$$

which just means that angular displacements must be zero at $t = 0$. The complexity of the expressions for the coefficients $f_{rr'}^j(t)$ that are obtained by solving this differential equation depend upon the presence or absence of symmetry in the diagonalized **R** tensor; for example, suppose that $\mathscr{R}_{xx} = \mathscr{R}_{yy}$. The differential equation that results is mathematically identical to the Schroedinger equation for the freely rotating symmetric top; if one defines

$$\begin{aligned} \mathscr{R} &= \tfrac{1}{2}(\mathscr{R}_{xx} + \mathscr{R}_{yy}) \\ \mathscr{R}_1 &= \mathscr{R}_{zz} - \tfrac{1}{2}(\mathscr{R}_{xx} + \mathscr{R}_{yy}) \end{aligned} \tag{3.63}$$

the results are

$$f^j_{rr'}(t) = \delta_{r,r'} \exp\left[-(j(j+1)\mathscr{R} + r^2\mathscr{R}_1)t\right] \tag{3.64}$$

For completeness, a few of the coefficients are listed in Table I for the case where $\mathscr{R}_{xx} \neq \mathscr{R}_{yy} \neq \mathscr{R}_{zz}$.[8,14,82,83]

One difficulty that arises when one attempts to apply these expressions in calculating correlation functions and correlation times for a molecule undergoing asymmetric random walk is that there is no prescription for estimating

TABLE I

Time-dependent Functions in the Expansion of $W(\delta\Omega)$ for a Molecule Undergoing Anisotropic Rotational Diffusion

j, r, r'	$f^j_{rr'}(t)$
0, 0, 0	1
1, 0, 0	$\exp(-2\mathscr{R}t)$
1, 0, ±1; 1, ±1, 0	0
1, ∓1, 1; 1, ±1, −1	$\dfrac{1}{2}\left\{\exp\left[-\left(2\mathscr{R}+\mathscr{R}_1+\dfrac{a}{\sqrt{3}}\right)t\right] \mp \exp\left[-\left(2\mathscr{R}+\mathscr{R}_1-\dfrac{a}{\sqrt{3}}\right)t\right]\right\}$
2, 0, 0	$\dfrac{1}{N^2}\{a^2 \exp(-F_1 t) + b^2 \exp(-F_2 t)\}$
2, ±1, 0; 2, 0, ±1	0
2, 0, ±2; 2, ±2, 0	$\dfrac{ab}{N^2\sqrt{2}}\{\exp(-F_1 t) - \exp(-F_2 t)\}$
2, ∓1, 1; 2, ±1, −1	$\tfrac{1}{2}\{\exp(-F_3 t) \mp \exp(-F_4 t)\}$
2, ±1, ±2; 2, ±2, ±1	0
2, ∓2, 2; 2, ±2, −2	$\dfrac{1}{2N^2}\{b^2 \exp(-F_1 t) + a^2 \exp(-F_2 t) \mp N^2 \exp(-F_5 t)\}$

$a = \sqrt{3}(\mathscr{R}_{xx} - \mathscr{R}_{yy})$

$\Delta = \left[\dfrac{a^2}{3} + (\mathscr{R}_{zz} - \mathscr{R}_{xx})(\mathscr{R}_{zz} - \mathscr{R}_{yy})\right]^{1/2}$

$b = 2(\mathscr{R}_1 + \Delta)$
$N = 2(\Delta b)^{1/2}$

$F_1 = 2(\mathscr{R}_{xx} + \mathscr{R}_{yy} + \mathscr{R}_{zz} + \Delta)$

$F_2 = 2(\mathscr{R}_{xx} + \mathscr{R}_{yy} + \mathscr{R}_{zz} - \Delta)$

$F_3 = 4\mathscr{R}_{xx} + \mathscr{R}_{yy} + \mathscr{R}_{zz}$

$F_4 = \mathscr{R}_{xx} + 4\mathscr{R}_{yy} + \mathscr{R}_{zz}$

$F_5 = \mathscr{R}_{xx} + \mathscr{R}_{yy} + 4\mathscr{R}_{zz}$

the elements of **R** for a molecule of arbitrary shape, and consequently, no method of determining the molecule-fixed coordinate system that diagonalizes this tensor. Of course, if an axis of symmetry is present, the principal z-axis of **R** can be taken to coincide with it and the symmetric diffuser results embodied in (3.64) can be used. Huntress[14] has briefly discussed how the differential equation for $W(\delta\Omega)$ might be solved for a nondiagonal **R**, and has given some results for a planar diffuser. We emphasize that the symmetry elements that are relevant to this problem have nothing to do with the inertial constants; rather, they refer to the variation in the intermolecular torque as one applies a symmetry operator to a molecule in the dense medium. To take a simple example, CH_4, CH_3D and $C_2H_2D_2$ are all characterized by interaction energy functions with tetrahedral symmetry and would thus be expected to be isotropic rotational diffusers. Only if inertial effects play an appreciable role in determining the motion would it be necessary to take note of the fact that these molecules are spherical, symmetric, and asymmetric tops, respectively, with different sets of principal inertial axes in the last two cases.

E. Rotational Diffusion Constants and Inertial Effects

We have also indicated that inertial effects are expected for sufficiently large values of **R**; specifically, one can generalize (3.46) and (3.50) slightly to find that the model will break down when any of the components of **R** is of the order $(1/2)(kT/I)^{1/2}$. In fact, it was realized some time ago that experimental rotational diffusion constants are often large enough to require consideration of inertial effects. Several workers have attempted to treat this problem;[84-89] recent developments lead to the conclusion that the best way to approach it is to utilize the theory of stochastic Liouville equations[47,90-94] to derive a memory function expression for $\mathcal{R}_{ii}$. In a later section, we will discuss some general aspects of this theory in connection with rotational motion; for the present, we will be content with a specific argument that leads to time-dependent diffusion constants $\mathcal{R}_{ii}(t)$.[85,94]

We start with the equation of motion for $W(\delta\Omega)$, the conditional distribution function for angular displacements. As with all statistical probability distributions, this function obeys the Liouville equation [(2.5) and (2.6)]. The formal closed solution of this equation can be written as

$$W(\delta\Omega; t) = \left\langle \exp\left[-i\int_0^t ds\,\mathcal{L}(s)\right]W(\delta\Omega; 0)\right\rangle \tag{3.65}$$

where we have explicitly indicated the times in both $W(\delta\Omega)$ and the Liouville operator $\mathcal{L}$, and the brackets indicate an average over the fluctuating angular momentum. We now assume that the components of the angular momentum

L_μ in (2.6) are Gaussian stochastic variables; Kubo has shown that an average over the fluctuations gives

$$W(\delta\Omega; t) = \exp\left[-i \int_0^t ds_1 \int_0^{s_1} ds_2\, \mathcal{O}(s_1, s_2) \right] W(\delta\Omega; 0) \qquad (3.66)$$

where the operator $\mathcal{O}$ can be written

$$\mathcal{O}(s_1, s_2) = \sum_i \frac{1}{I_{ii}^2} \langle L_i(s_1)L_i(s_2)\rangle \frac{\partial^2}{\partial\psi_i^2} \qquad (3.67)$$

where L_i is the component of angular momentum around the ith axis, and $d\psi_i$ is, as before, the infinitesimal angular displacement about that axis. In writing (3.67), it has been assumed that cross-correlations of the form $\langle L_i(s_1)L_j(s_2)\rangle$, $i \neq j$, are zero.

Equation (3.66) is now differentiated with respect to time to give

$$\frac{\partial W(\delta\Omega, t)}{\partial t} = \int_0^t ds\,\mathcal{O}(t, s)W(\delta\Omega, s) \qquad (3.68)$$

This integral equation for $W(\delta\Omega)$ can be solved by iteration if one introduces a model that will yield an expression for the correlation function of the angular momentum. The zeroth order iteration consists in assuming that $W(\delta\Omega, s) = W(\delta\Omega, t)$, on the right-hand side of (3.68). The resulting expression has the form of (3.60), but contains time-dependent rotational diffusion constants equal to[85]

$$\mathcal{R}_{ii}(t) = \frac{1}{I_{ii}^2} \int_0^t \langle L_i(0)L_i(s)\rangle\, ds \qquad (3.69)$$

F. Rotational Langevin Equation

In order to complete the calculation, it is necessary to insert an expression for the angular momentum correlation function. The most popular model that will yield a stochastic equation for angular momentum (or angular velocity) is the rotational analogue of the Langevin equation. In essence, we assume that the intermolecular torque $\mathbf{N}$ that appears in the Euler equations (2.1) can be written

$$\mathbf{N} = \underline{\xi}\cdot\boldsymbol{\omega} + \mathbf{N}(t) \qquad (3.70)$$

where $\underline{\xi}$ is a friction constant tensor, $\boldsymbol{\omega}$ is the (body-fixed) angular velocity, and $\mathbf{N}(t)$ is a fluctuating torque with the properties that

$$\langle \mathbf{N}(t)\rangle = 0$$
$$\langle N_i(t)N_j(t')\rangle = 2\xi_{ij}kT\delta(t - t') \qquad (3.71)$$

For simplicity, we consider spherical top molecules (for studies of the Langevin equation for symmetric tops, see Ref. 88); in this case, $\underline{\xi}$ is a diagonal tensor and the Langevin model leads to

$$\langle \omega_i(0)\omega_i(t) \rangle = \left(\frac{kT}{I} \right) \exp\left[-\frac{\xi_{ii} t}{I} \right] \tag{3.72}$$

Equation (3.69) becomes

$$\mathcal{R}_{ii}(t) = \frac{kT}{\xi_{ii}} \left(1 - \exp\left[-\frac{\xi_{ii} t}{I} \right] \right) \tag{3.73}$$

$$= \mathcal{R}_{ii}(\infty)\left(1 - \exp\left[-\frac{kTt}{I\mathcal{R}_{ii}(\infty)} \right] \right) \tag{3.74}$$

We have argued previously that inertial effects should become noticeable when $\mathcal{R}_{ii}(\infty) \simeq \frac{1}{2}(kT/I)^{1/2}$; the present model calculation further shows that the importance of the inertial effects depends upon the time variable as well as **R**. As expected, deviations of $\mathcal{R}_{ii}(t)$ from $\mathcal{R}_{ii}(\infty)$ are most noticeable at small t where the torques have not acted for sufficient time to alter the initial rotational velocity.

If (3.74) is now substituted into the rotational diffusion equation rather than $\mathcal{R}_{ii}(\infty)$, the resulting expression is still soluble for the symmetric diffuser; it can be shown[85] that (3.64) should be replaced by

$$f_{rr'}^j(t) = \delta_{rr'} \exp\left[-(\mathcal{R}_{xx}(j(j+1) - r^2) + r^2 \mathcal{R}_{zz})t \right] \tag{3.75}$$

where

$$\mathcal{R}_{ii} = \mathcal{R}_{ii}(\infty) + \frac{I\mathcal{R}_{ii}^2(\infty)}{tkT} \left\{ \exp\left[-\frac{kTt}{I\mathcal{R}_{ii}(\infty)} \right] - 1 \right\} \tag{3.76}$$

Although this derivation is primarily aimed at a calculation of small inertial corrections to the rotational diffusion theory, it is interesting to note that (3.75) gives Gaussian correlation functions in the limit of $kTt/I\mathcal{R}_{ii}(\infty)$ $= 0$, and that the coefficients in these Gaussians are identical to those derived previously from the free spherical top correlation functions [i.e., (2.20)]. The limitation of these results to small inertial effects arises from the well-known noncommutativity of the angles ψ_x, ψ_y, ψ_z for finite displacements. However, there is one important case where these solutions are more generally valid. This occurs when motion of the molecule about its own symmetry axis is inertial, but the axis itself rotates via small-step diffusion. Since the displacement angle ψ_z is identical to the Euler angle γ, it can be shown that (3.75) is a solution of the Langevin equations with $\mathcal{R}_{xx} = \mathcal{R}_{xx}(\infty) = \mathcal{R}_{yy}$, but $\mathcal{R}_{zz}$ given by (3.76).

In conclusion, we briefly discuss some of the attempts that have been made to evaluate rotational diffusion tensors. (We neglect inertial effects on **R** here, but simplify the notation by omitting the time argument.) Originally, it was believed[36,95] that the friction constant tensor $\underline{\xi}$ could be estimated from the solutions of the Stokes–Einstein equation[77] for the hydrodynamic drag on an ellipsoidal body rotating in a viscous medium. Perrin[77] has derived expressions for $\mathscr{R}_{ii}$; his results are most conveniently given in terms of the friction tensor $\underline{\xi}$; thus, one writes

$$\mathscr{R}_{ii} = \frac{kT}{\xi_{ii}} \tag{3.77}$$

If all three semi-axes of the ellipsoid are of different lengths a, b, c, the ξ_{ii} are functions of elliptic integrals; we do not repeat the expressions here.[80,96] For an axially symmetric ellipsoid where $c = b$, one finds

$$\xi_{zz} = \frac{32\pi\eta b^2(a^2 - b^2)}{6a - 3b^2 S} \tag{3.78}$$

$$\xi_{xx} = \xi_{yy} = \frac{32\pi\eta(a^4 - b^4)}{3(2a^2 - b^2)S - 6a} \tag{3.79}$$

with

$$S = \frac{2}{(a^2 - b^2)^{1/2}} \ln\left(\frac{a + (a^2 - b^2)^{1/2}}{b}\right), a > b$$

$$S = \frac{2}{(b^2 - a^2)^{1/2}} \tan^{-1}\left(\frac{(b^2 - a^2)^{1/2}}{a}\right), b > a \tag{3.80}$$

In (3.78) and (3.79), η is the shear viscosity of the liquid. There is now ample evidence that this theory overestimates the drag coefficients by an order of magnitude in many cases; the discrepancies are particularly noticeable when the rotating molecules are nearly spherical. A "microviscosity" theory has been proposed[97] for spherical molecules which gives smaller friction coefficients than the viscous drag model; it yields

$$\xi_{ii} = \frac{8\pi\eta a^3}{6} \tag{3.81}$$

which is smaller than the spherical limit of (3.78) and (3.79) by a factor of six.

However, we note that both of these theories contain a conceptual error which arises from the attempt to apply macroscopic hydrodynamics to a molecular problem. In the Newtonian mechanics that apply to a classical ensemble, the torque on a spherical molecule is zero. Since the rotational motion is free and $\underline{\xi}$ is zero, such a system will be described by one of the inertial models discussed in Section II rather than by rotational diffusion. Furthermore, it should not be surprising to find that the mean torques which are the determining factor for rotational drag are unrelated to the shear viscosity even when the molecules under consideration are nonspherical.

Recently, Hu and Zwanzig[98] have pointed out that a much improved hydrodynamic theory is obtained if one changes from the "stick" boundary conditions used by Perrin to "slip" boundary conditions. In the "slip" case, resistance to rotation is due entirely to the fact that a nonspherical object sweeps out a volume as it rotates or translates, thus displacing a certain amount of viscous fluid. For translation of a sphere of radius a, most of the viscous resistance is due to this displacement, as indicated by the values of the translational friction coefficients which go from $6\pi\eta a$ for the sticking boundary condition to $4\pi\eta a$ for the slipping boundary condition. Although general closed-form expressions were not obtained by Hu and Zwanzig for the friction coefficients of the symmetric ellipsoids treated, their numerical results are very much improved over those for the hydrodynamic theories with stick. In particular, the rotational friction coefficients vanish as the molecule approaches sphericity; this is clearly illustrated by the analytic expression obtained in the limit of small ellipticity $e = (a/c)^2 - 1$, where c is the length of the z semi-axis and a is the length of the x and the y semi-axes. Hu and Zwanzig showed that

$$\xi_{xx} = \frac{32\pi\eta a^3}{25} e^2 \left[1 - \frac{74}{25} e + \cdots \right] \tag{3.82}$$

A plot of their numerical results for the ratio of the slip friction coefficient to the stick value given by (3.79) and (3.80) is shown as a function of the axial ratio in Fig. 10. It is quite clear that this hydrodynamic model gives rotational friction coefficients that are an order of magnitude smaller than those for the stick boundary condition for axial ratios that are likely to correspond to real molecules, and that the results of the two calculations are not equivalent except for extremely nonspherical objects.

Although a calculation that relates the rotational friction tensor $\underline{\xi}$ to averages of an appropriate function of the intermolecular interaction law would appear to be the best method of solving this problem in the long run, very little quantitative work has been done along these lines to date. A crude argument has been presented[85] that relates the elements of the friction tensor to the angle dependence of the intermolecular potential $V(\mathbf{R}^N)$; since it seems likely that the memory function approach should lead to an improved derivation (and possibly an improved result as well), we note only that the final result is

$$\xi_{ii}^2 = \frac{2I}{\pi} \left\langle \frac{\partial^2 V(\mathbf{R}^N)}{\partial \psi_i^2} \right\rangle \tag{3.83}$$

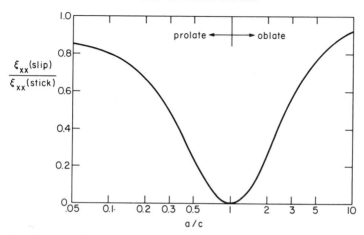

Fig. 10.   The ratio of the rotational friction constant ξ_{xx} for a symmetric rotator for slip and for stick boundary conditions at the surface of the rotator is shown as a function of the axial ratio c/a for the body. The friction constant refers to reorientation of the symmetry axis; in this model, the friction constant is zero for reorientation about the symmetry axis.

Qualitatively, this expression has the correct form in that it predicts that a component of the tensor will approach zero as the changes in the potential energy with angular changes about the axis of interest approach zero. However, no quantitative calculations based on (3.83) have been presented to date.

IV. EXPERIMENTAL EXAMPLES

No comprehensive review of the literature will be attempted here; rather, a relatively small number of systems will be chosen for discussion using the criteria that (a) the experimental data be extensive; (b) theoretical analysis of the data exist; (c) the results be illustrative of the models discussed in Sections II and III.

For example, the experimental studies and theoretical analyses of orientational relaxation in a class of relatively large and anisotropic molecules such as quinoline, nitrobenzene, and anisole[23,99] will not be discussed in detail here because it is now apparent that the behavior of these systems is too complex to be represented by any model of single-molecule rotation in the fluid. In general, two quite distinct relaxation times can be extracted from the depolarized light scattering for these liquids; often, it is found that the spectra are κ-dependent, which is interpreted as meaning that orientational relaxation can propagate as a wave, at least over a fraction of a wavelength. The physical picture deduced from these features (together with the detailed theoretical analyses[24]) is that the rotational motion of a molecule in these fluids is both strongly hindered and strongly coupled to that of its neighbors; consequently, a molecular theory for such fluids would have to deal explicitly with the correlated rotations that have been neglected in the models treated here.

Systems which can be adequately described by single-molecule rotation include methane, benzene, and possibly methyl iodide.

A. Methane

There is no doubt that the rotation of CH_4 and its deuterated isomers has been more extensively studied to date than any other case. Experimental data include: nmr T_1 values for all four species CH_iD_{4-i} in the liquid,[100,101] in the gas,[102,103] and in liquid[101] and gaseous mixtures;[104] inelastic neutron scattering for CH_4[105] and CD_4[106] in the liquid; infrared vibration–rotation bands for methane in the compressed gas, pure and in mixtures,[107,108] and for CH_4 and CD_4[109] as well as CH_3D and CHD_3[110] in the liquid, pure and dissolved in various solvents. Furthermore, extensive theoretical analyses of these data[100–115] have been presented which provide strong support for the conclusion that rotation of methane is only slightly hindered in the liquid or the compressed gas. In fact, the motion is sufficiently close to free rotation to require that an extended diffusion model be used rather than one of the stochastic models. Of course, an intermolecular potential function with tetrahedral symmetry would not be expected to cause much hindrance to rotation; on the other hand, the free rotation of CHD_3 or CH_3D occurs about a center-of-mass that is not coincident with the center of symmetry of the interaction, and thus might be expected to give rise to a larger "collision frequency" β^* than for CH_4 and CD_4, if the extended diffusion models are used to interpret the data.

Figure 11 shows experimental rotational correlation functions for CH_4 in the pure liquid obtained by calculating the cosine transforms of infrared and Raman spectra; the filled circles represent some relatively early work of Gordon,[114] and the open circles represent a recent calculation of McClung.[111] The differences between the two experimental $C_1(t)$ are minimal. Dashed lines show theoretical curves calculated from the J-diffusion model for two values of β^*; the M- and J-diffusion models do not differ appreciably at such small values of β^*. It is evident that the theoretical correlation functions are not very sensitive to the value of β^* chosen, and that no single value of β^* will reproduce both $C_1(t)$ and $C_2(t)$; nevertheless, $\beta^* = 0.75$ is a reasonably good choice for fitting both correlation functions.

The reduced zeroth moments of these correlation functions are listed as correlation times τ_l^* in Table II, together with values obtained by McClung from correlation functions for CH_4 in liquid solutions and CD_4, pure and in solution. It is of interest to note that the similarity of the *reduced* correlation

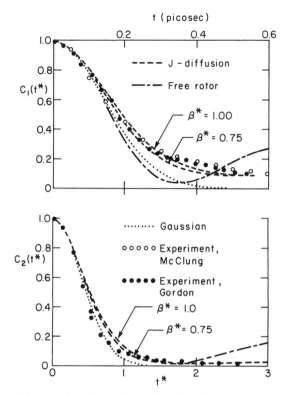

Fig. 11. Correlation functions for methane molecules in the liquid. The points are derived from experimental infrared and Raman bands; the lines shown for each value of *l* are given by various theoretical calculations, including exact (classical) free rotation; the Gaussian approximation to the exact curve; and two "best-fit" curves for the *J*-diffusion model. Reduced time t^* is shown on the bottom abcissa; time, in pico sec, is shown on the top.

times for CH_4 and CD_4 (in a given system) is an immediate indication that inertial motion is the dominant factor for the rotation of these molecules in the liquid. The moment of inertia which enters as part of the reducing factor plays no role in the purely stochastic models for reorientation, but gives reduced free rotor or inertial correlation functions that are invariant to changes in the moment of inertia for a spherical top.

McClung has fitted the experimental infrared correlation functions to the extended diffusion model, and has obtained good agreement. The values of β^* (or $1/\tau_L^*$, the reduced angular momentum correlation time) used are listed in Table II; the differences between his results and the first two entries in Table II for pure liquid CH_4 arise from the restriction that the first two correspond to a theory that fits $C_1(t)$ and $C_2(t)$ simultaneously. However,

TABLE II

Correlation Times, Collision Frequencies, and Orientational Correlation Times Calculated for Methanes in the Liquid from the Extended Diffusion Models

System	l	τ_l^*	$\beta^* = \tau_L^{*-1}$	$\pi\sigma_{CH_4-X}^2$ (Å^2)
CH_4, liquid, 90°K	1	1.3	0.75	8
CH_4, liquid, 98°K	2	0.5	0.75	8
CH_4, liquid, 117°K	1	1.3	1.2	12
CH_4, 15% in liquid Ar, 111°K	1	1.7	1.3	15
CH_4, 15% in liquid Kr, 135°K	1	1.8	1.4	18
CH_4, 15% in liquid Xe, 163°K	1	1.9	1.2	16
CD_4, liquid, 120°K	1	1.6	2.0	16
CD_4, 0.2% in liquid Ar, 91°K	1	2.0	2.2	20
CD_4, 0.2% in liquid Kr, 121°K	1	2.0	2.0	20
CD_4, 0.2% in liquid Xe, 163°K	1	1.6	1.4	16
CH_3D, in liquid CH_4, 91°K	1^a	1.2	1.1	
CH_3D, in liquid CH_4, 105°K	1^a	1.2	1.0	
CD_3H, in liquid CF_4, 135°K	1^a	1.4	1.0	

^{a} Reorientation of the C_3 axis.

the experimental curves can be reasonably well reproduced by either set of values. In general, it can be seen that the choices for the best-fit collision frequencies indicate that rotation in these liquids is only slightly hindered. This conclusion is supported by the fact that the free rotor and the Gaussian curves shown in Fig. 11 agree with experimental correlation functions for times up to $t^* \simeq 1$.

Equation (2.58) can now be used to estimate the "collision" cross-section for randomization of the angular momentum vector for a methane molecule. For an infinitely dilute solution of the spherical molecule, A dissolved in liquid B (also spherical molecules), (2.58) becomes

$$\beta_A^* = \rho_B \pi \sigma_{AB}^2 \left(\frac{8I_A}{\pi\mu_{AB}}\right)^{1/2} g(\sigma_{AB}) \qquad (4.1)$$

where μ_{AB} = the reduced mass of the AB pair and $\rho_B g(\sigma_{AB})$ is the number of B molecules per unit volume at a distance σ_{AB} from a molecule of A. Values of the collision cross-section were calculated for the solutions from (4.1) and for the pure liquids from (2.58), using $\rho_B g(\sigma_{AB}) = 0.1 \text{ Å}^{-3}$ in all cases. The results of this crude calculation are also listed in Table II, where it can be seen that the cross-section is roughly $8 \pm 2 \text{ Å}^2$, which is considerably smaller than the usual gas kinetic cross-section of $\sim 45 \text{ Å}^2$. Maryott, Malmberg, and Gillen[115] have analyzed the data for methane as well as a number of other spherical tops and have argued that this discrepancy indicates that a rotational collision number Z_{rot} should be introduced in these systems which is a measure of the ratio of the number of collisions per unit time needed to relax the angular momentum to the total number. Using somewhat different estimates of τ_L^* and σ, they obtain a collision number of 2.8 for methane; the analysis given here leads to a value of roughly 5. In view of the simplifications associated with the hard-sphere representation of methane molecules for the purposes of calculating "collision cross-sections" in the liquid, this difference is not significant; indeed, we will see that the cross-sections for gas-gas collisions are quite different from these estimates of the value in the liquid.

Figure 12 shows some infrared correlation functions for CH_4 plotted on a logarithmic scale; at long times, the curves show the exponential decay which

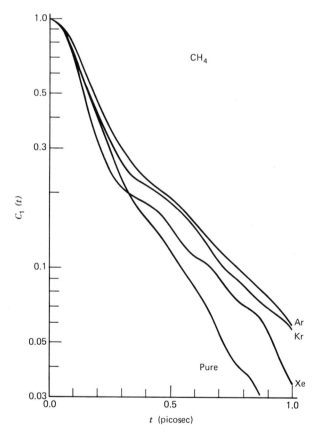

Fig. 12. Rotational correlation functions on a logarithmic scale for methane in the pure liquid and in several liquid rare gas solutions. All curves were obtained by Fourier transforming infrared bands.

is characteristic of a simple Markoffian mechanism for reorientation; on the other hand, the J-diffusion model also gives exponential correlation functions at long times, even for small values of β^*, so this observation is not in itself an indication that a stochastic model gives an adequate description of these systems.

An extensive analysis of both parallel and perpendicular infrared rotational bands for CHD_3 and CH_3D has been presented by Marsault, et al.,[110] who found the same general behavior for these symmetric top molecules in a variety of liquid solutions as that deduced for CH_4 and CD_4. The extended diffusion models gave curves that fit many of their experimental correlation functions with values of β^* of the order of unity. Even where detailed fit

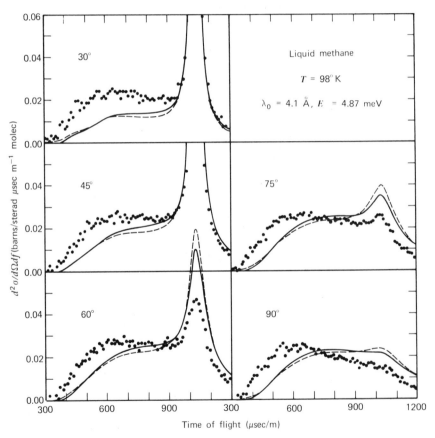

Fig. 13. Comparison of the observed and calculated time of flight distributions for neutrons of 4.1 Å wavelength scattered at various angles from liquid CH_4. The solid and dashed lines are different models of the translational motion (the dashed line is for simple diffusion); in both cases, the experimental correlation functions for $l = 1$ and 2 shown in Fig. 12 were used to calculate the rotational contribution to the scattering distribution.

of the extended diffusion models was not obtained, the general nature of the correlation functions and the values of the correlation times show quite clearly that rotation in these fluids is nearly free. A few of the correlation times obtained in this study are listed in Table II.

It is interesting to note that the inelastic neutron scattering data for liquid CH_4 and CD_4 has successfully been interpreted by substituting the *experimental* $C_1(t)$ and $C_2(t)$ given in Fig. 11 into (1.75). Figure 13 shows a comparison between the experimental inelastic scattering and two theoretical curves calculated by Sears[113] using different models for translation; the agreement is reasonably good, and can be noticeably improved if one includes the terms involving $C_3(t)$ and $C_4(t)$ in the rotational contribution; this has been done[106] by assuming that these functions are given by the Gaussian approximation shown for C_1, C_2 in Fig. 11 and for C_1 through C_4 in Fig. 1. In addition to improving the agreement for CH_4, this was the approach used to construct the theoretical curves shown for liquid CD_4 in Fig. 14. It can be seen that agreement between theory and experiment is quite good.

Arriving at a satisfactory understanding of the extensive measurements of nuclear spin relaxation times in the liquefied methanes has been a protracted process. In the first place, the early measurements of relaxation time were in error because of the strong effect of minute quantities of dissolved O_2 upon the magnetic relaxation.[100] Subsequent to this, interpretations of the data were put forth that relied in part upon some rather intuitive techniques for separating the contributions to $1/T_1$ due to intermolecular spin-spin motion (primarily due to translational displacements), to intramolecular spin-spin motion (reorientation) and to spin-rotation relaxation. Although an estimation of the translational term in $1/T_1$ could be obtained fairly reliably, either by dilution in a nonmagnetic solvent, or by using the standard theoretical expressions for this term, it proved to be considerably more difficult to sort out the intramolecular dipolar relaxation and the spin-rotational relaxation.

More recent studies have been aided by the fact that the elements of the spin-rotation tensor $\mathbf{C}$ are known for methane and for some of its deuterated isomers. If the tensor $\mathbf{C}$ is a scalar times $\mathbf{1}$, the spin-rotation contribution to the relaxation depends solely upon τ_L, the angular momentum relaxation time; one can then show that[103]

$$\frac{1}{T_1^{S-R}} = 0.056 T \tau_L \quad (CH_4) \tag{4.2}$$

$$= 0.040 T \tau_L \quad (CHD_3) \tag{4.3}$$

with τ_L in picoseconds. The presence of unequal diagonal elements in $\mathbf{C}$ adds extra terms to (4.2) and (4.3) that involve cross-correlations between angular momentum and reorientation such as those shown in (1.29); these

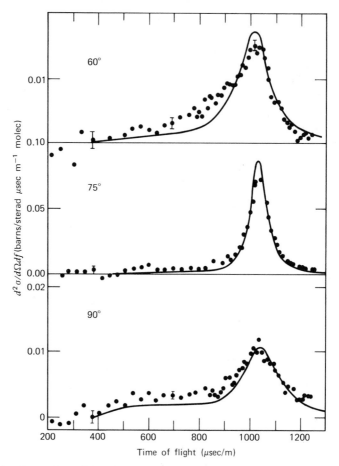

Fig. 14. Comparison of the observed and calculated time of flight distributions for liquid CD_4. The rotational contributions are calculated using the functions in Fig. 12, plus the Gaussian approximation for $l = 3$ and 4.

terms are generally approximated by decoupling the cross-correlations, an assumption which is valid only when τ_L^* is very short and the orientational correlation time τ_2^* is very long. Indeed, Hubbard[116] has noted that the rotational diffusion model and a decoupling assumption lead to

$$\tau_L^* \tau_2^* = \tfrac{1}{6} \qquad (4.4)$$

Although the extended diffusion models confirm this result at large β^*, explicit calculations show that (4.4) fails rather badly when β^* or $1/\tau_L^*$ is of the order of unity, as in the case of the liquid methanes. Consequently, a variety of experiments are needed to determine the spin-rotation and the rotational dipolar contributions to $1/T_1$ in these liquids.

However, there is one case where the spin-lattice relaxation is dominated by a single mechanism; this is deuteron relaxation, which is almost entirely due to the interaction of the nuclear quadrupole moment with the electric field gradient along a C—D bond. The deuteron relaxation times measured by deWit and Bloom[100] in liquid CD_4 and CHD_3 were found to be only slightly temperature-dependent and very nearly the same for the two molecules. For isotropic reorientation such as that expected in CD_4, (1.25) becomes

$$\left(\frac{1}{T_1^Q}\right) = \frac{3\pi^2}{2}\left(\frac{eqQ}{h}\right)^2 \tau_2 \tag{4.5}$$

At the time the experimental study was carried out, the quadrupole coupling constant eqQ/h was known only roughly; however, recent measurements on CHD_3[117] give 1.91×10^5 sec^{-1} for this constant; when this is combined with the experimental values of (T_1^Q) ranging from 10 sec at the melting point of $90°K$ to 12 sec at $110°K$, one finds

$$\tau_2 = 0.18 \text{ psec at } 90°K$$
$$= 0.15 \text{ psec at } 110°K$$

In reduced units, $\tau_2^* = 0.8$ at $110°K$. The curves relating correlation time to β^* for J-diffusion model given in Fig. 5 would indicate that τ_2^* should be 0.72 at $\beta^* = 2$ and 0.82 at $\beta^* = 3$, in agreement with the analyses of the infrared bands for this molecule that leads to a value of $\beta^* = 2$, as given in Table II.

With these indications that the proton relaxation time measurements are best interpreted in terms of the extended diffusion models rather than the rotation diffusion theory that has been so often applied to such data in the past, one can now attempt to estimate the spin-rotation and the intramolecular dipolar relaxation terms that contribute to the numbers obtained after the translational contribution has been subtracted away from the experimental $1/T_1$. Oosting and Trappeniers[103] have given a careful discussion of this problem and have interpreted their data in terms of an approximate inertial model[102] which is equivalent to taking only the zero-collision term in the extended diffusion models (but including quantization of rotational motion). As long as β^* or $1/\tau_L^*$ is small, the correlation times obtained from this truncated form of the theory will be reasonably close to those for the exact extended diffusion models, so that Oosting and Trappeniers' calculations should at least be indicative of the degree of consistency between the correlation times obtained from the proton relaxation data and the infrared band shapes. (These workers do show that an analysis based on the rotational diffusion models contains internal inconsistencies as well as yielding physically unrealistic values for the parameters of the model.)

Thus, we can concentrate on the inertial models. For the spherical top molecule CH_4, the intramolecular dipolar contribution to $1/T_1$ is then given by

$$\frac{1}{T_1^{DD}} = 7.7 \times 10^{10} \, \tau_2 \qquad (4.6)$$

We now have $(1/T_1^{DD})$ and $(1/T_1^{SR})$ (see (4.2), (4.3)) given in terms of two unknown correlation times; the third time that appears when the effects of asymmetry in the **C** tensor are included in (4.2) and (4.3) can be approximated by writing its reciprocal as $1/\tau_L + 1/\tau_2$. Oosting and Trappeniers estimated values of these two correlation times by assuming that the ratios of τ_L and of τ_2 for CH_4 and CHD_3 could be obtained from a theoretical argument; they then simultaneously fitted the theory to the data for both molecules. Their results for τ_2 and τ_L are of the correct order of magnitude for CH_4, but are somewhat more temperature-dependent than one would expect on the basis of the deuteron relaxation time in liquid CD_4. Therefore, we will present a slightly modified analysis that explicitly assumes that the extended diffusion models are valid in these fluids.

The first point to note is that $(1/T_1^{DD})$ is negligible in CHD_3 because of the weakness of the H—D spin–spin interaction. Thus, one can unambiguously extract $(1/T_1^{SR})$ for that molecule from the data. We will now attempt to predict the ratio of these relaxation times for CH_4 and CHD_3.

An approximation indicated by the extended diffusion models is to let $\tau_L \simeq \tau_2$ for the purpose of calculating the relatively small contribution of the anisotropic part to the spin-rotation relaxation. In that case, the expressions given by Oosting and Trappeniers become

$$\frac{1}{T_1^{SR}} = \frac{6.6 \times 10^8 \, T}{\beta_{CH_4}} \qquad \text{for } CH_4 \qquad (4.7)$$

$$= \frac{4.5 \times 10^8 \, T}{\beta_{CHD_3}} \qquad \text{for } CHD_3 \qquad (4.8)$$

where $\beta = 1/\tau_L$, as usual. In addition, the collision frequency given by (2.57) varies only with the square root of the mass of the methane molecules, so that

$$\frac{T_1^{SR}(CH_4)}{T_1^{SR}(CHD_3)} = 0.68 \left(\frac{m_{CHD_3}}{m_{CH_4}} \right)^{1/2} = 0.74 \qquad (4.9)$$

Figure 15 shows the experimental $1/T_1$ for liquid CH_4; the translational contribution calculated by Oosting and Trappeniers; the spin-rotation contribution, calculated from (4.9) and the spin-rotation times listed by

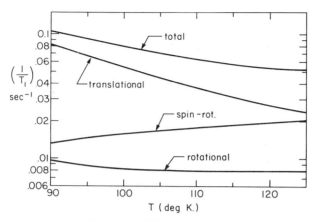

Fig. 15. Estimates of the contributions of different relaxation mechanisms to the proton spin-lattice relaxation time in liquid methane as a function of temperature.

Oosting and Trappeniers for CHD_3; and the rotational contribution to $1/T_1$ obtained by subtracting the translational and the spin-rotation terms from the total $1/T_1$.

The spin-rotation times shown in Fig. 15 are nearly a linear function of temperature T and give a collision frequency $\beta_{CH_4} = 4.1 \times 10^{12}$ sec^{-1} at 100°K, or $\beta^* = 0.8$. Considering the approximations made in obtaining this value, it is in reasonable agreement with McClung's choice of 1.2 to fit the infrared bands of liquid CH_4 at 111°K.

Finally, the rotational correlation times obtainable from the curve in Fig. 15 and (4.6) are also slowly varying functions of temperature, with $\tau_2 = 0.10$ psec, or $\tau_2^* = 0.5$. This value agrees reasonably well with the reduced correlation time for CD_4 deduced from the quadrupolar relaxation ($\tau_2^* = 0.8$) and extremely well with the correlation time obtained from the Raman spectrum of CH_4 (given as $\tau_2^* = 0.5$ in Table II).

The general conclusion to be drawn from this analysis seems clear: methane and its various deuterated isomers rotate relatively freely in the liquid. Analyses of the measured infrared and Raman spectra, of the inelastic neutron scattering, and of the proton and deuteron relaxation times all are consistent with this interpretation, and indicate that the angular momentum correlation decays with a characteristic time which is approximately equal to the time required for a 360° rotation of a free molecule. Alternatively, the rate at which collisions occur that lead to appreciable changes in angular momentum is approximately the same as the rate of free rotation; estimated values of the cross-section for these collisions are much smaller than the gas kinetic cross-section, indicating that only about one fourth of the actual binary encounters in the liquid affect the rotational motion of these molecules.

Analysis of the data concerned with the rotational motion of methane in the gas phase is noticeably simpler than in the case of the liquid data. In part, this is due to the fact that the binary collision model assumed in the extended diffusion theory becomes more realistic as the density decreases; indeed, the low-density data are quite clearly appropriate for the slightly perturbed free rotor that is generated by the extended diffusion models when β^* becomes small compared to unity. In addition, interpretation of the nuclear spin relaxation times is simplified because several of the contributions that are important in the liquid become so small in the gas that they can either be neglected, as in the case of the translational term in $1/T_1$ for all the methane molecules, or can be readily estimated to sufficient accuracy, as in the case of the rotational spin-spin dipolar relaxation term in $1/T_1$ for deuteron relaxation.

The general shape of the rotational correlation functions calculated from the Fourier transforms of a vibration–rotation band in methane gas mixed with various densities of nitrogen is shown in Fig. 16. The effect of an increase

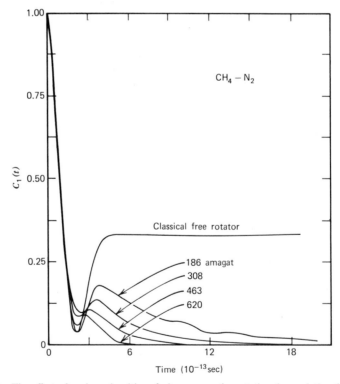

Fig. 16. The effect of various densities of nitrogen on the rotational correlation function of methane in the gas phase.

in the collision frequency at higher density is to produce an increase in the decay constants for the roughly exponential decays in the long-time tails of the correlation functions and is qualitatively quite similar to the theoretical curves shown in Figs. 2–4 for the extended diffusion models. Indeed, Eagles and McClung[112] have presented a detailed comparison of the band shapes for CH_4 in mixtures of He and N_2 with the semiclassical extended diffusion models. They find very good agreement at all densities and obtain reduced collision frequencies β^* from the fitting process shown in Fig. 17. It can be seen that the collision frequencies are linear functions of the density, as one might expect from a binary collision model. However, the best straight lines drawn through the points do not necessarily pass through the origin, as predicted from a binary collision model for β. In fact, the nonzero intercepts

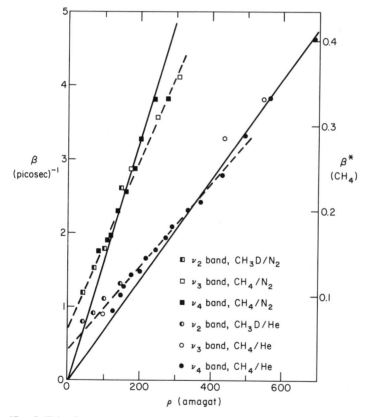

Fig. 17. Collision frequencies calculated from the best fit of spectra for the J-diffusion model to the experimental infrared bands for gaseous methane with added gases at room temperature and the indicated densities. Reduced collision frequencies are shown along the right-hand ordinate and unreduced β, along the left-hand.

for the dashed lines in Fig. 17 indicate that some source of broadening may be present in the vibration–rotation bands that (a) is independent of density and (b) has not been taken into account in the theoretical analysis. Some of the most likely sources of such broadening are: finite spectral slit widths in the spectrometer, hot bands, isotope effects on the vibrational frequency due to ^{13}C in natural abundance, and translational Doppler broadening. In fact, the net effect is probably a convolution of all these factors. This argument leads to the conclusion that the dashed lines in Fig. 17 may well be correct; if the slopes of these lines are used to calculate collision cross-sections, together with the assumption that $g(\sigma_{AB}) = 1$ in the gas, (2.57) gives

$$\pi(\sigma_{CH_4-N_2})^2 = 57 \text{ Å}^2$$
$$\pi(\sigma_{CH_4-He})^2 = 16 \text{ Å}^2$$

(Inasmuch as the results for CH_3D are also represented by these lines, the collision cross-sections include the CH_3D-N_2 and CH_3D-He pairs as well.)

The large difference between the cross-sections for collision with a nitrogen molecule and with a helium atom is unexpected; Lévi and Chalaye[107] have also determined β as a function of density for CH_3D in mixtures with CH_4, O_2, H_2, Ar, and Ne as well as N_2 and He. The collision cross-sections that can be estimated for these systems are:

$$\pi(\sigma_{CH_3D-Ne})^2 = 22 \text{ Å}^2$$
$$\pi(\sigma_{CH_3D-CH_4})^2 = 57 \text{ Å}^2$$
$$\pi(\sigma_{CH_3D-O_2})^2 = 52 \text{ Å}^2$$
$$\pi(\sigma_{CH_3D-Ar})^2 = 44 \text{ Å}^2$$

These values are consistent with the N_2 and He cross-sections, and indicate that moderate increases in the strength of the methane-foreign molecule interaction cause the cross-section to increase dramatically. The potentials for methane interacting with N_2, CH_4, O_2, and Ar are rather similar, and it is encouraging to see that the cross-sections are also similar for these pairs. Conceivably, a collision with a rare gas atom is less effective in altering the angular momentum of a methane molecule than in the case of a nonspherical collision partner of similar interaction strength. However, the uncertainties in the calculated cross-sections are too large to give more than a rough indication that such an effect may be present. Detailed computations of the dynamics of a tetrahedral methane colliding with a foreign molecule would be extremely helpful in rationalizing these values for the collision cross-sections.

It should be noted that the gas-phase binary cross-sections are larger than the ones deduced for the relaxation of angular momentum in the pure liquid. In part, this may be due to the choice of $g(\sigma)$ in the liquid, as mentioned previously. On the other hand, a binary collision model is not necessarily the best representation of the dynamics in a liquid. An alternative approach based on the ideas behind the rotational Langevin equation is perhaps more plausible. In this case, it is postulated that a fluctuating torque is exerted upon the central molecule which is the net torque due to all the intermolecular interactions. Since the molecules tend toward symmetrical distributions in a fluid in order to make the ensemble average torque on a molecule be zero, it can be argued that the frequency of occurrence of a large fluctuation in torque may actually be smaller in a liquid than in the gas. This implies that the "collision" model put forth as a justification for the extended diffusion theory may not be applicable in the liquid, even though the extended diffusion theory itself may still be applicable.

The nuclear magnetic relaxation times in gaseous methane yield constant ratios for β/ρ, and give values of the collision cross-sections that are consistent with those obtained from the analysis of the spectra. Consider first the deuteron relaxation times in CD_4 measured by Bloom, Bridges, and Hardy.[102] It is easy to demonstrate that the quadrupolar coupling between the deuteron spin and the electric field gradient along the C—D bond is the only important mechanism for relaxation in CD_4 gas. Thus, (4.5) yields the entire $1/T_1$; after substituting for the quadrupole coupling constant, one obtains

$$\left(\frac{1}{T_1}\right)_{CD_4\,gas} = (0.54 \times 10^{12}\ sec^{-2})\tau_2 \qquad (4.10)$$

We now wish to evaluate the correlation time τ_2 for the extended diffusion model in the limit of $\beta^* < 1$, since this is the range of interest (in the gas) according to Fig. 17. Thus, we are interested in the extreme left-hand side of Fig. 5. We limit the calculation to the spherical top molecule, for simplicity, and use the collisional expansion for the extended diffusion models that is given in (2.29). For sufficiently small β, the integral of the correlation function that gives the correlation time is determined almost entirely by a slowly decaying long-time tail of the correlation function which has been shown[46] to be (for the M- and J-diffusion)

$$C_l(t) = \frac{1}{\langle (d_{00}{}^l(x))^2 \rangle} \exp\{-\beta^* t^* [1 - \langle (d_{00}{}^l(x))^2 \rangle]\} \qquad (4.11)$$

where $x = \cos\theta_L$, with θ_L the polar orientation angle of the body-fixed angular momentum vector in a freely rotating symmetric top. In the limit of small β^*, the integral of (4.11) gives the correlation time τ_l^* as

$$\tau_l^* = \frac{\langle (d_{00}{}^l(x))^2 \rangle}{\beta^*[1 - \langle (d_{00}{}^l(x))^2 \rangle]} \qquad (4.12)$$

For a spherical top, $\langle (d_{00}{}^l(x))^2 \rangle = 1/(2l + 1)$, and (4.12) becomes

$$\tau_l^* = \frac{1}{2l\beta^*} \qquad (4.13)$$

(This result differs from that given by the model used by Bloom et al.,[102] the difference is due to the fact that all collisions contribute to τ_l^* even at small β^* (because the integration over t^* always includes times that are long compared to the inverse of the collision frequency β^*) whereas Bloom et al., only calculate the "zero-collision" part, that is, $\exp(-\beta^*t^*)$ times the free-rotor $C_l(t^*)$.) Since the experimental results can be expressed as

$$(T_1)_{CD_4gas} \text{ (sec)} = 0.1 \ \rho \text{ (amagat)}$$

the complete collisional theory gives

$$\beta \text{ (sec}^{-1}) = 1.4 \times 10^{10}\rho \text{ (amagat)} \qquad (4.14)$$

The rate of change of β with ρ calculated from Lévi and Chalaye's analysis[107] of the spectra for CH_3D/CH_4 mixtures is 1.2×10^{10} (amagat sec),$^{-1}$ in agreement with (4.14).

Proton relaxation times in gaseous CH_4 have also been determined.[102,103] Translational contributions to $1/T_1$ are negligible in this system, as is generally the case in gases. Equations (4.6) and (4.7) give $1/T_1$ for the two remaining mechanisms for relaxation in gaseous CH_4, and (4.13) gives the relation between τ_2 and β for a spherical top molecule in the gas. Thus,

$$\left(\frac{1}{T_1}\right)_{CH_4gas} = \frac{1.6 \times 10^{10}}{\beta} + \frac{20.0 \times 10^{10}}{\beta} \qquad (4.15)$$

at $T = 300K$, where both the rotational and the spin-rotation terms have been given explicitly in order to show that the spin-rotational relaxation mechanism is the dominant one in the gas (where β is small enough to allow one to use (4.13)). Bloom et al.[102] give

$$(T_1)_{CH_4gas} = 0.02\rho \text{ (sec/amagat)} \qquad (4.16)$$

at 300 K, which agrees well with Oosting and Trappeniers' measurements.[103] Equation (4.15) now yields

$$\beta \text{ (in sec}^{-1}) = 0.4 \times 10^{10} \ \rho \text{ (amagat)} \qquad (4.17)$$

for CH_4 colliding with CH_4. This result was also obtained by Oosting and Trappeniers by a slightly different argument; although it is of the correct order of magnitude, the slope of β versus ρ is too small by a factor of three relative to the two other estimates of this quantity. It is possible that the assumptions made in deriving (4.7) are in error; in particular, the contribution of the term in $(\Delta C)^2$ to the overall spin-rotation relaxation is somewhat uncertain, in part because of uncertainty in the experimental ΔC value for CH_4, and in part because of the difficulty of calculating the cross-correlation function for this term [shown in (1.29)]. However, McClung has performed the spherical top calculation for the extended diffusion models,[118] and has shown that the exact model is only slightly different from the approximate calculation that leads to (4.7), in the limit of small β. Consequently, the reason for the discrepancy must be sought elsewhere.

Up to this point, we have dealt only with the magnetic relaxation times for the methanes measured at a single temperature (300 K). In fact, experimental data cover the range from 110 to 300 K, and indicate that the CH_4 relaxation becomes considerably more efficient as the temperature decreases (at fixed density) whereas the deuteron relaxation is almost independent of T. The rise in T_1 for CH_4 gas at low temperature has been interpreted[102,103] as an indication that weak long-range interactions are effective in relaxing the angular momentum, especially when the molecules are moving slowly.[119] This could have a noticeable effect upon the time dependence of the spin-rotation term which is the dominant factor for CH_4 relaxation, but need not affect the reorientational motion that determines the correlation function for quadrupolar relaxation in CD_4. If these arguments are correct, it implies that the collisional description of the extended diffusion model needs to be modified somewhat to account for all the details of the gas phase magnetic and spectral data of these systems, including not only the temperature dependence but the values of the nuclear relaxation times obtained for the partially deuterated methanes.

B. Benzene

Our discussion of the rotational motion of methane has been greatly simplified by the assumption that both CH_4 and CD_4 are isotropic rotors, which is of course justified by consideration of both the moments of inertia and the symmetry of the pair-wise intermolecular interaction energy. Furthermore, the proton in CHD_3 and the deuteron in CH_3D lie on the inertial symmetry axes; we have avoided explicit discussion of other cases such as rotational relaxation of CH_2D_2, of the deuterons in CHD_3 or the protons in CH_3D which might be complicated by anisotropy in the reorientation of these molecules.

The benzene molecules C_6H_6 and C_6D_6 are examples of systems where anisotropy in the rotation might be expected to be an important factor. On the other hand, the shapes of these molecules are reasonably symmetric, so that one would not anticipate large hindrances to rotation in the liquids, particularly for rotation about the six-fold symmetry axes perpendicular to the molecular planes. It is obvious that there are two distinctly different kinds of molecule-fixed vectors in these systems:

1. Vector "p" perpendicular to the plane of the molecule; this includes the axis of permanent optical anisotropy, whose rotation is measured in depolarized light scattering experiments. In addition, the transition dipole moments or polarizability anisotropies associated with some of the normal modes of vibration are "p" vectors. In Herzberg's[120] notation, the selection rules for D_{6h} symmetry give vibrational bands corresponding to out-of-plane transition moments which are

$$\nu_4 \ (A_{2u}, \text{ infrared active})$$

$$\nu_{11} \ (E_{1g}, \text{ Raman active, depolarized})$$

2. Vector "i" lying in the molecular plane; this includes the vectors for magnetic relaxation, such as the magnetic dipole–dipole vector between pairs of protons that is the principal term in rotational proton relaxation; the dipole–dipole interaction between ^{13}C and an attached proton that is the primary mechanism for ^{13}C relaxation; the quadrupole-field gradient interaction vector along the C—D bond direction that causes deuteron relaxation; and a number of spectral bands corresponding to in-plane vibrational transition moments, including those associated with the normal modes ν_1, ν_2 (A_{1g}, Raman active, polarized), ν_{12}, ν_{13}, and ν_{14} (E_{1u}, infrared active), and ν_{15}, ν_{16}, ν_{17}, and ν_{18} (E_{2g}, Raman active, depolarized).

The pertinent nuclear relaxation times have long been known for liquid benzene; indeed, one of the first indications[121] that the standard Bloembergen–Purcell–Pound theory of proton relaxation was in error because of its use of the Stokes–Einstein theory for estimating the rotational diffusion constant arose from the anomalously long proton relaxation times for benzene, especially those measured at infinite dilution in a nonmagnetic solvent after careful removal of dissolved oxygen. In addition, Raman bands and depolarized light scattering have been measured, as well as the inelastic neutron scattering from both C_6H_6 and C_6D_6.

It turns out that the spectral band widths are too small for these molecules in the liquid to allow one to obtain accurate rotational correlation functions for this molecule. The moment of inertia and/or the hindrance to rotation is large enough to yield bands that are only a few cm^{-1} in width (or a few meV, in the case of neutron scattering); over such narrow frequency ranges, a variety of other comparable sources of broadening are present which complicate the extraction of rotational data from the measurements. After careful analysis, it has proved to be possible to estimate the rotational line broadening; however, at this level of accuracy, one is realistically limited to calculations of the rotational correlation times, which are equal to the reciprocal of the rotational half-widths at half-height, measured in units of radians/sec. We have argued that two types of correlation time may be present: those for a vector perpendicular to the molecular plane ($\tau_l^{(p)}$), and those for an in-plane vector ($\tau_l^{(i)}$). Assuming only that the symmetry of the molecule is sufficient to give equivalent rotational motion about any axis lying in the plane of the molecule, (1.21) can be evaluated for $\eta = 90°(i)$ and for $\eta = 0°(p)$ to give

$$\tau_1^{(i)} = \tau_{11} \tag{4.18}$$

$$\tau_2^{(i)} = \tfrac{1}{4}\tau_{20} + \tfrac{3}{4}\tau_{22} \tag{4.19}$$

$$\tau_1^{(p)} = \tau_{10} \tag{4.20}$$

$$\tau_2^{(p)} = \tau_{20} \tag{4.21}$$

where

$$\tau_{lm} = \int_0^\infty \langle D_{mm}^{\,l}(\delta\Omega) \rangle \, dt$$

Evidently, explicit evaluation of the τ_{lm} will require that a model be introduced; however, it is helpful to list some of the values for the correlation times that have been obtained experimentally.

The rotational part of the proton spin-lattice relaxation time has been determined by measuring T_1 for C_6H_6 at infinite dilution in C_6D_6;[122] a value of 60 sec at 20 C obtained in early work has been superceded by the more recent measurements which give 103 sec at 25 C. Both results indicate that T_1 in pure C_6H_6 is dominated by translational relaxation, since the measured $T_1 = 20$ sec.[123,124] Assuming that spin-rotation contributions to $(T_1)_{rot}$ are negligible, one finds

$$\frac{1}{103} = 3\frac{\gamma^4\hbar^2}{b^6}\tau_2^{(i)} \tag{4.22}$$

where it has been assumed that the only important proton–proton interactions are with the pair of nearest neighbors to a proton in the C_6H_6 ring. Since this gives the separation distance $b = 2.47$ Å, one finds $\tau_2^{(i)} = 1.28$ psec from the proton spin relaxation data. The ^{13}C spin-relaxation time is 29 sec,[124,125] and measurements of the nuclear Overhauser effect show that the dipole–dipole contribution to $(T_1)_{13C}$ is 37 sec. Thus, one can write

$$\left(\frac{1}{T_1^{DD}}\right)_{13C} = 2.2 \times 10^{10}\tau_2^{(i)} \tag{4.23}$$

or $\tau_2^{(i)} = 1.2$ psec. The deuteron relaxation time C_6D_6 is 1.6 sec,[126,127] and is related to the correlation time by

$$\left(\frac{1}{T_1^Q}\right)_D = \frac{3\pi^2}{2}\left(\frac{eqQ}{h}\right)^2 \tau_2^{(i)} \tag{4.24}$$

If one takes $eqQ/h = 1.8 \times 10^5$ Hz as a typical value for CD bonds, one finds $\tau_2^{(i)} = 1.3$ psec. It seems that a value of 1.2 ± 0.1 psec is the best estimate for $\tau_2^{(i)}$ at the present time.

It is possible that the correlation times for C_6H_6 and C_6D_6 will differ because of mass differences; however, even in the inertial models where this effect is a maximum, the correlation times should vary as the ratio of the square roots of the moment of inertia, and this is negligibly different from unity, for the purposes of this analysis.

Reorientation of the axis perpendicular to the plane of the molecule has been studied by measuring the depolarized light scattering[128–131] and the shape of the v_2 Raman band for C_6H_6[127,132] and for C_6D_6.[132]

Figure 18 shows the v_2 band for C_6D_6 together with contributions due to a hot band, and an isotopically split vibrational mode. After subtracting this intensity the remaining polarized intensity $I_{pol}(\omega)$ can be written

$$I_{pol}(\omega) = I_{iso}(\omega) + \tfrac{4}{3}I_{aniso}(\omega) \tag{4.25}$$

The depolarized intensities $I_{dep}(\omega)$ for the v_2 band are directly equal to $I_{aniso}(\omega)$. This intensity is a convolution of the vibrational line shape, the spectral slit function, and the orientational spectrum; since $I_{iso}(\omega)$ is a convolution of the vibrational line and the spectral slit function (translational broadening is negligible on the frequency scale of these experiments), one can combine the data for the polarized and depolarized bands to extract sufficient information to estimate at least the orientational line width. In this way, Gillen and Griffiths found[127]

$$\tau^{(p)} = 3.1 \text{ psec} \quad (C_6D_6 \text{ at } 25 \text{ C})$$
$$= 2.8 \text{ psec} \quad (C_6H_6 \text{ at } 25 \text{ C})$$

In calculating $\tau^{(p)}$ from the experimental band widths, it has been assumed that the effects of collision-induced depolarized light scattering upon the band shape can be neglected; since this contribution to the scattering generally occurs with spectral half-widths of 20 to 40 cm^{-1}, the true reorientation times $\tau^{(p)}$ may be somewhat longer than the values quoted.

Depolarized light scattering measurements have yielded values of $\tau^{(p)}$ equal to 3.1,[129] 3.3,[130] and 3.8[131] psec (uncorrected for collision-induced scattering) and 2.6 psec, based on a decomposition of the frequency-dependent scattering into two Lorentzians, and association of the narrow one with reorientational

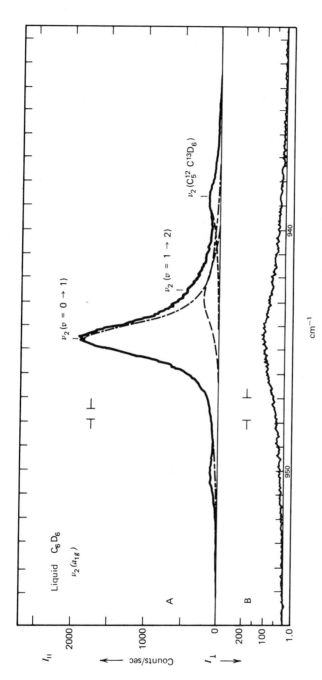

Fig. 18. Photon-counting intensities of the polarized (I_{11}) and the depolarized (I_1) Raman spectra for C_6D_6 at room temperature; the estimated hot band and isotopic band intensities are sketched; spectral slit widths for the two spectra are also indicated.

motion.[128] Consequently the best estimate of $\tau^{(p)}$ at present is 2.8 ± 0.2 psec for C_6H_6 in the liquid at 25 C; the time for C_6D_6 is perhaps 10% longer.

The values of $\tau^{(i)}$ and $\tau^{(p)}$ can now be used to estimate

$$\tau_{20} = 2.8 \pm 0.2 \text{ psec}$$
$$\tau_{22} = 0.8 \pm 0.3 \text{ psec} \tag{4.26}$$

for C_6H_6 in the liquid at $25°C$. The distinction between the times for C_6D_6 and C_6H_6 is not quantitatively significant at the present level of uncertainty and will be ignored.

The correlation times given in (4.26) indicate that the reorientation of benzene in the liquid is clearly anisotropic, with the rapid reorientation associated with rotation about the molecular symmetry axis, as might be expected.

In order to progress further, it is now necessary to introduce a model; in the absence of detailed experimental correlation functions, the criteria for a choice are almost entirely those of physical intuition. However, intuition is aided by a calculation of the reduced correlation times:

$$\tau_{20}^* = \tau_{20}\left(\frac{kT}{I}\right)^{1/2} = 4.7 \pm 0.3$$

$$\tau_{22}^* = \tau_{22}\left(\frac{kT}{I_{zz}}\right)^{1/2} = 1.0 \pm 0.3 \tag{4.27}$$

$(I_{zz} = 2I = 2.92 \times 10^{-38}$ g cm^2). The value of τ_{20}^* is large enough to suggest that small-step rotational diffusion of the molecular symmetry axis is occurring in the liquid; the correlation times for $l = 2$ shown in Fig. 5 for the extended diffusion models indicate that τ_{20}^* is well into the small-step region. Although the extended diffusion models could be used to analyze the benzene data in principle, they are presently incapable of representing anisotropic reorientation. The small-step model can be used, but it is necessary at least to include possible inertial effects upon rotation about the symmetry axis. Consequently, we will employ the small-step model in a form that allows for a time-dependent rotational diffusion constant; in the case where only rotation about the z-axis is fast enough for this to be important, (3.75) and (3.76) give

$$\langle D_{00}{}^2(\delta\Omega)\rangle = \exp\left[-6\mathcal{R}_{xx}t\right] \tag{4.28}$$

$$\langle D_{22}{}^2(\delta\Omega)\rangle = \exp\left[-(2\mathcal{R}_{xx} + 4\overline{\mathcal{R}}_{zz})t\right] \tag{4.29}$$

$$\overline{\mathcal{R}}_{zz} = \mathcal{R}_{zz} + \frac{I_{zz}\mathcal{R}_{zz}{}^2}{tkT}\left\{\exp\left[-\frac{kTt}{I_{zz}\mathcal{R}_{zz}}\right] - 1\right\} \tag{4.30}$$

(We omit the (∞) notation here for simplicity.) The correlation time τ_{20} is now directly given by

$$\tau_{20} = \frac{1}{6\mathscr{R}_{xx}} \qquad (4.31)$$

so that $\mathscr{R}_{xx} = 6 \times 10^{10} \text{ sec}^{-1}$ for benzene in the liquid at 25 C. The integration of (4.29) to obtain τ_{22} must be done numerically; τ_{22}^* is plotted in Fig. 19 as a function of $\mathscr{R}_{zz}^* = \mathscr{R}_{zz}(I_{zz}/kT)^{1/2}$ for the experimental $\mathscr{R}_{xx}^* = 0.05$ ($= \mathscr{R}_{xx}(I_{zz}/kT)^{1/2}$). The correlation time calculated in the asymptotic limit for small $\mathscr{R}_{zz}^*$ is also shown; it is calculated from

$$\tau_{22}^* = \frac{1}{2\mathscr{R}_{xx}^* + 4\mathscr{R}_{zz}^*} \qquad (4.32)$$

The limiting value of the inertial-corrected τ_{22}^* is 0.6, and corresponds to the case of the Gaussian approximation to free rotation about the z-axis combined with small-step reorientation of the z-axis with $\mathscr{R}_{xx}^* = \mathscr{R}_{yy}^* = 0.05$. Although this limiting value is too small to be in perfect agreement with experiment, it is nearly within the estimated uncertainties of the experimental

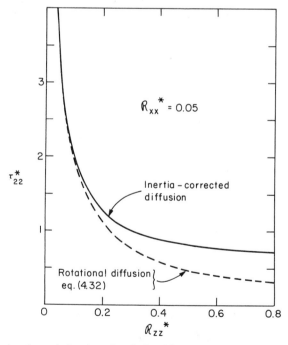

Fig. 19. Reduced correlation time τ_{22}^* calculated from (4.29) for a model of inertial rotation about the symmetry axis and small-step reorientation of the axis with a reduced diffusion constant $\mathscr{R}_{xx}^*$.

τ_{20}^* and τ_{22}^*, and is consistent with the physical picture of anisotropic rotation for these molecules in the liquid, with only slight hindrance to rotation about the molecular symmetry axis.

Two studies of the inelastic neutron scattering in liquid benzene have been reported;[133,134] in both cases, the intensity and line width of the "quasielastic" scattering were measured. The term "quasielastic" means the scattering at low-energy transfers due to translational and rotational motions and excluding scattered neutrons whose energy loss or gain is due to interaction with an internal vibration of the molecule. The dependence of the intensity upon the scattering parameter κ is of little interest in the present discussion because one learns only about static correlations from such data.[135] (In general, one has

$$\int_{\text{band}} I(\omega, \kappa)\, d\omega = \frac{1}{\pi} \int_{-\infty}^{\infty} \int_{\text{band}} C(t, \kappa) e^{i\omega t}\, dt\, d\omega \qquad (4.33)$$

$$= C(0, \kappa)$$

where $C(t, \kappa)$ is an unnormalized correlation function whose value at $t = 0$ is proportional to the equilibrium correlations present in the fluid.)

On the other hand, line width measurements can be quite useful, although they obviously do not contain as much information as a complete line shape study. For a Lorentzian line (in the absence of information to the contrary, this assumption is as good as any), the width at half-height $E_{1/2}$ is related to the molecular dynamics by

$$E_{1/2} = \frac{2\hbar}{\tau_N(\kappa)} \qquad (4.34)$$

just as in other band width studies; in fact, the features that distinguish neutron scattering widths from light scattering are primarily (a) a difference in selection rules; (b) the fact that light scattering is in practice limited to $\kappa = 0$ because the wavelength of visible radiation is large compared to correlation distances in fluids that are not in their critical region.

It is well known that the neutron scattering from hydrogeneous molecules is dominated by the incoherent proton scattering. If the coherent scattering in benzene is neglected, the nuclear correlation time in liquid C_6H_6 becomes

$$\tau_N(\kappa) = \int_0^\infty F_s(\kappa, t)\, dt \qquad (4.35)$$

where $F_s(\kappa, t)$ is given by (1.74); furthermore, the incoherent nuclear rotational correlation function defined in (1.75) can be written explicitly as

$$C_{\text{inc}}(t, \kappa) = [j_0(\kappa R_H)]^2 + 3[j_1(\kappa R_H)]^2 \langle D_{11}{}^1(\delta\Omega)\rangle$$
$$+ \tfrac{5}{4}[j_2(\kappa R_H)]^2 [\langle D_{00}{}^2(\delta\Omega)\rangle + 3\langle D_{22}{}^2(\delta\Omega)\rangle] + \cdots \qquad (4.36)$$

In the benzene experiments, $R_H = 2.47$ Å, the distance between a proton and the benzene center-of-mass. The correlation function in (4.36) can be calculated on the assumption that the model that was produced to explain the magnetic relaxation and the spectral data is correct; (3.75) and (3.76) then give the requisite time-dependent functions as

$$\langle D_{rr}^{\,l}(\delta\Omega)\rangle = \exp\left\{-\mathcal{R}_{xx}^*[l(l+1) - r^2]t^* - \frac{r^2 t^{*2}}{2}\right\} \qquad (4.37)$$

An expression for the translational correlation function is also required; here we use[136]

$$\langle e^{i\kappa\cdot\delta\mathbf{r}_{cm}(t)}\rangle = e^{-\kappa^2\rho(t)} \qquad (4.38)$$

$$\rho(t) = D\left[t - \frac{1}{\zeta\sqrt{3}}e^{-\zeta t}\sin\sqrt{3}\zeta t\right] \qquad (4.39)$$

where $\zeta = kT/2Dm$ and D is the macroscopic diffusion constant for benzene $(2.0 \times 10^{-5}$ cm^2/sec at 20°C) Linewidths calculated as a function of κ from (4.36)–(4.39) and (1.74) are compared with the data for C_6H_6 at room temperature in Fig. 20. There are two particularly noteworthy features to be seen: first, rotational motion of the protons has a large effect upon the width of the quasielastic scattering at all but the smallest κ; secondly, a relatively large number of terms must be included in order to calculate $C_{inc}(t, \kappa)$ for $\kappa > 1$ Å^{-1}; that is, (4.36) must be extended to at least the $l = 4$ term and possibly further. In their analysis of the C_6H_6 data, Winfield and Ross[133] give a clear demonstration that the reason for the discrepancy between computed curves similar to the one in Fig. 20 and the experimental data in the region where $\kappa = 1$ to 2 Å^{-1} arises from the neglect of coherent scattering in the calculation of the total cross-section; therefore, it may be concluded that the model that gives rise to (4.37) is quite adequate to describe the reorientation of benzene in the liquid at room temperature, if one uses a value for $\mathcal{R}_{xx}$, the small-step diffusion constant for reorientation of the symmetry axis, equal to $(6 \pm 1) \times 10^{-10}$ sec^{-1}.

It is of interest to compare the rotational diffusion constant deduced from experiment with that calculated from the Stokes–Einstein equation for an oblate spheroid undergoing Brownian rotation in a viscous liquid; we will estimate values using both the "stick" and the "slip" boundary conditions. Based on intermolecular distances in the crystal and van der Waals radii of the atoms, it seems reasonable to take the lengths of the molecular major and minor axes to be 7.4 and 3.7 Å, respectively. Equations (3.79) and (3.80) then give $\xi_{xx} = \xi_{yy} = 3.7 \times 10^{-23}$ g cm^2/sec for "stick" boundary conditions. Consequently, (3.77) gives $\mathcal{R}_{xx} = \mathcal{R}_{yy} = 0.9 \times 10^{10}$ sec^{-1}, which is a

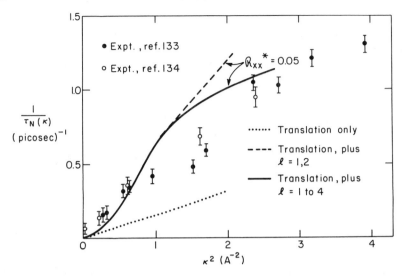

Fig. 20. Reciprocals of the quasielastic neutron scattering correlation time calculated from the spectral half-widths observed for liquid C_6H_6 are plotted as a function of the scattering parameter $\kappa = (4\pi/\lambda) \sin (\theta/2)$. The points show the experimental data; the curves are calculated for: translation only; translation plus rotational contributions for $l = 1$ and 2 for rotation about the symmetry axis plus small-step reorientation of the axis; translation plus the same model for the rotational terms but including all l values of 4 or less.

factor of 7 smaller than the value deduced from experiment. If it is assumed that "slip" rotation in a viscous fluid is a better model for rotational diffusion, one can use the curve in Fig. 10 together with an axial ratio of 2 for an oblate symmetric rotor to calculate $\mathcal{R}_{xx} = \mathcal{R}_{yy} = 3 \times 10^{10} \ \text{sec}^{-1}$, which is still too small by a factor of 2. This is a considerable improvement over the "stick" model; inasmuch as the calculated values vary with the cube of the molecular dimensions, a relatively minor change in the estimated size parameters can give complete agreement between the experimental value and the theory for a slip boundary condition.

The fact that the relaxation time τ_{20} can be interpreted in terms of a theory which makes it proportional to the viscosity of the fluid suggests a method for determining the importance of correlated rotations in this system.[129] If the depolarized light scattering is measured at varying benzene concentrations in an optically isotropic solvent (so that the solvent does not contribute to the scattered intensity), the orientational relaxation times obtained from the band width in the usual fashion will vary with concentration if benzene–benzene correlated rotations are present in addition to the expected linear variation with solution viscosity. In fact, the data show that changes in the benzene concentration do not affect the rotational correlation times, which is a strong indication that single-molecule models give an adequate description of rotation in this fluid.

C. Methyl Iodide

So far, we have discussed the reorientation of methane, a spherical top molecule with a weakly angle-dependent intermolecular potential function, and benzene, an oblate symmetric top with intermolecular interactions that lack dipolar terms and show almost no hindrance to rotation about the C_6 symmetry axis. In contrast, methyl iodide is a highly prolate symmetric top with an appreciable permanent dipole moment term in its intermolecular potential. It also happens that the studies of its reorientations in the liquid are among the most detailed and painstaking of any published to date. Even though inelastic neutron scattering data appear to be lacking in this case, measurements and analysis of nuclear spin relaxation times together with the infrared and Raman band shapes provide a relatively complete picture of the nature of the reorientation of this molecule. Some pertinent molecular parameters are listed in Table III.

TABLE III
Molecular Parameters for Methyl
Iodide

$$I = \begin{cases} 112 \times 10^{-40} \text{ g cm}^2 \text{ (CH}_3\text{I)} \\ 139 \times 10^{-40} \text{ g cm}^2 \text{ (CD}_3\text{I)} \end{cases}$$

$$I_{zz} = \begin{cases} 5.5 \times 10^{-40} \text{ g cm}^2 \text{ (CH}_3\text{I)} \\ 10.8 \times 10^{-40} \text{ g cm}^2 \text{ (CD}_3\text{I)} \end{cases}$$

C—H distance = 1.11 Å
C—I distance = 2.14 Å
H—H distance = 1.84 Å
H—I distance = 2.69 Å
HCH angle = 111.4°
ICH angle = 107.5°
eqQ/h (for deuterium) = 1.80×10^5 Hz

$$\begin{matrix} \sqrt{I/kT} \\ (300 \text{ K}) \end{matrix} = \begin{cases} 0.52 \text{ psec (CH}_3\text{I)} \\ 0.58 \text{ psec (CD}_3\text{I)} \end{cases}$$

Methyl iodide possesses three vibrational modes of symmetry A_1 and three doubly degenerate modes of E symmetry, and all are active in both the Raman and infrared. According to the theory summarized in Section I, Fourier transforms of the rotational spectral bands give correlation functions that can be denoted by $C_l^{(A_1)}(t)$ and $C_l^{(E_j)}(t)$ where $l = 1$ for infrared and 2 for

Raman bands and j denotes the jth degenerate mode. These functions are related to molecular reorientations by

$$C_1^{(A_1)}(t) = \langle D_{00}{}^1(\delta\Omega)\rangle \qquad (4.40)$$

$$C_1^{(E_j)}(t) = \langle D_{11}{}^1(\delta\Omega)\rangle \qquad (4.41)$$

$$C_2^{(E_j)}(t) = \frac{|\alpha_{xy}^{(E_j)}|^2\langle D_{11}{}^2(\delta\Omega)\rangle + |\alpha_{xz}^{(E_j)}|^2\langle D_{22}{}^2(\delta\Omega)\rangle}{|\alpha_{xy}^{(E_j)}|^2 + |\alpha_{xz}^{(E_j)}|^2} \qquad (4.42)$$

where $\alpha_{xy}^{(E_j)}$, $\alpha_{xz}^{(E_j)}$ are the derivatives of the polarizability elements with respect to the normal coordinate associated with the jth mode of E symmetry. In view of the lack of information concerning the values of the polarizabilities, it is hard to see how one can gain quantitative information from the Raman E bands. However, the interpretation of the correlation functions obtained from the A_1 Raman bands and all the infrared bands should be straight-forward. Of course, calculations of the "true" rotational bands require corrections for spectral slit width, hot bands, vibrational widths, and collision-induced absorption, as has already been discussed in connection with the benzene experiments. Several studies of the spectra have appeared[20,137-142] that have taken some or all of these factors into account in calculations of the correlation functions or correlation times. For example, Raman spectra of CH_3I are shown in Fig. 21; it is evident that there is considerable overlap between bands in the depolarized spectrum in some cases (especially between v_1 and v_4), and that accurate estimations of the rotational band widths are far from trivial in this system. On the other hand, the E bands are noticeably broader than the A_1 bands, which shows that the correlation times associated with the E bands are shorter than those for the A_1 bands. In this way, one can state qualitatively at least that rotation about the molecular axis is noticeably faster than reorientation of this axis. However, more quantitative measures of these rates can be obtained from other experiments, so only the A_1 Raman correlation times will be listed in the summary in Table IV. The value obtained from the v_1 band is quite uncertain; indeed, Fig. 21 indicates that v_3 is by far the strongest of the depolarized A_1 bands and thus should yield the most precise band width data. According to (4.40) and Table IV, it can be concluded that

$$\tau_{20} = 1.5 \pm 0.2 \text{ psec} \qquad (4.43)$$

for CH_3I at room temperature; it also appears that this time is about 10% longer for CD_3I than for CH_3I.

Measurements of the depolarized Rayleigh light scattering[20,132] for CH_3I at room temperature are sufficiently detailed to give the correlation

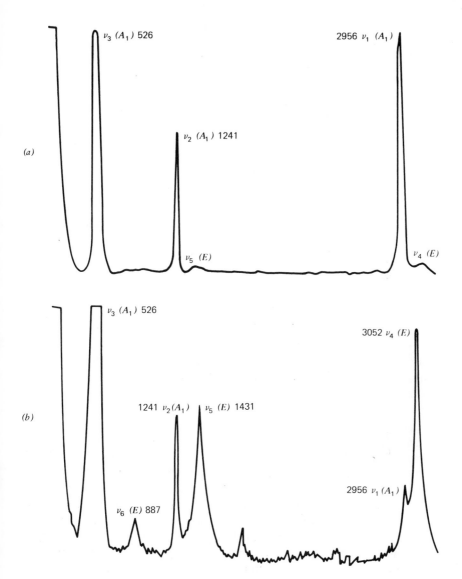

Fig. 21. Raman spectra of liquid CH_3I: (*a*) polarized, (*b*) depolarized. Although the scales of the spectra are different, they are related by the fact that the polarized *E* bands are 4/3 as intense as the corresponding depolarized bands. The numbers denote band center positions in cm^{-1}.

TABLE IV

Correlation Times for Methyl Iodide at 300°K (psec)

Method	CH_3I	CD_3I
v_1, Raman	0.8^{139}	1.2^{139}
v_2, Raman	$1.1,^{139}$ 1.4^{132}	1.8^{139}
v_3, Raman	$1.4,^{139}$ $1.6,^{140}$	1.5^{139}
	$1.7,^{141}$ $1.6,^{142}$	
Depolarized scattering	$1.6,^{20}$ 1.6^{132}	
v_3, infrared	1.0^{137}	1.5^{137}
v_6, infrared	0.4^{137}	0.5^{137}
Dipolar absorption	3.1^{143}	
T_1, deuterium		0.40^{144}
T_1, ^{13}C	See text	
T_1, proton rotations	See text	

function as well as the correlation time. The resulting $C_2(t)$ is shown in Fig. 22, and can be seen to be reasonably exponential for $t^* > 1$ and quite different from the free-rotor curve for $t^* > 0.5$. The correlation times obtained from these experiments are listed in Table IV, and agree well with the Raman results. Thus, these data lead one to conclude that correlated rotational motions are unimportant in this system, at least in regard to their effect on τ_{20}.

Correlation functions and correlation times have also been determined from an A_1 and an E band in the near infrared,[137] and from pure rotational absorption in the far infrared.[143] Values of $C_1(t)$ calculated from the A_1 band are shown in Fig. 22 and correlation times are listed in Table IV. The behavior of these correlation functions is consistent with $C_2(t)$ with respect to their linearity on a logarithmic scale at long time and their deviation from the free-rotor curve. However, a comparison of the correlation times for $l = 1$ and $l = 2$ (which are essentially equal to the inverses of the slopes of the linear portions of the curves shown in Fig. 22) gives the first indication of the presence of some serious unresolved problems in the interpretation of the data for methyl iodide.

We first note that the experimental reduced correlation times are

$$\tau_{20}^* = 3; \qquad \tau_{10}^* = 2 \qquad (4.44)$$

The magnitudes of these times (plus the linearity of $\ln C_l(t)$) indicates that reorientation of the symmetry axis is governed by a Markoffian mechanism. However, if the molecule is undergoing small-step diffusion, as has often been assumed in the published work on this system, the ratio τ_{20}^*/τ_{10}^* should

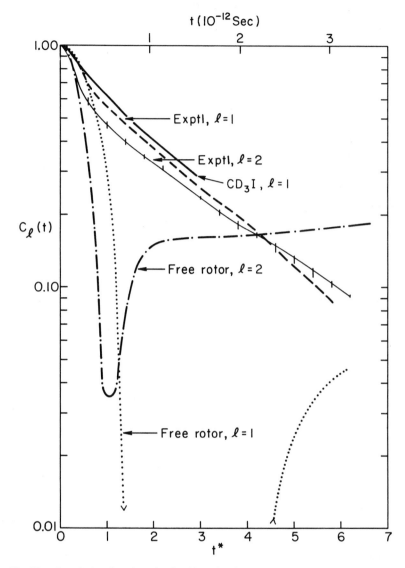

Fig. 22. Correlation functions for liquid methyl iodide compared with free-rotor curves. (The free-rotor curve for $l = 1$ goes negative between $t^* = 1.5$ and 4.5.) Experimental curves for $l = 1$ were obtained from the corrected infrared spectral band shapes; intensities of depolarized light scattering were used to calculate $C_2(t)$.

87

be $\frac{1}{3}$ rather than $\frac{3}{2}$. (A possible alternative is to invoke a large-step jump model such as that proposed by Ivanov.)

On the other hand, the "pure" dipolar absorption spectrum in the far infrared[143] gives a correlation time τ_{10} that differs by a factor of three from the value obtained from the infrared bands, and which is much closer to the predicted small-step diffusion value of 4.5 psec. The most obvious explanation for the differences in τ_{10} obtained in the two experiments is that correlated motions can contribute to the far infrared absorption, but are absent from the vibration–rotation spectra. However, the ratios obtained from infrared data for τ_{20}/τ_{10} in dilute solutions of CH_3I are also close to 3, and correlation effects should be negligible in those systems; furthermore, agreement between τ_{20} values obtained from the pure rotational Raman and the vibration-rotational Raman is also inconsistent with an explanation based on correlated motion.

Perhaps the most likely explanation for the discrepancy arises from the problem of eliminating the effects of nonrotational sources of line broadening upon the measured correlation times, especially for infrared band shape analyses. In particular, relaxation of the vibrating molecule occurs due to energy transfer and to vibrational dephasing caused by collisions in the fluid. For vibration–rotation band shape analyses, the total correlation time $\tau_{lm}(\text{expt})$ can be written as

$$\frac{1}{\tau_{lm}(\text{expt})} = \frac{1}{\tau(\text{vib})} + \frac{1}{\tau_{lm}(\text{rot})} \tag{4.45}$$

As already noted, $\tau(\text{vib})$ can be measured in the Raman by determining the widths of the polarized bands. In fact, the value for methyl iodide is known to be 2.3 ± 0.2 psec from measurements of the A_1 bands widths. If it is assumed that this time is also applicable to the pure vibrational relaxation in the infrared band of methyl iodide, $\tau_{lm}(\text{rot})$ can be calculated from (4.45). After correcting the infrared correlation functions for hot bands and variations of dielectric constant through the band,[20] one obtains a correlation time $\tau_{10}(\text{expt}) = 1.4 \pm 0.1$ psec (given by the slope of the linear portion of the dashed curve in Fig. 22); in this case, (4.45) gives $\tau_{10}(\text{rot}) = 3.6 \pm 0.4$ psec. This calculation shows that most, if not all, of the apparent inconsistency between infrared and Raman correlation times is due to the omission of vibrational broadening in the analysis of the infrared band shapes.

In addition, all of these spectral band shapes are affected by collision-induced absorption[145] or scattering;[19] not enough is known about this phenomenon for nonspherical molecules to allow one to estimate its relative importance in the various measured spectra, but it has been estimated[20] that the fraction of the total intensity of depolarized light scattering due to collision-induced anisotropy in the polarizability is of the order of 0.1; inasmuch as this scattering occurs at larger frequency shifts than that due to rotational motion, it gives additional "nonrotational" intensity in the wings of the band. These effects appear in polarized Raman bands which lends an additional element of uncertainty to the estimation of the nonrotational part of the depolarized bands. It is well known that they also give rise to easily measurable far infrared absorption in nonpolar fluids in the region around 100 cm^{-1},[146] and consequently, one suspects that they may not be negligible in the infrared experiments.

A simple but general method exists for determining if appreciable non-rotational contributions are present in a given spectral band. This is based on Gordon's[1] proof that the second moment M_2 of the rotational spectrum must be given by

$$M_2(l) = l(l + 1)\left(\frac{kT}{I}\right) \tag{4.46}$$

for a spectrum associated with a vector or tensor component that is parallel to the molecular symmetry axis, independent of the detailed rotational dynamics. The second moment is calculated by evaluating

$$M_2(l) = \frac{\displaystyle\int_{\text{band}} \omega^2 I_{\text{rot}}(\omega)\,d\omega}{\displaystyle\int_{\text{band}} I_{\text{rot}}(\omega)\,d\omega}$$

In those cases where second moments were evaluated for experimental methyl iodide bands,[20,137,138,142] they differ considerably from the pre-dicted values, often even after apparent elimination of "nonrotational" effects. Thus, it appears that the spectral correlation times listed in Table IV should be regarded as subject to revision when better methods of estimating nonrotational spectral intensity have been developed for nonspherical molecules.

Nuclear spin relaxation times have also been determined for methyl iodide. Since the interpretation of these data is free from many of the ex-traneous factors that appear to be present in the spectral measurements, one might anticipate that a clearer picture would emerge from an analysis of the rotational contributions to the spin-lattice relaxation times T_1. We first consider the deuteron quadrupolar relaxation in CD_3I; (1.21) and (1.25) can be used to express the relaxation time in terms of the correlation times τ_{2k}. Since this molecule can be assumed to be a symmetric rotor without loss of generality, one can use the coupling constant and the ICH angle listed in Table III to show that

$$\left(\frac{1}{T_1^Q}\right)_D = (0.48 \times 10^{-12} \text{ sec}^{-2})\tau_D \tag{4.47}$$

$$\tau_D = 0.13\tau_{20} + 0.25\tau_{21} + 0.62\tau_{22} \tag{4.48}$$

It can be seen that the τ_D (which is listed in Table IV) is dominated by τ_{22}, as one would anticipate for a vector that is nearly perpendicular to the molecular axis. Other T_1 values measured for CH_3I include the intra-molecular part of the proton relaxation time[144] (obtained by extrapolation to infinite dilution in a nonmagnetic solvent) and the ^{13}C time.[144,147] Measurements of nuclear Overhauser effect[147] show that the ^{13}C T_1 is affected by a nondipolar relaxation mechanism which is concluded to be spin-rotational in origin. It is then not surprising to find that spin rotation also plays an important role in determining the intramolecular proton relaxation. Thus, we write

$$\left(\frac{1}{T_1}\right)_{^{13}C} = \left(\frac{1}{T_1^{DD}}\right)_{^{13}C} + \left(\frac{1}{T_1^{SR}}\right)_{^{13}C} \tag{4.49}$$

and an analogous equation for $(1/T_1)_H$. The fact that the ^{13}C—H vector which dominates the ^{13}C dipolar relaxation has the same orientation in the molecular frame as the C—D vector gives rise to a correlation time that is identical to the deuteron time. If a small ^{13}C—I contribution to the relaxation time is omitted, one can write

$$\left(\frac{1}{T_1^{DD}}\right)_{^{13}C} = 5.5 \times 10^{10}(\text{sec}^{-2})\tau_D \tag{4.50}$$

After inserting the numerical value of τ_D (and estimating the ^{13}C—I term), one finds

$$(T_1^{DD})_{^{13}C} = 48 \pm 2 \text{ sec} \tag{4.51}$$

Equation (4.51) agrees very well with the direct determination of this time from spin-lattice relaxation combined with a nuclear Overhauser effect measurement, which gives

$$(T_1^{DD})_{^{13}C} = 52 \pm 6 \text{ sec} \tag{4.52}$$

One thus has

$$(T_1^{SR})_{^{13}C} = \begin{cases} 18 \pm 2 \text{ sec } (\tau_D = \tau_{^{13}C}) \\ 14 \pm 2 \text{ sec (NOE)} \end{cases} \tag{4.53}$$

In the absence of experimental values of the spin-rotation tensor **C** for methyl iodide, it is difficult to extract dynamical information from (4.53).

(We will not discuss the efforts made to estimate **C** and thus to deduce correlation times from (4.53); these calculations[144,147] are based on approximate dynamical models wherein the angular momentum and the orientational relaxations have been decoupled, and thus compound the uncertainty introduced by the lack of knowledge concerning **C**.) We note only that the most reasonable choices for spin-rotation correlation times and coupling constants give rise to a physical picture for the rotational dynamics that is consistent with the one to be developed here. In particular, Gillen, Schwartz, and Noggle estimate that

$$\left(\frac{1}{T_1^{DD}}\right)_H = 0.020 \pm 0.005 \text{ sec}^{-1} \tag{4.54}$$

with a large uncertainty arising from the fact that the spin-rotational term dominates the total relaxation process so that $(1/T_1^{DD})_H$ is the difference of two nearly equal numbers. After inserting the H—H distance given in Table III, (1.20) gives

$$\left(\frac{1}{T_1^{DD}}\right)_H = (4.3 \times 10^{10} \text{ sec}^{-2})(\tfrac{1}{4}\tau_{20} + \tfrac{3}{4}\tau_{22}) \tag{4.55}$$

or, using (4.54),

$$\tau_{20} + 3\tau_{22} = (1.8 \pm 0.5) \text{ psec.} \tag{4.56}$$

Up to this point, we have not restricted our analysis of the data by introducing a specific model for rotational motion in this system. On the other hand, the experimental results do not seem to yield unequivocal values of the correlation times, primarily because of doubts concerning the times obtained from the infrared and Raman spectra. If we ignore the inconsistencies, it would appear that the correlation times for methyl iodide at room temperature are given by (4.43) for τ_{20}; (4.56) would then indicate that τ_{22} is 0.1 psec within an uncertainty of at least 0.3 psec; and (4.48) finally yields τ_{21} to be between 0.5 and 1 psec. Studies of the temperature dependence[140,142,144] of τ_{20} indicate that

$$1/\tau_{20} = Ae^{-E_a/RT} \tag{4.57}$$

with an activation energy $E_a = 1.9$ kcal/mole. Because of the similarity between this temperature variation and that of the viscosity, it has been argued that the reorientation of the symmetry axis is governed by small-step diffusion. (This conclusion can only be accepted at the cost of ignoring the fact that the spectral τ_{20}/τ_{10} differs greatly from the expected value.) Proceeding in this way, one finds that

$$\mathscr{R}_{xx} = \mathscr{R}_{yy} = 0.11 \text{ psec}^{-1} \tag{4.58}$$

at room temperature (or $\mathscr{R}_{xx}^* = 0.06$, a value which is not too different from the reduced rotation diffusion constant for benzene). We now assume that rotation about the molecular symmetry axis is nearly free so that the decay of correlations due to this motion is given by the Gaussian model. Thus,

$$\tau_{2k}^* = \int_0^\infty \exp\left[-\left(\mathscr{R}_{xx}^* t^*(6 - k^2) + \frac{k^2 t^{*2}}{2} \right) \right] dt^* \qquad (4.59)$$

A numerical integration yields

$$\tau_{21} = 0.5 \text{ psec}$$
$$\tau_{22} = 0.3 \text{ psec} \qquad (4.60)$$

which are consistent with the experimental results. Although Griffith's analysis[140] differs in some minor ways from that given here, he finds that the temperature dependence of the relaxation times for rotation about the symmetry axis is quite small, which is the expected result for inertial motion.

Inasmuch as there is some reason to believe that the model of small-step diffusion for reorientation of the symmetry axis plus nearly free rotation about the axis may be adequate to describe methyl iodide in the liquid, it is of interest to calculate $\mathscr{R}_{xx}$ from the Stokes–Einstein model. If we take the major axis of the molecule to be the sum of the iodine and the methyl van der Waals radii (2.15 and 2.00 Å, respectively[148]) plus the bond length, we find $a = 6.3$ Å; furthermore, the length of the minor axis b can be set equal to 4.15 Å, the average of the methyl and iodine diameters. Equations (3.77), (3.79), and (3.80) give $\mathscr{R}_{xx} = 0.022 \text{ psec}^{-1}$ which is, as usual, much smaller than the experimental value. However, the friction coefficients shown in Fig. 10 for the slip boundary condition give $\mathscr{R}_{xx} = 0.18 \text{ psec}^{-1}$ for an axial ratio of 0.65, which is acceptable, considering the uncertainties in the experimental value of 0.11 psec^{-1} and in our estimates of the lengths of the molecular axes.

V. SUMMARY

In the three experimental examples discussed, it emerges that rotational motion about one or more molecular axes can be quite free even in dense systems if the symmetry of the intermolecular potential is sufficiently high. When rotation is nearly free, the extended diffusion model or the Gaussian model appears to give an adequate method at least for calculating the correlation time, and, in the case of the J-diffusion model, for calculating the entire correlation function for isotropic reorientation. However, rotation in liquid benzene and in methyl iodide is noticeably anisotropic, and appears to be described best by a combination of small-step reorientation of the symmetry axes accompanied by almost free rotation about the symmetry axes.

A large literature exists that we have not mentioned. Other small molecules such as methyl cyanide[132,149] and chloroform[20,132,150] have been studied in detail. Gillen and Noggle[151] have analyzed much of the data for these and several other molecules using an approach similar to that discussed here for CH_3I, and have concluded that CH_3CN and $CHCl_3$ are also among the molecules that exhibit small-step diffusion of the symmetry axis plus nearly free rotation about that axis. Maryott, Malmberg, and Gillen[115] have collected and analyzed the spin-rotation data for a number of tetrahedrally symmetric (spherical top) molecules; their findings indicate that nearly free rotation in such liquids is more nearly the rule than it is the exception; consequently, the extended diffusion models give a much better description of these systems than the small-step rotational diffusion theory which has so often been invoked in previous studies. Models for describing molecules with internal rotations as well as external have been developed, and comparisons with experiment have been made.[152] The discrete-site jump models discussed in Section III have long been used to treat reorientational motions in plastic crystals and in fluids with strong forces hindering rotation (such as hydrogen-bonded liquids). Molecular dynamics computer studies of some simple model molecules have begun to appear[47,92,153] which should provide the basis for tests of truly molecular theories of rotation.

Some unanswered questions remain concerning even the systems discussed here: is there some way to justify the fact that the experimental rotational diffusion constants agree with the values calculated from a model of rotation in a viscous fluid with slip boundary conditions? Do long-range angle-dependent terms in the intermolecular potential function (such as dipole-dipole interactions) affect the dynamics and if not, why not? How does one rationalize the application of the extended diffusion model, which appears to be based on a binary collision picture, to fluids as dense as normal liquids, and is there a way to rationalize the values obtained for the fitting parameter, whether one calls it β, the collision frequency, or $1/\tau_L$, the angular momentum correlation time?

A. Memory Functions

Even though the tentative answers to these and many other questions concerning rotational dynamics in dense phases are subject to revision at present, we will conclude by describing a general approach to this problem which shows promise of generating a number of new insights in the near future. This is the technique of memory functions. Inasmuch as the general principles involved have been reviewed recently,[92,154] we will give a specific discussion of the possible applications to rotational problems[38,155-158] and, as an illustration, show how some of the models already described can be obtained as special cases of the general theory. (In fact, some tentative

steps in this direction have already been taken in connection with the derivation given in Section III for the time-dependent rotational diffusion constant.)
   For simplicity, we consider only

$$C_l(t) = \langle D_{00}{}^l(\delta\Omega) \rangle \tag{5.1}$$

and the memory functions associated with it. (To include correlation functions for $D_{k,m}^l(\delta\Omega)$ with $k, m \neq 0$ would merely complicate the notation.) The memory function $K_l(\tau)$ for $C_l(t)$ can be defined by the equation

$$\frac{d}{dt} C_l(t) = - \int_0^t K_l(\tau) C_l(t - \tau) \, d\tau \tag{5.2}$$

Of course, little progress is made by expressing the unknown function $C_l(t)$ in terms of an integral equation involving a second unknown function. The advantage of this approach lies primarily in the discovery that the behavior of $C_l(t)$ over the entire time regime can be calculated if one knows $K_l(\tau)$ at short times only. In order to develop the theory more fully, some additional mathematical formalism is necessary. We define $\mathscr{I}_l(\omega)$ and $\hat{K}_l(\omega)$ to be the complex Laplace transforms of $C_l(t)$ and $K_l(t)$, respectively:

$$\mathscr{I}_l(\omega) = \mathscr{I}_l'(\omega) + i\mathscr{I}_l''(\omega) = \int_0^\infty C_l(t) e^{i\omega t} \, dt \tag{5.3}$$

$$\hat{K}_l(\omega) = \hat{K}_l'(\omega) + i\hat{K}_l''(\omega) = \int_0^\infty K_l(t) e^{i\omega t} \, dt \tag{5.4}$$

where $\mathscr{I}_l'(\omega)/\pi$, the real part of $\mathscr{I}_l(\omega)/\pi$, is equal to the spectral intensity $\hat{I}_l(\omega)$ at angular frequency ω. Note also that

$$\int_0^\infty \frac{d}{dt} C_l(t) e^{i\omega t} \, dt = i\omega \mathscr{I}_l(\omega) - 1 \tag{5.5}$$

and

$$\hat{\mathscr{G}}_l(\omega) = \int_0^\infty \mathscr{G}_l(t) e^{i\omega t} \, dt = \omega^2 \mathscr{I}_l(\omega) + i\omega \tag{5.6}$$

where $\mathscr{G}_l(t)$ is a new correlation function defined as

$$- \mathscr{G}_l(t) = \frac{d^2}{dt^2} C_l(t) \tag{5.7}$$

$$= \left\langle \sum_m \frac{d}{dt} D_{m0}^{*\,l}(\Omega(0)) \frac{d}{dt} D_{m0}{}^l(\Omega(t)) \right\rangle \tag{5.8}$$

(In deriving (5.8), we have assumed that $C_l(t)$ is calculated for a classical ensemble and thus is a real even function of t.

We can now take the Laplace transform of (5.2) to find

$$i\omega \mathscr{I}_l(\omega) = 1 - \hat{K}_l(\omega)\mathscr{I}_l(\omega) \tag{5.9}$$

Thus,

$$I_l(\omega) = \frac{\hat{K}'_l(\omega)}{\hat{K}'_l(\omega) + [\omega + \hat{K}''_l(\omega)]^2} \tag{5.10}$$

Also, a time differentiation of (5.2) gives

$$\mathscr{G}_l(t) = K_l(t) + \int_0^t d\tau \int_0^\tau d\tau' K_l(\tau)K_l(\tau')C_l(\tau - \tau') \tag{5.11}$$

It can now be seen that $\mathscr{G}_l(t)$ and $K_l(t)$ are equal at short times; the Laplace transforms of these functions are related by

$$\hat{K}_l(\omega) = \hat{\mathscr{G}}_l(\omega)\left[1 + \frac{\hat{\mathscr{G}}_l(\omega)}{\omega}\right]^{-1} \tag{5.12}$$

Although the relation between $K_l(t)$ and the dynamical variables of the system is not explicit, one can gain some insight concerning the memory function by its relation to $\mathscr{G}_l(t)$, which does have an explicit form. In fact, several authors[49,50] have suggested that experimental $\mathscr{G}_l(t)$ contain useful information about the decay of angular momentum correlations. We can use (5.7) to obtain this correlation function from experimental spectra, since

$$\mathscr{G}_l(t) = \frac{I}{2\pi} \int_{-\infty}^\infty \omega^2 I_l(\omega)e^{-i\omega t} d\omega \tag{5.13}$$

for symmetrical band shapes. However, detailed calculations[10] show that $\mathscr{G}_l(t)$ can be identified with the angular momentum correlation function only at short times. Indeed, an explicit evaluation of the derivatives in (5.8) gives

$$\mathscr{G}_l(t) = \frac{l(l + 1)}{2I^2} \sum_{n, n' = -1, 1} (2\delta_{nn'} - 1)\langle L_n^*(0)L_{n'}^*(t)D_{-nn'}(\delta\Omega)\rangle \tag{5.14}$$

where

$$L_n^* = LD_{n, 0}^1(\phi_L, \theta_L)$$

with L, ϕ_L, θ_L the spherical polar components of the body-fixed angular momentum of a symmetric top molecule. If the angular displacement $\delta\Omega$ is negligibly small over the time for decay of the correlation for angular momentum (5.14) takes on the approximate form used previously:

$$\mathscr{G}_l(t) \simeq \frac{l(l + 1)}{2}\langle \boldsymbol{\omega}_\perp(0) \cdot \boldsymbol{\omega}_\perp(t)\rangle \tag{5.15}$$

where $\boldsymbol{\omega}_\perp$ is the body-fixed angular velocity perpendicular to the molecular symmetry axis. However, when $\delta\Omega$ is appreciable, $\mathscr{I}_l(t)$ involves cross-correlations between angular velocity and orientation angles.

We have seen that the memory function $K_l(t)$ can be equated to the angular velocity correlation function only at short times. At longer times, it is still mathematically well defined by (5.12), (5.6), and (5.14); however, its physical significance is no longer clear.

We now show how two of the models discussed previously can be derived by assuming a particular form for $K_l(t)$. First, we consider the J-diffusion model,[158] which can be obtained as a consequence of assuming that

$$K_l(t) = K_l^{(fr)}(t)e^{-\beta t} \tag{5.16}$$

where $K_l^{(fr)}(t)$ is the free-rotor memory function. The Laplace transform of (5.16) is

$$\hat{K}_l(\omega) = \hat{K}_l^{(fr)}(\omega - i\beta) \tag{5.17}$$

According to (5.9),

$$\hat{K}_l^{(fr)}(\omega - i\beta) = -i\omega - \beta + \frac{1}{\mathscr{I}_l^{(fr)}(\omega - i\beta)} \tag{5.18}$$

In order to evaluate $\mathscr{I}_l^{(fr)}(\omega - i\beta)$, we turn to (2.40) and remember that $C_l(t, 0)$ is a free-rotor correlation function; thus,

$$\mathscr{I}^{(fr)}(\omega - i\beta) = \langle\mathscr{C}_l(\omega)\rangle + i\langle\mathscr{S}_l(\omega)\rangle \tag{5.19}$$

Since

$$\mathscr{I}_l(\omega) = \frac{1}{\hat{K}_l(\omega) + i\omega} \tag{5.20}$$

one can use (5.17), (5.18), and (5.19) to show that

$$\mathscr{I}_l(\omega) = \frac{1}{[\langle\mathscr{C}_l(\omega)\rangle + i\langle\mathscr{S}_l(\omega)\rangle]^{-1} - \beta} \tag{5.21}$$

A calculation of the real part of $\mathscr{I}_l(\omega)/\pi$ now gives an expression for $\hat{I}_l(\omega)$ that is identical to (2.48), and completes the proof that the memory function of (5.16) generates J-diffusion spectral densities and correlation functions.

This calculation also suggests a method of deriving the isotropic small-step diffusion model. We have already seen that the J-diffusion model approaches rotational random walk as the frequency factor β becomes large; according to (5.16), an approach to this limit causes the memory function to decay more and more rapidly until it begins to take on the appearance of a Dirac delta function. Inasmuch as the integral of $\exp(-\beta t)$ is $1/\beta$, we can write

$$K_l(t) = K_l(0)\beta^{-1}\delta(0) \tag{5.22}$$

for large β, where (5.15) and (5.11) show that

$$K_l(0) = l(l + 1)\left(\frac{kT}{I}\right) \tag{5.23}$$

The integral in (5.2) can be performed with the aid of (5.22) to give

$$-\frac{d}{dt} C_l(t) = l(l + 1)\mathscr{R}C_l(t) \tag{5.24}$$

where $\mathscr{R} = kT/I\beta$. A trivial integration of (5.24) gives (3.47), which is a special case of the general result of (3.64).

More generally, one can use (5.3) and (5.9) to obtain the memory function for any system with a known correlation function; thus, both experimental and model memory functions can be calculated. If the features of $K_l(t)$ can be determined in this way for some specific cases, the information gained will be helpful in generating approximate forms which may be useful in calculations of $C_l(t)$ and $I_l(\omega)$ for a range of conditions. Figs. 23 and 24

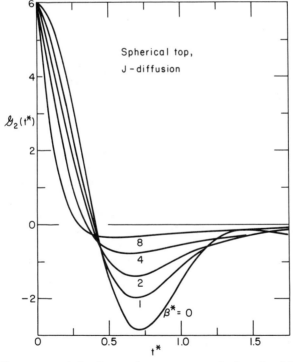

Fig. 23.   The cross-correlation for angular momentum and orientation defined by (5.7) is shown as a function of reduced time for various values of the reduced collision frequency. The model was taken to be J-diffusion for a spherical top.

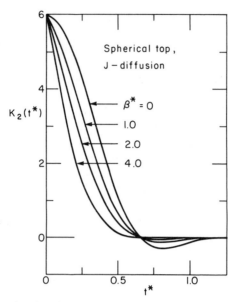

Fig. 24. The memory functions for $l = 2$ are shown here for the same models that gave the correlation functions of Fig. 23.

show $K_2(t)$ and $\mathscr{G}_2(t)$ for a spherical top rotating according to the J-diffusion model. Reduced units are used such that

$$K_2(0) = \mathscr{G}_2(0) = l(l + 1) \tag{5.25}$$

and a range of reduced β values are given, including free rotation ($\beta^* = 0$). It can be seen that the memory function decays rapidly with time, and that the rate of decay increases as β^* increases. On the other hand, Fig. 9 shows that the decay of $C_2(t)$ becomes slower for large β^*. This suggests that an accurate representation of the memory function at small t^* may be capable of generating accurate correlation functions even at large t^* even when the explicit J-diffusion model is not invoked. This idea is the basis for calculations of correlation functions and correlation times from the exact values of the first few moments of $K_l(t)$;[155,156] for example, Keyes[155] notes that

$$\tau_l = 1/K_l(0) \tag{5.26}$$

(A direct result of (5.20) and the fact that $\tau_1 = \pi \hat{I}_l(0)$, if $I_l(\omega)$ is normalized to unit area.) If one approximates $K_l(t)$ by

$$K_l(t) = \mu_l^{(0)} \exp\left[-\frac{1}{2} t^2 \frac{\mu_l^{(2)}}{\mu_l^{(0)}} \right] \tag{5.27}$$

where $\mu_l^{(n)}$ is the nth moment of $K_l(t)$, it can be shown that

$$\frac{1}{\tau_l^*} = \frac{l(l+1)}{2} \left[\frac{2\pi}{\dfrac{\langle \mathbf{N} \cdot \mathbf{N} \rangle}{(kT)^2} + s} \right]^{1/2} \tag{5.28}$$

where $s = 3$ and 11 for $l = 1$ and 2, respectively, and $\langle \mathbf{N} \cdot \mathbf{N} \rangle$ is the mean square torque on the molecule. It is not clear that the Gaussian form for the memory function is correct even for rapidly decaying $K_l(t)$, but this approximation can in principle be tested against experiment.

Another approach to the generation of correlation functions is based on the fact that a heirarchy of memory functions exists[154,155,157] which are defined by

$$\frac{dK_l^{(n-1)}}{dt} = - \int_0^t K_l^{(n)}(\tau) K_l^{(n-1)}(t-\tau) \, d\tau \tag{5.29}$$

So far, we have been discussing the lowest-order equation, with $C_l \equiv K_l^{(0)}$, $K_l \equiv K_l^{(1)}$. It has been suggested[155] that the time scales for the decay of these functions become shorter and shorter as the index increases; thus, a short-time approximation for the nth memory function may be adequate to generate accurate lower-order functions even at long times. The initial tests of this hypothesis[153,157] indicate that this is indeed the case.

At present, the principal defect of this method of computing rotation dynamics is the relatively tenuous connection between the memory functions and explicit molecular properties (such as torques, velocities, etc.). Nevertheless, a number of calculations based on these ideas have already appeared. The derivation of (3.69) for the relationship between rotational diffusion constant and the autocorrelation function of the molecular angular velocity is based on the memory function formalism, plus an assumption concerning the time scale for reorientation relative to that for the change of angular velocity. One might guess that it will be possible to use this approach to deduce a method of calculating β or τ_L, the parameter of the extended diffusion models, which does not involve "collision" frequencies in dense liquids. An improved theory relating rotational diffusion constant to molecular properties is also desirable, and it seems that this might be obtainable from the hierarchy of memory functions for such systems.

References

1. R. G. Gordon, *Adv. Mag. Res.*, **3**, 1 (1968).
2. W. A. Steele, in *Transport Phenomena in Fluids*, H. J. M. Hanley, Ed., Dekker, New York, 1969, Chap. 8.
3. M. E. Rose, *Elementary Theory of Angular Momentum*, Wiley, New York, 1957.
4. A. R. Edmonds, *Angular Momentum in Quantum Mechanics*, Princeton University Press, Princeton, 1957.

5. J. O. Hirschfelder, C. F. Curtiss, and R. B. Bird, *Molecular Theory of Gases and Liquids*, Wiley, New York, 1954, Chap. 12, Appendix B.

6. A. Abragam, *The Principles of Nuclear Magnetism*, Oxford University Press, London, 1961, Chap. VIII, Sec. IIE.

7. R. L. Hilt and P. S. Hubbard, *Phys. Rev.*, **134**, A392 (1964); L. K. Runnels, *Phys. Rev.*, **134**, A28 (1964).

8. W. T. Huntress, Jr., *Adv. Mag. Res.*, **4**, 2 (1970).

9. Ref. 6, Chap. VIII, Sec. IIF.

10. A. G. St. Pierre and W. A. Steele, *J. Chem. Phys.*, **62**, 2286 (1975).

11. P. W. Atkins, in *Electron Spin Relaxation in Liquids*, L. T. Muus and P. W. Atkins, Eds., Plenum Press, New York, 1972.

12. M. Bloom, F. Bridges, and W. N. Hardy, *Can. J. Phys.*, **45**, 3533 (1967).

13. P. S. Hubbard, *Phys. Rev.*, **131**, 1155 (1963); *Phys. Rev. A*, **9**, 481 (1974).

14. W. T. Huntress, *J. Chem. Phys.*, **48**, 3524 (1968); C. H. Wang, D. M. Grant, and J. R. Lyerla, Jr., *J. Chem. Phys.*, **55**, 4674 (1971).

15. D. Kivelson, in *Electron Spin Relaxation in Liquids*, L. T. Muus and P. W. Atkins, Eds., Plenum Press, New York, 1972.

16. R. G. Gordon, *J. Chem. Phys.*, **42**, 3658 (1965); **43**, 1307 (1965).

17. R. Pecora and W. A. Steele, *J. Chem. Phys.*, **42**, 1872 (1965).

18. S. Bratos and E. Marechal, *Phys. Rev. A*, **4**, 1078 (1971); E. Brindeau, S. Bratos, and J. Leickman, *Phys. Rev. A*, **6**, 2007 (1972); F. J. Bartoli and T. Litovitz, *J. Chem. Phys.*, **56**, 404, 413 (1972); L. A. Nafie and W. L. Peticolas, *J. Chem. Phys.*, **57**, 3145 (1972).

19. J. A. Bucaro and T. A. Litovitz, *J. Chem. Phys.*, **54**, 3846 (1971); **55**, 3585 (1971); H. B. Levine and G. Birnbaum, *J. Chem. Phys.*, **55**, 2914 (1971); W. M. Gelbart, *J. Chem. Phys.*, **57**, 699 (1972); M. Thibeau and B. Oksengorn, *Mol. Phys.*, **15**, 579 (1968); *J. Phys. (Paris)*, **30**, 47 (1969).

20. P. van Konynenburg and W. A. Steele, *J. Chem. Phys.*, **56**, 4776 (1972).

21. Ref. 2, Appendix. However, (8.A.37) and (8.A.38) are incorrect as they stand; the proper form is obtained by taking complex conjugates of all D_{km}^2 shown there.

22. However, see T. Keyes and D. Kivelson, *J. Chem. Phys.*, **56**, 1057 (1972); T. Keyes, *Mol. Phys.*, **23**, 737 (1972).

23. G. I. A. Stegeman and B. P. Stoicheff, *Phys. Rev. A*, **7**, 1160 (1973); G. R. Alms, D. R. Bauer, J. I. Brauman, and R. Pecora, *J. Chem. Phys.*, **59**, 5304 (1973); M. D. Levenson and A. L. Schawlow, *Optics Comm.*, **2**, 192 (1970).

24. H. C. Andersen and R. Pecora, *J. Chem. Phys.*, **54**, 2584 (1971); **55**, 1496 (1971); T. Keyes and D. Kivelson, *J. Chem. Phys.*, **54**, 1786 (1971); **56**, 1876 (1972); C. H. Chung and S. Yip, *Phys. Rev. A*, **4**, 928 (1971); A. Ben Reuven and N. D. Gershon, *J. Chem. Phys.*, **54**, 1049 (1971); N. D. Gershon and I. Oppenheim, *Physica*, **64**, 247 (1973); N. K. Ailawadi, B. J. Berne, and D. Forster, *Phys. Rev. A*, **3**, 1472 (1971); V. Volterra, *Phys. Rev.*, **180**, 156 (1969).

25. However, see D. A. Condiff and J. S. Dahler, *J. Chem. Phys.*, **44**, 3988 (1966); H. Brenner, *J. Colloid Sci.*, **20**, 104 (1965).

26. W. M. Lomer and G. G. Low, in *Thermal Neutron Scattering*, P. Egelstaff, Ed., Academic, New York, 1965; S. Yip, in *Spectroscopy in Biology and Chemistry*, Academic, New York, 1974; K. E. Larsson, *Ber. Bunsenges.*, **75**, 352 (1971).

27. J. G. Powles, *Adv. Phys.*, **22**, 1 (1973).

28. V. F. Sears, *Can. J. Phys.*, **44**, 1279 (1966).

29. W. A. Steele and R. Pecora, *J. Chem. Phys.*, **42**, 1863 (1965).

30. A. DasGupta, S. I. Sandler, and W. A. Steele, *J. Chem. Phys.*, **62**, 1769 (1975); K. E. Gubbins, C. G. Gray, P. A. Egelstaff, and M. S. Anath, *Mol. Phys.*, **25**, 1353 (1973).

31. T. J. Krieger and M. S. Nelkin, *Phys. Rev.*, **106**, 290 (1957); A. Rahman, *J. Nucl. Energy*, **A13**, 128 (1961); M. Antonini, P. Ascarelli, and G. Caglioti, *Phys. Rev.*, **136**, A 1280 (1964); A. K. Agrawal and S. Yip, *Phys. Rev. A*, **171**, 263 (1968); K. E. Larsson and L. Bergstedt, *Phys. Rev.*, **151**, 117 (1966); K. E. Larsson, *Phys. Rev. A*, **3**, 1006 (1971); A. K. Agrawal and S. Yip, *J. Chem. Phys.*, **46**, 1999 (1967); C. T. Chudley and R. J. Elliott, *Proc. Phys. Soc.*, **77**, 353 (1961).

32. R. G. Gordon, *J. Chem. Phys.*, **43**, 1307 (1965).

33. S. Bratos, J. Rios, and Y. Guissani, *J. Chem. Phys.*, **52**, 439 (1970); S. Bratos and J. Rios, *Compt. Rend.*, **269**, 90 (1969).

34. H. Dardy, V. Volterra, and T. A. Litovitz, *Symp. Faraday Soc.*, **6**, 71 (1972).

35. R. L. Fulton, *J. Chem. Phys.*, **55**, 1386 (1971); B. Crawford, Jr., A. C. Gilbey, A. A. Clifford, and T. Fujiyama, *Pure Appl. Chem.*, **18**, 373 (1969) and references therein.

36. P. Debye, *Polar Molecules*, Chemical Catalogue Co., New York, 1929; A. Fröhlich, *Theory of Dielectrics*, Oxford University Press, Oxford, 1949.

37. U. M. Titulaer and J. M. Deutch, *J. Chem. Phys.*, **60**, 1502 (1974); E. Fatuzzo and P. R. Mason, *Proc. Phys. Soc.*, **90**, 741 (1967); R. H. Cole, *Mol. Phys.*, **26**, 269 (1973).

38. T.-W. Nee and R. Zwanzig, *J. Chem. Phys.*, **52**, 6353 (1970).

39. P. Bordewijk, *Adv. Mol. Relaxation Processes*, **5**, 285 (1973).

40. H. Goldstein, *Classical Mechanics*, Addison-Wesley, Reading, Mass., 1950.

41. A. G. St. Pierre and W. A. Steele, *Phys. Rev.*, **184**, 172 (1969).

42. A. K. Agrawal and S. Yip, *Phys. Rev.*, **171**, 263 (1968).

43. R. G. Gordon, *J. Chem. Phys.*, **44**, 1830 (1966).

44. R. E. D. McClung, *J. Chem. Phys.*, **51**, 3842 (1969); **54**, 3248 (1971); **55**, 3459 (1971); **57**, 5478 (1972).

45. M. Fixman and K. Rider, *J. Chem. Phys.*, **51**, 2425 (1969).

46. A. G. St. Pierre and W. A. Steele, *J. Chem. Phys.*, **57**, 4638 (1972).

47. G. D. Harp and B. J. Berne, *Phys. Rev. A*, **2**, 975 (1970).

48. W. Feller, *An Introduction to Probability Theory and Its Applications*, Vol. I, 3rd ed., Wiley, New York, 1950, Chap. VI.

49. J. E. Anderson and R. Ullman, *J. Chem. Phys.*, **55**, 4406 (1971); A. Gerschel, I. Darmon, and C. Brot, *Mol. Phys.*, **23**, 317 (1972).

50. J. Kushick and B. Berne, *J. Chem. Phys.*, **59**, 4486 (1973).

51. D. Enskog, *Kgl. Svenska Vetenskapakad. Handl.*, **64**, No. 4 (1922).

52. T. Einwohner and B. J. Alder, *J. Chem. Phys.*, **49**, 1458 (1968).

53. P. van Konynenburg and W. A. Steele, *J. Chem. Phys.*, **62**, 2301 (1975).

54. M. Theodosopulu and J. S. Dahler, *J. Chem. Phys.*, **60**, 4048 (1974).

55. D. E. O'Reilly, *J. Chem. Phys.*, **57**, 885 (1972).

56. H. D. Dardy, V. Volterra, and T. A. Litovitz, *J. Chem. Phys.*, **59**, 4491 (1973); W. M. Madigosky and T. A. Litovitz, *J. Chem. Phys.*, **34**, 489 (1961).

57. E. N. Ivanov, *Sov. Phys. I.E.T.P.*, **18**, 1041 (1964).

58. For instance, see V. I. Smirnov, *Linear Algebra and Group Theory*, transl. and ed. by R. A. Silverman, McGraw-Hill, New York, 1961, Sect. 17.

59. Ref. 48, Chap. XV, Sec. 2.

60. J. D. Hoffman, *J. Chem. Phys.*, **23**, 1331 (1955).

61. Ref. 48, p. 377, 434.

62. J. I. Lauritzen, *J. Chem. Phys.*, **28**, 118 (1958); J. D. Barnes, *J. Chem. Phys.*, **58**, 5193 (1973).

63. J. I. Lauritzen and R. Zwanzig, *Adv. Mol. Relaxation Processes*, **5**, 339 (1973).

64. D. C. Look and I. J. Lowe, *J. Chem. Phys.*, **44**, 3437 (1966).

65. C. Brot and I. Darmon, *Mol. Phys.*, **21**, 785 (1971).

66. C. Thibaudier and F. Volino, *Mol. Phys.*, **26**, 1281 (1973); S. Alexander, A. Baram, and Z. Luz, *Mol. Phys.*, **27**, 441 (1974).
67. R. I. Cukier and K. Lakatos-Lindenberg, *J. Chem. Phys.*, **57**, 3427 (1972).
68. J. D. Hoffman, *Archiv. Sci.*, **12**, 36 (1959).
69. J. D. Hoffman and B. M. Axilrod, *J. Res. Natl. Bur Stds.*, **58**, 61 (1957).
70. A. Ben Reuven and E. Zamin, *J. Chem. Phys.*, **55**, 475 (1971).
71. S. Chandrasekhar, *Rev. Mod. Phys.*, **15**, 1 (1943).
72. D. E. O'Reilly, *J. Chem. Phys.*, **49**, 5416 (1968); *Ber. Bunsenges. Phys. Chem.*, **75**, 208 (1971); D. E. O'Reilly and E. M. Peterson, *J. Chem. Phys.*, **55**, 2155 (1971); **56**, 2262 (1972).
73. J. E. Anderson and R. Ullman, *J. Chem. Phys.*, **47**, 2178 (1967); J. E. Anderson, *J. Chem. Phys.*, **47**, 4879 (1967); R. Ullman, *J. Chem. Phys.*, **49**, 831 (1968).
74. S. Glarum, *J. Chem. Phys.*, **33**, 639 (1960).
75. T. A. Litovitz and G. McDuffie, *J. Chem. Phys.*, **39**, 729 (1963); D. A. Pinnow, S. J. Candau, and T. A. Litovitz, *J. Chem. Phys.*, **49**, 347 (1968).
76. Ref. 10, eq. (A.6).
77. F. Perrin, *J. Phys. Radium*, **5**, 497 (1934); **7**, 1 (1936).
78. L. D. Favro, *Phys. Rev.*, **119**, 53 (1960).
79. K. A. Valiev and M. M. Zaripov, *Sov. Phys.*, *J.E.T.P.*, **15**, 353 (1962).
80. D. E. Woessner, *J. Chem. Phys.*, **37**, 647 (1962).
81. W. A. Steele, *J. Chem. Phys.*, **38**, 2411 (1963).
82. R. Pecora, *J. Chem. Phys.*, **50**, 2650 (1969).
83. W. A. Steele, *J. Chem. Phys.*, **43**, 2598 (1965).
84. E. P. Gross, *J. Chem. Phys.*, **23**, 1415 (1955).
85. W. A. Steele, *J. Chem. Phys.*, **38**, 2404 (1963).
86. H. Shimizu, *Bull. Chem. Soc. Japan*, **39**, 2385 (1966).
87. P. W. Atkins, A. Loewenstein, and Y. Margalits, *Mol. Phys.*, **17**, 329 (1969).
88. K. Mishima, *J. Phys. Soc. Japan*, **31**, 1796 (1971); K. Mishima and Y. Tanaka, *J. Phys. Soc. Japan*, **32**, 581 (1972).
89. P. S. Hubbard, *Phys. Rev. A*, **6**, 2421 (1972); **8**, 1429 (1973); *Phys. Rev. A*, **9**, 481 (1974).
90. R. Kubo, *Adv. Chem. Phys.*, **16**, 101 (1969).
91. H. Mori, *Progr. Theor. Phys. (Kyoto)*, **33**, 423 (1965); **34**, 399 (1965).
92. B. J. Berne and G. D. Harp, *Adv. Chem. Phys.*, **17**, 63 (1970).
93. R. Kubo, *J. Math. Phys.*, **4**, 174 (1963); *J. Phys. Soc. Japan*, **17**, 1100 (1962).
94. Ref. 38, Appendix B.
95. N. Bloembergen, E. M. Purcell, and R. V. Pound, *Phys. Rev.*, **73**, 679 (1948).
96. H. Shimizu, *J. Chem. Phys.*, **37**, 765 (1962); **40**, 754 (1964).
97. A. Gierer and K. Wirtz, *Z. Naturforsch.*, **8a**, 532 (1951); N. E. Hill, W. E. Vaughn, A. H. Price, and M. Davies, *Dielectric Properties and Molecular Behavior*, van Nostrand, New York, 1969.
98. C.-M. Hu and R. Zwanzig, *J. Chem. Phys.*, **60**, 4354 (1974).
99. A. Szöke, E. Courtens, and A. Ben Reuven, *Chem. Phys. Letters*, **1**, 87 (1967); V. S. Starunov, E. V. Tiganov, and I. L. Fabelinskii, *J.E.T.P. Letters*, **4**, 176 (1966); **5**, 260 (1967); H. C. Lucas, D. A. Jackson, J. G. Powles, and B. Simic-Glavaski, *Mol. Phys.*, **18**, 505 (1970).
100. M. Bloom and H. S. Sandu, *Can. J. Phys.*, **40**, 289 (1962); G. A. deWit and M. Bloom, *Can. J. Phys.*, **43**, 986 (1965).
101. J. H. Rugheimer and P. S. Hubbard, *J. Chem. Phys.*, **39**, 522 (1963).
102. M. Bloom, F. Bridges, and W. H. Hardy, *Can. J. Phys.*, **45**, 3533 (1967).
103. G. J. Gerritsma, P. H. Oosting, and N. J. Trappeniers, *Physica*, **51**, 381 (1971); P. H. Oosting and N. J. Trappeniers, *Physica*, **51**, 395, 418 (1971).

104. K. Lalita, *Can. J. Phys.*, **52**, 876 (1974).
105. B. A. Dasannacharya and G. Vekataraman, *Phys. Rev.*, **156**, 196 (1967); references to previous work are given therein.
106. G. Venkataraman, B. A. Dasannacharya, and R. K. Rao, *Phys. Rev.*, **161**, 133 (1967).
107. G. Lévi and M. Chalaye, *Chem. Phys. Letters*, **19**, 263 (1973).
108. R. L. Armstrong, S. M. Blumenfeld, and C. G. Gray, *Can. J. Phys.*, **46**, 1331 (1968).
109. A. Cabana, R. Bardoux, and A. Chamberland, *Can. J. Chem.*, **47**, 2915 (1969).
110. J. P. Marsault, F. Marsault-Herail, and G. Lévi, *Mol. Phys.*, **26**, 997 (1973).
111. R. E. D. McClung, *J. Chem. Phys.*, **55**, 3459 (1971).
112. T. E. Eagles and R. D. McClung, *J. Chem. Phys.*, **59**, 435 (1973); **61**, 4070 (1974).
113. V. F. Sears, *Can. J. Phys.*, **45**, 237 (1967).
114. R. G. Gordon, *J. Chem. Phys.*, **42**, 3658 (1965); **43**, 1307 (1965).
115. A. A. Maryott, M. S. Malmberg, and K. T. Gillen, *Chem. Phys. Letters*, **25**, 169 (1974).
116. P. S. Hubbard, *Phys. Rev.*, **131**, 1155 (1963).
117. S. C. Wofsy, J. S. Munter, and W. Klemperer, *J. Chem. Phys.*, **53**, 4005 (1970).
118. R. E. D. McClung, *J. Chem. Phys.*, **51**, 3842 (1969) (Equations (52), (53), (61) and (62)).
119. J. A. Courtney and R. L. Armstrong, *J. Chem. Phys.*, **52**, 2158 (1970); references to other work are contained therein.
120. G. Herzberg, *Infrared and Raman Spectra of Polyatomic Molecules*, van Nostrand, New York, 1945, pp. 118, 2533, 63, 364.
121. W. B. Moniz, J. A. Dixon, and W. A. Steele, *J. Chem. Phys.*, **38**, 2418 (1963).
122. R. W. Mitchell and M. Eisner, *J. Chem. Phys.*, **33**, 86 (1960); M. Eisner and R. W. Mitchell, *Bull. Am. Phys. Soc.*, **6**, 363 (1961); J. G. Powles and R. Figgins, *Mol. Phys.*, **10**, 155 (1966).
123. G. W. Nederbragt and C. A. Reilly, *J. Chem. Phys.*, **24**, 1110 (1956); J. G. Powles, *Ber. Bunsenges. Phys. Chem.*, **67**, 328 (1963).
124. W. Haeberlen, H. W. Speiss, and D. Sweitzer, *J. Mag. Res.*, **6**, 39 (1972).
125. G. C. Levy, *J. C. S. Chem. Comm.*, **1972**, 47; L. M. Jackman, E. S. Greenberg, N. M. Szeverenyi, and G. K. Schnorr, *J. C. S. Chem. Comm.*, **1974**, 141.
126. J. G. Powles, M. Rhodes, and J. H. Strange, *Mol. Phys.*, **11**, 515 (1966).
127. K. T. Gillen and J. E. Griffiths, *Chem. Phys. Letters*, **17**, 359 (1972).
128. D. A. Jackson and B. Simic-Glavaski, *Mol. Phys.*, **18**, 393 (1970).
129. G. R. Alms, D. R. Bauer, J. I. Brauman, and R. Pecora, *J. Chem. Phys.*, **58**, 5570 (1973).
130. I. L. Fabelinskii, *Trudy Fiz. Inst. Akad. Nauk SSSR*, **9**, 181 (1958).
131. V. S. Starunov, *Opt. Spektrosk.*, **18**, 2 (1965).
132. F. J. Bartoli and T. A. Litovitz, *J. Chem. Phys.*, **56**, 413 (1972).
133. D. J. Winfield and D. K. Ross, *Mol. Phys.*, **24**, 253 (1972).
134. V. Trepădus, S. Răpeau, I. Pădureanu, V. A. Parfenov, and A. G. Novikov, *J. Chem. Phys.*, **60**, 2832 (1974).
135. J. G. Powles, *Adv. Phys.*, **22**, 1 (1973).
136. L. Glass and S. Rice, *Phys. Rev.*, **165**, 186 (1968).
137. T. Fujiyama and B. Crawford, Jr., *J. Phys. Chem.*, **73**, 4040 (1969); C. E. Favelukes, A. A. Clifford, and B. Crawford, Jr., *J. Phys. Chem.*, **72**, 962 (1968).
138. W. G. Rothschild, *J. Chem. Phys.*, **53**, 3265 (1970).
139. H. S. Goldberg and P. S. Pershan, *J. Chem. Phys.*, **58**, 3816 (1973).
140. J. E. Griffiths, *Chem. Phys. Letters*, **21**, 354 (1973).
141. M. Constant and R. Fauquembergue, *J. Chem. Phys.*, **58**, 4030 (1973); M. Constant, M. Delhaye, and R. Fauquembergue, *Compt. Rend. B*, **271**, 1117 (1970).
142. R. B. Wright, M. Schwartz, and C. H. Wang, *J. Chem. Phys.*, **58**, 5125 (1973).
143. N. Constant and R. Fauquembergue, *Compt. Rend. B*, **272**, 1293 (1971).
144. K. T. Gillen, M. Schwartz, and J. Noggle, *Mol. Phys.*, **20**, 899 (1971).

145. C. G. Gray, *J. Phys. B*, **4**, 1661 (1971); H. B. Levine and G. Birnbaum, *J. Phys. Rev.*, **154**, 86 (1967); H. B. Levine, *Phys. Rev.*, **160**, 159 (1967); I. Ozier and K. Fox, *J. Chem. Phys.*, **52**, 1416 (1970).

146. N. W. B. Stone and D. Williams, *Mol. Phys.*, **10**, 87 (1965); L. Marabella and G. E. Ewing, *J. Chem. Phys.*, **56**, 5445 (1972); W. Ho, G. Birnbaum, and A. Rosenberg, *J. Chem. Phys.*, **55**, 1028 (1971); G. Birnbaum, W. Ho, and A. Rosenberg, *J. Chem. Phys.*, **55**, 1039 (1971); L. Mannik, A. R. W. McKellar, N. Rich, and J. C. Stryland, *Can. J. Phys.*, **48**, 95 (1970).

147. J. R. Lyerla, D. M. Grant, and C. H. Wang, *J. Chem. Phys.*, **55**, 4676 (1971); J. R. Lyerla, D. M. Grant, and R. K. Harris, *J. Phys. Chem.*, **75**, 585 (1971).

148. L. Pauling, *The Nature of the Chemical Bond*, 2nd ed., Cornell Univ. Press, Ithaca, 1948, p. 189.

149. L. Yarwood, *Adv. Mol. Relaxation Processes*, **5**, 375 (1973); J. E. Griffiths, *J. Chem. Phys.*, **59**, 751 (1973); T. E. Bull and J. Jonas, *J. Chem. Phys.*, **53**, 3315 (1970); T. T. Bopp, *J. Chem. Phys.*, **47**, 3621 (1967); D. E. Woessner, B. S. Snowden, Jr., and E. T. Strom, *Mol. Phys.*, **14**, 265 (1968).

150. D. L. Vander Hart, *J. Chem. Phys.*, **60**, 1858 (1974); G. R. Alms, D. R. Bauer, J. I. Brauman, and R. Pecora, *J. Chem. Phys.*, **59**, 5310 (1973); I. Laulicht and S. Meirman, *J. Chem. Phys.*, **59**, 2521 (1973); D. L. Hogenboom, D. E. O'Reilly, and E. M. Peterson, *J. Chem. Phys.*, **52**, 2793 (1970); R. R. Shoup and T. C. Farrar, *J. Mag. Res.*, **7**, 488 (1972); W. T. Huntress, *J. Phys. Chem.*, **73**, 103 (1969).

151. K. T. Gillen and J. H. Noggle, *J. Chem. Phys.*, **53**, 801 (1970).

152. D. Wallach, *J. Chem. Phys.*, **47**, 5258 (1967); **50**, 1219 (1969); D. E. Woessner, *J. Chem. Phys.*, **36**, 1 (1962); C. P. Smyth, in *Molecular Relaxation Processes*, Academic Press, New York, 1966, p. 1; Y. K. Levine, P. Partington and G. C. K. Roberts, *Mol. Phys.*, **25**, 497 (1973); S. J. Seymour and J. Jonas, *J. Mag. Res.*, **8**, 376 (1972); Y. K. Levine, N. J. M. Birdsall, A. G. Lee, J. C. Metcalfe, P. Partington, and G. K. C. Roberts, *J. Chem. Phys.*, **60**, 2890 (1974); D. Wallach, *J. Phys. Chem.*, **73**, 307 (1969); C. P. Smyth, *Adv. Mol. Relaxation Processes*, **1**, 1 (1967); J. Crossley, *Adv. Mol. Relaxation Processes*, **6**, 39 (1974).

153. J. Barojas, D. Levesque, and B. Quentrec, *Phys. Rev. A*, **7**, 1092 (1973).

154. B. J. Berne, *Physical Chemistry: An Advanced Treatise*, Vol. VIII B, D. Henderson, Ed., Academic, New York, 1971.

155. D. Kivelson and T. Keyes, *J. Chem. Phys.*, **57**, 4599 (1972); T. Keyes, *J. Chem. Phys.*, **57**, 767 (1972).

156. F. Bliot, C. Abbar, and E. Constant, *Mol. Phys.*, **24**, 241 (1972); G. W. Parker and F. Lado, *Phys. Rev. B*, **8**, 3081 (1973).

157. B. Quentrec and B. Bezot, *Mol. Phys.*, **21**, 879 (1974).

158. D. Chandler, *J. Chem. Phys.*, **60**, 3508 (1974).

ROLES OF REPULSIVE AND ATTRACTIVE FORCES IN LIQUIDS: THE EQUILIBRIUM THEORY OF CLASSICAL FLUIDS

HANS C. ANDERSEN*

*Department of Chemistry,
Stanford University,
Stanford, California*

DAVID CHANDLER*

*School of Chemical Sciences,
University of Illinois,
Urbana, Illinois*

JOHN D. WEEKS

*Bell Laboratories,
Murray Hill, New Jersey*

CONTENTS

* Alfred P. Sloan Fellow

Abstract

The attractive and the repulsive intermolecular forces play very different roles in forming the equilibrium structure of a dense liquid. The harsh repulsive forces (which fix the *shape* of the molecules) essentially determine the high-density structure. The effect of the attractive forces on the structure can often be ignored or treated by perturbation theory. This idea, which goes back to the time of van der Waals, forms the basis for a quantitative theory of the equilibrium structural and thermodynamic properties of liquids which is reviewed in this article. The structure due to the repulsive forces alone is related to that of the hard sphere model fluid by using an accurate perturbation technique called the "blip function" method. A very simple and accurate theory for the thermodynamic properties of simple dense liquids follows using only the first term in the high-temperature (weak interaction) expansion of the partition function in powers of the attractive interaction.

The effect of the attractive forces on the structure (important at lower density) can be calculated using a generalization of the Mayer cluster theory for ionic solutions called the optimized cluster theory (OCT). By a diagramatic summation, the bare attractive potential is replaced by a renormalized potential. The renormalized potential becomes progressively weaker as the density is increased and the repulsive molecular cores are packed more closely together, thus screening out most of the effects of the attractive forces. A theoretical analysis of the diagrams summed in the OCT as well as numerical results are presented. Some of the implications of these ideas to a number of areas of interest in the liquid state are discussed; for example, the theory of freezing, the static structure of complex molecular liquids and liquid crystals, and dynamical properties of liquids. The relationship between the OCT and the well-known mean-spherical-model integral equation and the diagrams it sums is also given.

I. INTRODUCTION

This article is a review of a theory of liquids that has been developed during the past few years.[1-5] The principal physical concepts and major simplifying feature in this theory originated with the work of van der Waals long ago. It is the idea that for a dense fluid the harsh repulsive forces (which are nearly hard core interactions) dominate the liquid structure. This means the *shape* of molecules determines the intermolecular correlations. Attractive forces, dipole–dipole interactions, and other slowly varying interactions play a minor role in the structure. Their effect on the thermodynamic properties of a fluid is essentially that of a mean-field or uniform background potential. As a result, if a dense liquid is composed of spherical (or nearly spherical) molecules, the intermolecular structure should be very similar to that of a fluid made up of hard spheres.

In recent years, an important use of the historic van der Waals idea was the theory of freezing developed by Longuet-Higgins and Widom.[6] At about the same time, Reiss[7] concluded on the basis of the scaled particle theory that a liquid has "its volume or density determined by the soft [attractive] part of the intermolecular potential. Once this volume is established one may consider the liquid as a hard sphere fluid confined within it." Other works that were based on the van der Waals concept and set the foundation for the theory reviewed herein are the perturbation theory of liquids developed by Barker and Henderson[8] and Verlet's[9] hard sphere model for the structure of simple liquids.

A qualitative explanation of why the repulsive intermolecular forces dominate the structure of most dense fluids follows from a description of the environment of a particle in a liquid. For simplicity, consider an atomic liquid. The phase diagram is shown in Fig. 1. High density corresponds to thermodynamic states at which $\rho^{-1/3} \lesssim r_0$, where ρ is the average particle density, and r_0 is the location of the minimum in the intermolecular pair potential. A glance at the phase diagram shows that "high density" characterizes most of the liquid phase outside of the critical region. Note that $\rho^{-1/3}$ provides an estimate of the average separation between nearest neighbors. Thus, in a dense liquid, nearest neighbors are crushed extremely close to one another. Any displacement of a particle will cause a large change in the energy associated with the interparticle repulsions. The change in energy associated with the attractions will be relatively small because these interactions are not quickly varying functions of the interparticle separation. As a result, the high-density structure is determined mainly by the repulsive forces.

If the attractions were strong and quickly varying, they too would play an important role in the liquid structure. However, there are few one-component fluids for which this is the case. Liquid water is one of these few exceptions to the van der Waals idea. The hydrogen bond interactions between water molecules are as quickly varying as most of the repulsive forces. Thus, the hydrogen bond plays a crucial role in determining the local tetrahedral arrangement of the water molecules.

Fortunately, the van der Waals idea is correct for nearly all nonhydrogen-bonding dense fluids. In order to demonstrate its quantitative validity, one must be able to calculate the intermolecular correlations that are produced

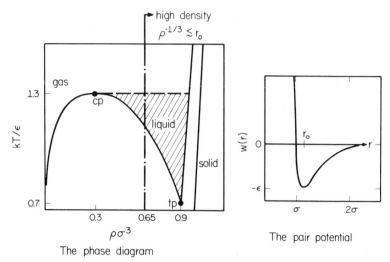

Fig. 1. Phase diagram and intermolecular pair potential for the Lennard–Jones fluid. For particular choices of the length and energy parameters, σ and ε, the Lennard–Jones system is a qualitatively accurate model for several liquids. (See Ref. 23). Notice that the density at the critical point (cp) is roughly one third the density at the triple point (tp); and the critical temperature is about twice the triple temperature. This approximate scaling holds for nearly all nonhydrogen bonding liquids that are composed of molecules that are relatively small (having $\lesssim 10$ atoms). Thus, the phase diagram shown here is a qualitative representation of the phase diagram for many one-component fluids. The "high-density" region comprises thermodynamic states for which the particle density, ρ, is greater than about twice the critical density, ρ_c.

by the repulsive interactions. This is the subject of Section II where the blip function theory is discussed. The blip function theory is derived by expanding the properties of a general repulsive force system about those of a hard sphere fluid. The theory provides a rigorous relationship between the equilibrium properties of the hard sphere fluid and the properties of fluids with realistic repulsive forces. This produces a major simplification of the classical many-body problem since the configurational properties of the hard sphere fluid are independent of temperature and scale according to one length, the hard sphere diameter. Furthermore, since the properties of the hard sphere system are known from the results of computer simulations,[10] the relationship allows one to calculate equilibrium properties of systems with continuous repulsive forces.

The blip-function method is used in Section III to calculate the equilibrium pair correlation function produced by the repulsive forces in the Lennard–Jones fluid (a model for atomic liquids). The repulsive force structure is then compared with the structure produced by the full Lennard–Jones potential. The comparison demonstrates that at high densities the pair correlations in the Lennard–Jones fluid are indeed dominated by the repulsive forces. The thermodynamic ramifications of this structural phenomenon are also discussed in Section III.

At low and moderate densities, molecules in a fluid are not very close together, and the attractive forces do play a significant role in the intermolecular structure. A theory to describe the role of attractive forces is discussed in Section IV. The principal procedure in this theory is a rearrangement of the Mayer cluster expansion which allows one to replace the attractive interactions with a renormalized potential.[5] This renormalized (or screened) potential embodies the repulsive force screening which is present in dense fluids. "Repulsive force screening" denotes the mechanism for the reduction of the effect of attractions (and other slowly varying forces) as the density is increased; at high density the particles are so close to one another that the repulsive forces form a structure (essentially due to excluded volume effects) which is not appreciably changed by the attractive interactions. At low densities, repulsive force screening does not exist, and the renormalized potential becomes, in the limit of zero density, the bare attractive interaction.

The principal results obtained from the renormalized cluster expansion are called the optimized cluster theory (OCT). This theory can be used to calculate equilibrium properties of liquids at moderate and low, as well as high, densities. The results of some of these calculations are presented in Section IV.

There are two basic theoretical techniques that are discussed in Sections II through IV. These are the blip function method to describe the effects of repulsive forces, and the OCT to describe the effects of attractions. Together they provide a unified theory for understanding the equilibrium properties of many gases and liquids. However, they are not without limitations. The blip function method is useful only when the molecules in a liquid are fairly spherical. The OCT is reliable only when the density times the renormalized potential is small. The limitations are discussed more fully in Sections II and IV.

There are many important implications to the theory reviewed in this article. They are outlined in Section V. Some of the topics included in the discussion are as follows: the theory of freezing; molecular motion in liquids (i.e., transport coefficients, rotational relaxation times; and time correlation functions); and the static structure of complex molecular liquids.

Appendix A contains a summary of some terminology concerning cluster diagrams. This terminology is used in Section IV. Appendix B presents the diagramatic formulation of the mean-spherical-model integral equation. This equation has been the focus of great interest in recent years and it is intimately related to the theory developed in Section IV.

In closing this Introduction we note four points concerning the format of this article. First, it is not a comprehensive review of all the modern theories of liquids. Rather, the scope is limited to emphasizing the different roles of repulsive and attractive forces in liquids. For comprehensive reviews the reader is referred to the articles by Barker and Henderson[11] and Andersen.[12] Second, the reader will find that the theory required to describe the role of attractive forces on the liquid structure is inherently more complicated then the formalism needed to describe the effects of the repulsions. Furthermore, the short-ranged harsh repulsions are usually all that one needs to study in order to understand the equilibrium intermolecular structure. For these reasons, the article has been constructed so that Sections II, III, and V can be read independently of Section IV. Third, we consider explicitly simple classical fluids only. However, the theory we present is generalizable to more complex systems. Some applications and extensions of the theory which we will not refer to directly but may be of interest to the reader are the theory of liquid mixtures by Lee and Levesque,[13] studies of molecular fluids by Steele and Sandler[14] and by Sandler et al.,[15] the theory of quantum fluids by Kalos et al.,[16] Shiff's work on astrophysics,[17] and the theory of polar liquids by Stell et al.[18] Finally, we have tried not to simply duplicate the material found in our earlier publications on the blip function expansion and the OCT. Rather, we present in this article relatively different ways of developing those theories. As a result, this review should be regarded as a supplement to Refs. 1–5 and not as a replacement.

II. REPULSIVE FORCES AND THE HARD SPHERE MODEL

Although many workers from the time of van der Waals on realized the the importance of repulsive forces in determining a dense liquid's structure, this insight did not immediately yield a quantitative theory of liquids. It proved very difficult to obtain an accurate description of even the simplest model for a repulsive force system, the hard sphere fluid, so verification and further development of these ideas were not possible.

Thus it is hard to overestimate the importance of the molecular dynamics and Monte Carlo "computer experiments" on the hard sphere fluid which became feasible around 1960[19] and the derivation[20] and the exact solution [21,22] of the Percus–Yevick equation for hard spheres. These gave us for the first time accurate structural and thermodynamic data for a model fluid system over the entire fluid phase. The liquid state was less well understood than the gaseous or solid state primarily because of the lack of a model system, comparable to the ideal gas or the harmonic solid. The modern work on the hard sphere fluid has provided a similarly useful, though necessarily more complicated, starting point for theories of the liquid state.

Analytical formulae are now available which accurately reproduce the results of machine computations on the hard sphere fluid both for the thermodynamic properties and for the pair correlation functions.[10,23] If these hard sphere results are in fact to be useful for quantitative calculations on real fluids, we must relate the properties of the idealized hard sphere fluid to those of realistic fluids with their smoothly varying repulsive forces. A relationship can be established by considering functional Taylor expansions for the equilibrium properties of a general repulsive force fluid.[4] Articles by Barker[24] and Gubbins et al.[25] provide an alternative approach that could be used to arrive at the same final results [(2.11), (2.16) and (2.17)].

We discuss here the simplest case where the repulsive interactions in the fluid can be represented by "soft sphere" pair potentials $u_R(r)$ which depend only on the scalar distance r between pairs of molecules. Although the smooth $u_R(r)$ may seem to be very different from the discontinuous hard sphere potential $u_d(r)$, which is 0 for $r > d$ and infinite for $r < d$ (their difference being infinite for $r < d$), the thermodynamic and structural properties of both the hard and soft sphere systems are determined by the Mayer cluster functions $f_d(r) = e^{-\beta u_d(r)} - 1$ and $f_R(r) = e^{-\beta u_R(r)} - 1$, respectively, which are very similar to each other [see (2.2) and (2.6)]. Indeed, for a reasonable choice of hard sphere diameter d, the f functions differ from each other only over a small region of space if $u_R(r)$ is a harshly repulsive potential (see Fig. 2). Here $\beta = 1/k_B T$ where T is the temperature, k_B is Boltzmann's constant and the subscript R denotes the soft sphere repulsive force system and d denotes a hard sphere system with diameter d.

We can exploit this similarity by considering a "test system" where the f function $f_\mu(r)$ gradually changes from that of the hard sphere system to that of

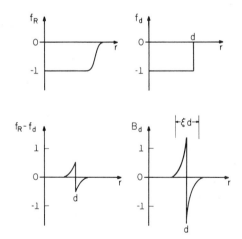

Fig. 2. Schematic plots of some functions considered in the blip function method: (*a*) the Mayer f function for a soft-sphere repulsive potential $f_R(r)$; (*b*) the hard sphere f function $f_d(r)$; (*c*) difference between (*a*) and (*b*); (*d*) blip function $B_d(r)$, showing the significance of the parameter ξ. According to (2.9), the hard sphere diameter d is chosen to make the net area under $r^2 B_d(r)$ equal to zero.

the soft sphere system by means of a coupling parameter μ:

$$f_\mu(r) = f_d(r) + \mu \Delta f(r) \qquad 0 \leq \mu \leq 1$$

where

$$\Delta f(r) = f_R(r) - f_d(r) \tag{2.1}$$

The negative of the dimensionless excess Helmholtz free energy density $\mathscr{A}_\mu$ (with respect to an ideal gas at the same volume V, temperature T, and number density $\rho = N/V$) of a system of N test particles is given by

$$\mathscr{A}_\mu \equiv \frac{-\beta \Delta A_\mu}{V} = V^{-1} \ln Q_\mu \tag{2.2}$$

where

$$Q_\mu = V^{-N} \int \mathbf{dr}^N \prod_{i<j=1}^{N} [1 + f_\mu(r_{ij})] \tag{2.3}$$

Here ΔA_μ is the excess Helmholtz free energy and $\int \mathbf{dr}^N$ denotes an integration over all positions of the N particles in the volume V. Using the familiar coupling parameter technique, we differentiate and then integrate (2.2) with respect to μ and find the exact result

$$\mathscr{A}_R - \mathscr{A}_d = \frac{\rho^2}{2} \int_0^1 d\mu \int \mathbf{dr} \, \Delta f(r) y_\mu(r) \tag{2.4}$$

Here $\mathscr{A}_R$ is the quantity $\mathscr{A}$ for the soft sphere system ($\mu = 1$), $\mathscr{A}_d$ that of the hard sphere system ($\mu = 0$) and

$$y_\mu(r_{12}) = \frac{\dfrac{N(N-1)}{\rho^2} \int \mathbf{dr}^{(N-2)} \prod_{\substack{i<j \\ (i,j) \neq (1,2)}} [1 + f_\mu(r_{ij})]}{\int \mathbf{dr}^N \prod_{i<j} [1 + f_\mu(r_{ij})]} \tag{2.5}$$

Because of the integrations in (2.5), $y_\mu(r)$ is a continuous function of r even when, as in the present case, $f_\mu(r)$ is not. Also $y_\mu(r)$ is positive and is simply related to the pair correlation function $g_\mu(r)$ by

$$[1 + f_\mu(r)] y_\mu(r) = g_\mu(r) \tag{2.6}$$

$g_\mu(r)$ is also called the radial distribution function and is defined so that $4\pi \rho r^2 g_\mu(r) \, dr$ gives the average number of molecules at a distance between r and $r + dr$ from a central molecule.[26]

We anticipate that for a reasonable choice of hard sphere diameter d, the fractional change in $y_\mu(r)$, denoted by

$$\delta y_\mu(r) \equiv \frac{y_\mu(r) - y_d(r)}{y_d(r)} \tag{2.7}$$

will be small. Hence we separate out this fractional change in (2.4):

$$\mathscr{A}_R - \mathscr{A}_d = \frac{\rho^2}{2} \int d\mathbf{r}\, B_d(r) + \frac{\rho^2}{2} \int_0^1 d\mu \int d\mathbf{r}\, B_d(r)\delta y_\mu(r) \qquad (2.8)$$

where $B_d(r)$ is the "blip function" given by

$$B_d(r) \equiv y_d(r)\Delta f(r) \qquad (2.9)$$

Note that $B_d(r)$, and not simply $\Delta f(r)$, appears naturally. The product $y_d(r)\Delta f(r)$ contains the physical effect of the medium. As the density increases, the medium makes it more likely that a pair of particles will be separated by distances at which $\Delta f(r)$ is nonzero. As a result, the effects of $\Delta f(r)$ grow as the density increases.

Both $\Delta f(r)$ and $B_d(r)$ are nonzero over only a small range of r near $r = d$. Let ξd denote the range of $B_d(r)$. Then the dimensionless parameter ξ can be defined as

$$\xi \equiv \frac{1}{d}\int_0^\infty |B_d(r)|\, dr$$

$$\cong \frac{1}{4\pi d^3}\int |B_d(r)|\, d\mathbf{r} \qquad (2.10)$$

For harshly repulsive potentials ξ, the "softness" parameter, is much less than unity (see Fig. 2). It is zero only when the continuous repulsive potential becomes the hard sphere potential.

We want to relate the soft sphere thermodynamic properties to those of a hard sphere system with diameter d chosen to make the correction terms on the right-hand side of (2.8) as small as possible. A particularly appealing choice for d is one that makes the first term, which is apparently of order ξ, vanish. That is, we pick d such that

$$\int B_d(r)\, d\mathbf{r} = \int [e^{-\beta u_R(r)} - e^{-\beta u_d(r)}]y_d(r)\, d\mathbf{r} = 0 \qquad (2.11)$$

This is indeed a felicitous choice for d, since we will show that the complicated second term in (2.8) is then of order ξ^4 and thus can frequently be neglected entirely. Furthermore $\delta y_\mu(r)$ is of order ξ^2 with this same choice of d.

To show these facts, differentiate and integrate (2.5) with respect to μ. After integration over irrelevant coordinates the result can be written in the general form

$$y_\mu(r) = y_d(r)\left[1 + \int_0^\mu d\mu' \int d\mathbf{s}\, K_{\mu'}(\mathbf{r}, \mathbf{s})B_d(s)\right] \qquad (2.12)$$

or, using (2.7), and integrating over the angles of **s**,

$$\delta y_\mu(r) = \int_0^\mu d\mu' \int_0^\infty ds \, s^2 \overline{K}_{\mu'}(r, s) B_d(s) \tag{2.13}$$

where

$$\overline{K}_\mu(r, s) = \int_0^\pi \sin \theta_s \, d\theta_s \int_0^{2\pi} d\varphi_s K_\mu(\mathbf{r}, \mathbf{s}) \tag{2.14}$$

Here $\overline{K}_\mu(r, s)$ is a complicated function expressible in terms of integrals over higher-order distribution functions.* Fortunately for the present purposes we need know only some of its general properties. In an isotropic fluid $\overline{K}_\mu(r, s)$ can depend only on the magnitudes of the vectors **r** and **s** since there is no distinguishable direction left after the angular integration in (2.14). This integration also smooths out the jump discontinuities which can occur in the full kernel $K_\mu(\mathbf{r}, \mathbf{s})$ and leaves $\overline{K}_\mu(r, s)$ a continuous function of s. Recalling (2.11), we may rewrite (2.13) as

$$\delta y_\mu(r) = \int_0^\mu d\mu' \int_0^\infty ds \, s^2 [\overline{K}_{\mu'}(r, s) - \overline{K}_{\mu'}(r, d)] B_d(s) \tag{2.15}$$

$B_d(s)$ is nonzero only for a small range of s near $s = d$. This range has width of approximately ξd. In this region the term in square brackets in (2.15) is $0(\xi)$, since $\overline{K}_\mu$ is a continuous function of s. The term $B_d(s)$ provides another factor of $0(\xi)$ in the integrand, by (2.10). Hence δy_μ is $0(\xi^2)$ for all r. Then for $\mu = 1$,

$$y_R(r) = y_d(r)[1 + 0(\xi^2)]$$

or

$$g_R(r) \equiv e^{-\beta u_R(r)} y_R(r) = e^{-\beta u_R(r)} y_d(r)[1 + 0(\xi^2)] \tag{2.16}$$

Similarly, the last term in (2.8) can be written as

$$\frac{\rho^2}{2} \int_0^1 d\mu \int d\mathbf{r} \, B_d(r)[\delta y_\mu(r) - \delta y_\mu(d)]$$

Using the same argument as before, we would expect this integral to be $0(\xi^2)$ if $\delta y_\mu(r)$, like $\overline{K}_\mu(r, s)$, were of order unity. However we have already shown that $\delta y_\mu(r)$ is $0(\xi^2)$. Thus the integral actually is $0(\xi^4)$ and we have

$$\mathscr{A}_R = \mathscr{A}_d + 0(\xi^4) \tag{2.17}$$

Equation (2.11) gives us a criterion for the choice of a temperature- and density-dependent hard sphere diameter d associated with a system of soft

---

* $K_d(\mathbf{r}, \mathbf{s})$ can be easily derived from the explicit expressions given by Andersen et al.[4]

spheres with potential u_R. Then both the thermodynamics and structure of the soft sphere system can be approximated very simply using (2.16) and (2.17) and the known results for hard sphere systems. The thermodynamic relationship (2.17) is inherently more accurate than the structural one (2.16).

The associated hard sphere diameter d calculated using (2.11) is a decreasing function of temperature and density, in accordance with physical intuition. As the temperature is increased at constant density, the soft sphere molecules have more kinetic energy and can approach each other more closely before finally being repelled by $u_R(r)$. Hence the associated hard sphere diameter decreases. Also when the density is increased at constant temperature, molecules are squeezed closer together so the associated diameter should decrease. The density dependence is much less than the temperature dependence, however.

The correction term of $0(\xi^4)$ to the thermodynamics, given explicitly in (2.8), is also a function of temperature and density and must vanish identically in the low-density limit since $\delta y_\mu(r)$ is $0(\rho)$. The criterion (2.11) assures that (2.17) gives the correct second virial coefficient, and the correct low-density form for $g(r)$ is obtained from (2.16). Thus the lowest-order theory ((2.16) and (2.17)) is exact at low density and when ξ is small should remain accurate even at high density.

Equations (2.16) and (2.17) have been compared to computer "experimental" results for two soft-sphere systems.[4] In the first, $u_R(r)$ is the inverse twelfth-power potential, $\epsilon(\sigma/r)^{12}$ and ξ is rather large ($\xi = 0.35$ at $\rho\sigma^3 = 0.8$ and $k_B T/\epsilon = 1$). In the second, $u_R(r)$ is the repulsive part of the Lennard–Jones potential (see (3.4)) and $\xi = 0.14$ at $\rho\sigma^3 = 0.8$ and $k_B T/\epsilon = 0.8$, so this is a more harshly repulsive potential. For both systems the thermodynamic relationship (2.17) is quantitatively accurate when $\rho d^3 \lesssim 0.93$ (at higher densities the hard sphere fluid is metastable; the hard sphere fluid freezes at $\rho d^3 = 0.93$). ξ is small enough in the repulsive force Lennard–Jones system that both the structural relation (2.16) and the thermodynamic relation (2.17) are very accurate. This system forms the basis of our discussion of the Lennard-Jones liquid in the next section.

When ξ is large, the structural relation (2.16) would be expected to fail at high density. This is the case for the inverse twelfth-power potential,[4,27] and the correction term given in (2.12) is important. We have written down explicitly[4] the correction term to the next order in ξ, which arises when $\overline{K}_\mu(r, s)$ is approximated by $\overline{K}_d(r, s)$ in (2.13). This term can be calculated using presently available data on the hard sphere fluid.[11] This process could in principle be continued with further terms in the Taylor series expansion of $\overline{K}_\mu(r, s)$ about $\mu = 0$ but the additional correction terms require complicated averages over high-order hard sphere distribution functions. What is needed is an approximation for the entire integral over μ in (2.13), that is, a summation to all orders of ξ of the most important terms in the Taylor series. It may be possible to develop such a theory using the types of techniques to be discussed later in Section IV.

Fortunately, for many applications the correction terms are not needed. Further, when studying particular systems one can determine *a priori* whether the corrections are needed for accurate results. This is done by testing the internal consistency of the simple theory given in (2.16) and (2.17). The thermodynamic properties calculated from (2.17) can be compared with those calculated from (2.16) using the energy or virial equations (see (3.9) and (3.10)). The results of the two separate calculations differ to order ξ^2. If the results agree, then ξ is small enough that both equations are accurate. This has been the case in a number of applications.

The blip function expansion can be easily generalized to multicomponent systems and molecular systems with nonspherical pair potentials, requiring essentially only a change of notation.[28] For example, if the molecular pair potential is of the form $u_R(\mathbf{r}_i, \mathbf{r}_j, \mathbf{\Omega}_i, \mathbf{\Omega}_j)$, where $\mathbf{\Omega}_i$ and $\mathbf{\Omega}_j$ are Euler angles giving the orientation of molecules i and j, then an approximation for the pair correlation function is

$$g_R(\mathbf{r}_i, \mathbf{r}_j, \mathbf{\Omega}_i, \mathbf{\Omega}_j) \cong e^{-\beta u_R(\mathbf{r}_i, \mathbf{r}_j, \mathbf{\Omega}_i, \mathbf{\Omega}_j)} y_d(|\mathbf{r}_i - \mathbf{r}_j|) \tag{2.18}$$

where d is chosen by a generalization of (2.11) which includes an integration over the Euler angles. However, this simple approach is accurate only when the molecules are fairly spherical so that ξ is small.

Thus the blip function method has wide applicability and is very simple to use in practice. In particular it permits us to give a quantitative discussion of the properties of simple liquids, the subject of the next section.

III. ROLE OF REPULSIVE FORCES IN LIQUIDS

Now that we can deal accurately with the repulsive forces in a fluid, we will apply the van der Waals ideas discussed in the Introduction to calculate the thermodynamic and structural properties of realistic fluids, which have attractive as well as repulsive forces.[2,3] As an example, we consider the Lennard–Jones fluids, where the pair potential energy is given by

$$w(r) = 4\epsilon \left[\left(\frac{\sigma}{r} \right)^{12} - \left(\frac{\sigma}{r} \right)^6 \right] \tag{3.1}$$

Here σ has dimensions of length and ϵ of energy. This model fluids gives a fairly accurate description of the properties of rare gas liquids and has been extensively studied by computer simulations to which we can compare our results.[29,30] The main purpose of this comparison is to gain confidence in the validity of these ideas before applying them to more complicated systems.

A. Separation of the Pair Potential

As discussed in Section I, the structure of a liquid at high density should be determined mainly by the repulsive forces. Thus we separate the intermolecular potential into a *reference* part $u_0(r)$ containing all the repulsive forces and a *perturbation* part $u(r)$ containing all the attractive forces:

$$w(r) = u_0(r) + u(r) \tag{3.2}$$

Note that the force determined from (3.1) is repulsive for all $r < r_0$. Here $r_0 = 2^{1/6}\sigma$ is the distance at which the potential reaches its minimum value. If $u_0(r)$ is to contain all these repulsive forces, no other repulsions and no attractive forces we require

$$\frac{du_0(r)}{dr} = \frac{dw(r)}{dr} \qquad r \leq r_0$$
$$u(r) = 0 \qquad r > r_0 \tag{3.3}$$

Equation (3.3) and (3.2) then determine $u_0(r)$ and $u(r)$ uniquely as

$$\begin{aligned} u_0(r) &= w(r) - w(r_0), & r &\leq r_0 \\ &= 0, & r &> r_0 \end{aligned} \tag{3.4a}$$

and

$$\begin{aligned} u(r) &= w(r_0), & r &\leq r_0 \\ &= w(r), & r &> r_0 \end{aligned} \tag{3.4b}$$

where $w(r_0) = -\epsilon$ is the minimum value of the potential (3.1). This potential separation is shown in Fig. 3.

With this separation, the physical arguments of Section I suggest that at high densities the effects of the attractive interactions $u(r)$ on the structure

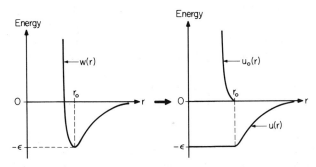

Fig. 3. The unique separation of the Lennard–Jones potential, $w(r)$, into a part $u_0(r)$ containing all the repulsive interactions in $w(r)$ (and no attractions), and a part $u(r)$ containing all the attractive interactions in $w(r)$ (and no repulsions).

of the liquid should be small. In the simplest approximation we ignore them entirely and get

$$g(r) \cong g_0(r) \tag{3.5}$$

Thus the structure of the liquid should be very similar to that of the hypothetical reference fluid which has the same repulsive forces and no attractive forces.

B. Thermodynamic Perturbation Theory

The thermodynamic consequences of this postulated structural behavior are easy to calculate, using coupling parameter methods similar to those in Section II. Consider a "λ system" with pair potential

$$w_\lambda(r) = u_0(r) + \lambda u(r) \qquad 0 \le \lambda \le 1 \tag{3.6}$$

Differentiating and integrating the canonical partition function with respect to λ, we find the exact result

$$\mathscr{A} - \mathscr{A}_0 = -\frac{\beta \rho^2}{2} \int_0^1 d\lambda \int d\mathbf{r} \, u(r) g_\lambda(r) \tag{3.7}$$

Here $g_\lambda(r)$ is the pair correlation function when the pair potential is $w_\lambda(r)$, $\mathscr{A}$ is the negative of the dimensionless excess free energy density for the Lennard–Jones system ($\lambda = 1$) and $\mathscr{A}_0$ that for the reference system ($\lambda = 0$). If attractive forces have little effect on the fluid's structure, then $g_\lambda(r) \cong g_0(r)$ and the λ integration in (3.7) is trivial. We then get the high-temperature approximation (HTA),

$$\mathscr{A} \cong \mathscr{A}_0 - \frac{\beta \rho^2}{2} \int d\mathbf{r} \, u(r) g_0(r) \tag{3.8}$$

so named because the attractions would be expected to be of little importance at high enough temperatures. However, the arguments given in the Introduction suggest that (3.5) and hence (3.8) should also be accurate at high density even when the temperature is not high. To test the accuracy of (3.5) and (3.8) one must be able to evaluate $g_0(r)$ and $\mathscr{A}_0$. This is easily done by applying (2.16) and (2.17), respectively, to the Lennard–Jones repulsion $u_0(r)$. The results reported below are calculated in just this way.

Equation (3.8) is usually given as the first term in the high-temperature (weak interaction) series[31,32] which can be derived by expanding $\mathscr{A}$ in powers of the perturbation potential. However, the higher-order terms are so complicated it is difficult to tell whether the weak interaction expansion converges at liquid temperatures or when the first-order result (3.8) will be accurate. The closed expression (3.7) shows it is the effect of the perturbation potential on the liquid's structure that determines the accuracy of the first approximation (3.8). Thus it is very important to choose a potential separation where $g_0(r) \cong g(r)$ if a simple first-order thermodynamic perturbation theory is to be accurate.

C. Results

Earlier perturbation theories of liquids[8,33] proposed different separations of the potential into reference and perturbation parts. These earlier reference systems differ considerably from the repulsive force reference system, and the perturbation parts contain some strong and rapidly varying forces. As a result, the effect of the perturbation potential on the fluid's structure is large and (3.5) and (3.8) are not accurate for these separations at liquid temperatures and densities.

Figure 4 gives $g(r)$ for the Lennard–Jones liquid at a state near the triple point. This is compared with the $g_0(r)$ for the repulsive force reference system and with the references systems used in the earlier thermodynamic perturbation theories of McQuarrie and Katz[33] ($u_0^{MK}(r) = 4\epsilon(\sigma/r)^{12}$) and Barker and Henderson[8] ($u_0^{BH}(r)$ is the positive part of the Lennard–Jones potential). As expected, (3.5) is a poor approximation in the latter theories and thus large

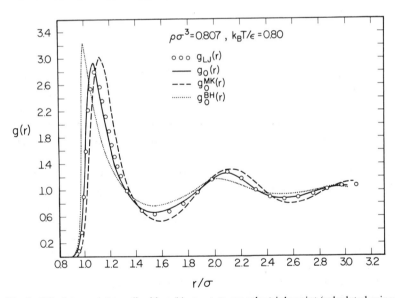

Fig. 4. The Lennard–Jones liquid $g_{LJ}(r)$ at a state near the triple point (calculated using the optimized cluster theory) compared with the repulsive force reference system $g_0(r)$ (calculated using the blip function method) and the correlation functions for the reference systems proposed by Barker and Henderson, $g_0^{BH}(r)$ (calculated using the blip function method), and McQuarrie and Katz, $g_0^{MK}(r)$ (results taken from Monte Carlo calculations of Hansen and Weis[27]). The calculated correlation functions are essentially identical to those that would be given by "exact" Monte Carlo calculations of those functions.

errors result from the use of the HTA ("first-order perturbation theory") for these theories at liquid temperatures and densities.[2,3,10] In contrast, the repulsive force reference system $g_0(r)$ is seen to be very close to the Lennard–Jones $g(r)$. This provides a striking confirmation of the correctness of the physical picture introduced by van der Waals long ago. Figure 4 is typical of the agreement we find between $g(r)$ and $g_0(r)$ for all high-density states.

The main difference between $g_0(r)$ and $g(r)$ is that the first peak in $g(r)$ is moved slightly outward from that of $g_0(r)$. This is easily understood: when attractions are present the neighboring molecules would like to sit in the potential minimum at r_0 for energetic reasons, though only a very small shift is possible because of the packing of the repulsive cores. This small difference in $g(r) - g_0(r)$ is sufficient to cause the use of $g_0(r)$ in the pressure (virial) equation,

$$\frac{\beta P}{\rho} = 1 - \frac{\beta \rho}{6} \int r \frac{dw(r)}{dr} g(r)\, \mathbf{dr} \tag{3.9}$$

to give too large results[10,23] because $dw(r)/dr$ weights the shifted region very heavily. More accurate thermodynamic results arise when $g_0(r)$ is used in the internal energy equation,

$$\frac{\beta \Delta E}{N} = \frac{\beta \rho}{2} \int w(r) g(r)\, \mathbf{dr} \tag{3.10}$$

since $w(r)$ weights the shifted region less strongly than does $dw(r)/dr$. Best of all, however, is the use of the approximation $g_0(r) \cong g_\lambda(r)$ in (3.7) since the smooth $u(r)$ is practically constant over the shifted region. Thus the HTA (3.8) can be very accurate, and quantitative results for the pressure and internal energy of dense liquids can be obtained[3,10,23] by differentiating (3.8) with respect to ρ or β. Table I gives some representative results.

This happy state of affairs extends over that part of the temperature-density plane where (3.5) holds. This includes trivially all high-temperature states ($k_B T/\epsilon \gtrsim 3$) of any density, where $\beta u(r)$ is small, and all high-density states ($\rho \sigma^3 \gtrsim 0.65$) even when the temperature is low and $\beta u(r)$ is large. Thus a large part of the entire phase diagram can be understood using these simple physical ideas, including the dense liquid region where most other approaches have failed.

At lower density and low temperatures and particularly as we approach the critical point the physical ideas suggesting (3.5) and (3.8) no longer hold and we must consider explicitly the effects of the attractive forces on $g(r)$. (This is necessary even at high density if the virial pressures are to be accurate.) The most straightforward approach would be to take the first-order correction to $g_\lambda(r)$ in a power series about $\lambda = 0$, as is done by Barker and Henderson.[11] However, like the μ expansion discussed in Section II, the first-order

TABLE I
Pressure and Internal Energy for the Lennard–Jones
Fluid for Several Representative High-Density States
Calculated from (3.8)

		$\beta P/\rho$		$-\Delta E/N\epsilon$	
$\rho\sigma^3$	$k_B T/\epsilon$	HTA[a]	MD[b]	HTA[a]	MD[b]
0.88	0.94	2.82	2.77	6.03	6.04
0.85	2.202	4.22	4.20	4.77	4.76
0.85	1.128	2.78	2.78	5.69	5.69
0.85	0.76	0.74	0.82	6.06	6.07
0.75	1.304	1.55	1.61	4.99	5.02
0.75	1.071	0.76	0.89	5.15	5.17
0.75	0.84	0.38	0.37	6.01	6.04
0.65	1.585	1.19	1.25	4.20	4.23

[a] Obtained by numerical differentiation of (3.8).
[b] Molecular dynamics calculations summarized in Verlet and Weis.[10] The uncertainty in these results is ± 0.05.

term in λ for $g_\lambda(r)$ involves complicated averages over three- and four-particle correlation functions, and when $\beta u(r)$ is large, it is not clear why just one more term in λ will be sufficient. In the next section we discuss a renormalization procedure which provides a general treatment of the attractive forces. The renormalization of the perturbation gives us a mathematical basis with which we can understand, in part at least, why the repulsive forces at high density prove so effective in screening out the effects of the attractions. However, before turning to this problem, we consider the relationship between the basic theory described in this section—the HTA—and the van der Waals equation, which in a sense, motivated the present theory.

D. Van der Waals Equation

The van der Waals equation for a one-component fluid is

$$P = P_d - a\rho^2 \tag{3.11}$$

where a is a positive constant and P_d denotes the pressure of the hard sphere fluid. With the appropriate choice for a and d, this equation provides a remarkably accurate description of the equation of state for a dense fluid.*

* Van der Waals actually assumed that $P_d = k_B T\rho(1 - b\rho)^{-1}$, where b is a positive constant which describes the volume of a molecule. This particular form for P_d is correct only for a one-dimensional hard rod fluid. Results are considerably improved when more accurate results for P_d are used. See Widom for an excellent discussion.[34]

This fact can be understood within the framework of the theory developed in this section. By employing (2.16) and (2.17) of the blip function expansion, (3.8) can be expressed as

$$\frac{A}{Nk_B T} \cong \frac{A_d}{Nk_B T} - \frac{\rho a(\rho)}{k_B T} \tag{3.12}$$

where

$$\rho a(\rho) = -\frac{1}{2} \int_d^\infty 4\pi \rho r^2 g_d(r) u(r) \, dr \tag{3.13}$$

The only important approximation involved in this result is the assumption that the structure of the fluid is determined by the repulsive forces. It has been demonstrated that this is an excellent approximation at high densities, i.e., $0.65 \lesssim \rho d^3 \lesssim 0.9$.

Strictly speaking, $a(\rho)$ is also a function of temperature. This arises from the temperature dependence of the hard sphere diameter d. However, for realistic harshly repulsive interactions, d is only a weak function of T. Thus to a good approximation the temperature dependence of $a(\rho)$ can be neglected.

Over the limited region of high density, a linear interpolation of the density dependence of $\rho a(\rho)$ should be fairly accurate, that is,

$$\rho a(\rho) \cong a_0 + \rho a \tag{3.14}$$

where a_0 and a are constants chosen to give the best fit to $\rho a(\rho)$ over the high-density region.* Equation (3.14) may seem even more plausible when one realizes that for short-ranged potentials (such as the Lennard–Jones potential) the major contribution to the integral in (3.13) comes for those values of r in the first coordination shell. Since $4\pi \rho r^2 g_d(r) \, dr$ is the number of hard spheres in a shell between r and $r + dr$ that surround a central hard sphere, the density dependence of the integral in (3.13) is essentially the same as the density dependence of the number of nearest neighbors in a hard sphere fluid. This should vary smoothly with density in the high-density region.

Equations (3.14) and (3.12) can be combined and then differentiated with respect to ρ to yield the pressure. Provided one neglects the density dependence in d (an approximation that is even better than (3.14)), one obtains

$$\frac{\beta P}{\rho} = \frac{\beta P_d}{\rho} - \beta a \rho \tag{3.15}$$

which is van der Waals' equation.

* This choice of the constants a_0 and a may differ considerably from values chosen to fit low-density results. The familiar evaluation of the van der Waals parameters in terms of critical point data[32] is not justified since the HTA and hence the van der Waals equation is not accurate in this low-density regime.

Thus, the basic approximations which justify the van der Waals' equation for real dense fluids are as follows: (1) the structure of the fluid is determined by the repulsive forces; and (2) the number of nearest neighbors to a particle in a dense liquid is approximately a linear function of the density. Approximation (2) is reasonable, but not so accurate as approximation (1).

There exists a well-defined, but certainly unrealistic, class of models for which the van der Waals equation is the exact equation of state.[35] These models are classical fluids for which the total pair potential is $u_d(r) + u_K(r)$, where $u_K(r)$ is a Kac potential. The general form for a Kac potential is

$$u_K(r) = \gamma^v F(\gamma r) \tag{3.16}$$

where v denotes the dimensionality of the system,

$$\int \mathbf{dx}\, F(x) = \alpha \tag{3.17}$$

is a finite constant, and γ tends to zero. Thus, $u_K(r)$ is both infinitely weak and infinitely long ranged. It is intuitively obvious that such a weak and slowly varying interaction will have no effect of the fluid structure, and that its effect on the thermodynamics will be one of a simple mean field or background potential.

To study the effect of $u_K(r)$ mathematically one may start from (3.7). For the model system defined above, we have the exact result:

$$\mathscr{A}(T, \rho; u_d + u_K) = \mathscr{A}_d - \frac{\beta\rho^2}{2} \int_0^1 d\lambda \int d\mathbf{r}\, g_\lambda(r) u_K(r)$$

$$= \mathscr{A}_d - \frac{\beta\rho^2}{2} \alpha - \frac{\beta\rho^2}{2} \int_0^1 d\lambda \int \mathbf{dr}[g_\lambda(r) - 1] u_K(r) \tag{3.18}$$

Since $[g_\lambda(r) - 1]$ decays to zero at large r in a one-phase system, the last integral in (3.18) vanishes as γ tends to zero (this is called the Kac limit). As a result,

$$\mathscr{A}(T, \rho; u_d + u_K) = \mathscr{A}_d - \frac{\beta\rho^2}{2} \alpha \tag{3.19}$$

By differentiating this expression for the free energy one obtains van der Waal's equation with $a = -\alpha/2$ in the one-phase region. In the two-phase region, the thermodynamic properties are obtained by a Maxwell construction.

The Kac potential produces no forces between molecules. As a result, it rigorously has no effect on the fluid structure, and as we have shown, the van der Waals equation is exact for the Kac model. However, this development of the van der Waals equation does not justify the use of the equation for describing real fluids since the Kac potential is clearly unphysical. The utility of the van der Waals equation arises from the fact that it is a fairly accurate equation even when one considers fluids with realistic pair potentials. In the preceding discussion, we have described the physical reasons for this accuracy.

IV. EFFECT OF ATTRACTIVE FORCES ON LIQUID STRUCTURE

We have seen above that for dense monatomic liquids the structure of the liquid is dominated by the repulsive part of the interatomic potential. The effect of the attractions is much smaller than we might have expected on the basis of the magnitude of the dimensionless parameter, $\epsilon/k_B T$, which characterizes their strength. This is reminiscent of (and, as we shall see below, formally analogous to) the fact that in fluids of charged particles the range and strength of the interparticle correlations is much smaller than we might have expected on the basis of the range and the strength of the Coulomb potential. This latter fact is sometimes called "screening" or "shielding" of the Coulomb potential. Its earliest theoretical description was in the pioneering work of Debye and Hückel[36] on the subject of dilute ionic solutions. They found that although the interionic potential (divided by $k_B T$) for two ions separated by a distance r is approximately

$$\frac{\beta Z_i Z_j e^2}{\epsilon r} \tag{4.1}$$

at large distances, the strength of their correlations (i.e., the pair correlation functions minus unity) is approximately

$$\frac{-\beta Z_i Z_j e^2}{\epsilon} \frac{e^{-\kappa r}}{r} \tag{4.2}$$

Here Z_i and Z_j are the valences of the ions, e is the magnitude of the electronic charge, ϵ is the dielectric constant of the solvent, and κ is the inverse of the "Debye screening length." Its formula is

$$\kappa^2 = \frac{4\pi\beta e^2}{\epsilon} \sum_i Z_i^2 \rho_i \tag{4.3}$$

The exponential factor in (4.2) guarantees that the correlations are weaker and of shorter range than the potential. Mayer[37] extended the work of Debye and Hückel by using the formalism of cluster expansions for fluids. In his theory the function given by (4.2) appears as a screened or renormalized potential energy of interaction among ions. It described a many-body effect which tends to counteract and cancel the direct interactions among the ions. Some of the Debye–Hückel results were obtained and corrections were expressed in terms of the renormalized potential rather than the original unscreened potential in (4.1). In this section we shall discuss how the ionic cluster theory method of Mayer can be applied to the problem of the structure of dense liquids like the Lennard–Jones fluid. We will define a renormalized potential which plays a role analogous to the Debye–Hückel screened potential. At high densities this renormalized potential is much weaker than the actual attractive potential and seems to describe a many-body screening effect in which the short-ranged repulsive forces tend to counteract and cancel the effect of the attractive forces. We shall call this *repulsive force screening* to distinguish it from the Coulomb screening in ionic solutions.

We will first give a brief review of the cluster theory for the pair correlation function of fluids. The Mayer ionic solution theory will be outlined, including the graphical definition of the screened potential. Then the case of dense fluids will be discussed, showing how the concepts of a renormalized potential can be generalized to cover this situation. The consequences of this for the theory of liquid structure and thermodynamics will then be developed.

A. Cluster Theory of a Fluid of Attracting Hard Spheres

The model fluid of interest has an interatomic potential which is the sum of a hard sphere part plus a perturbation. (We will discuss later the case where there are soft sphere rather than hard sphere repulsive forces.) If $w(r)$ is the interatomic potential, then

$$
\begin{aligned}
w(r) &= \infty & r < d \\
&= u(r) & r > d
\end{aligned}
\tag{4.4}
$$

where $u(r)$ is the perturbation.

Cluster theories of fluids are often expressed in terms of the Mayer f function for the interaction, defined by

$$
f(r) = \exp\left[-\beta w(r)\right] - 1
$$

$$
= f_d(r) + \left[1 + f_d(r)\right] \sum_{n=1}^{\infty} \frac{1}{n!} (\phi(r))^n
\tag{4.5}
$$

where $f_d(r)$ is the Mayer f function for the hard sphere potential, and

$$
\phi(r) = -\beta u(r)
\tag{4.6}
$$

This separation of f into various parts is of use because we expect that in some sense the effect of ϕ on the structure is small compared with the effect of the hard sphere repulsions represented by f_d.

The Mayer cluster theory[38] provides a formula for the pair correlation function of a fluid in terms of an infinite series of diagrams. Each diagram represents an integral whose integration variables represent particle positions and whose integrands contain factors of the Mayer f function of the relative positions of various particles. These diagrams can be expressed in terms of f_d and ϕ, and by a process of topological reduction[38b] they can be expressed in terms of the functions $h_d(r)$ and $\phi(r)$, where

$$h_d(r) = g_d(r) - 1 \tag{4.7}$$

and $g_d(r)$ is the pair correlation function for hard spheres at the density of interest. The resulting expression for $\ln g(r)$ is a convenient starting point for our discussion. It is

$$
\begin{aligned}
\ln g(r) = \ &\ln g_d(r) + \text{sum of all connected diagrams with} \\
&\text{two root points (labeled 1 and 2 and separated} \\
&\text{by a distance } r\text{), any number of field points, at} \\
&\text{most one } h_d \text{ bond and any number of } \phi \text{ bonds} \\
&\text{connecting any two points, at least one } \phi \text{ bond,} \\
&\text{no articulation points, and no reference pair of} \\
&\text{articulation points, such that the diagram does} \\
&\text{not become disconnected if the two roots are} \\
&\text{removed} \tag{4.8}
\end{aligned}
$$

Appendix A contains a brief review of some of the graph theoretic terms used here. Figure 5 shows various examples of these diagrams. For the present discussion it is necessary to understand only the following features of these diagrams: (1) The points 1 and 2 are represented as open circles and should be imagined as being a distance r apart (see Fig. 5). (2) The field points are represented as closed circles. (3) Each bond in the diagram is represented as a line connecting a pair of points. (4) There are two types of bonds, ϕ bonds (represented by solid lines) and h_d bonds (represented by dashed lines). (5) The value of each diagram is a certain multidimensional integral over the positions of the field points. (6) There is a well-defined prescription for writing down the integral corresponding to any particular diagram.

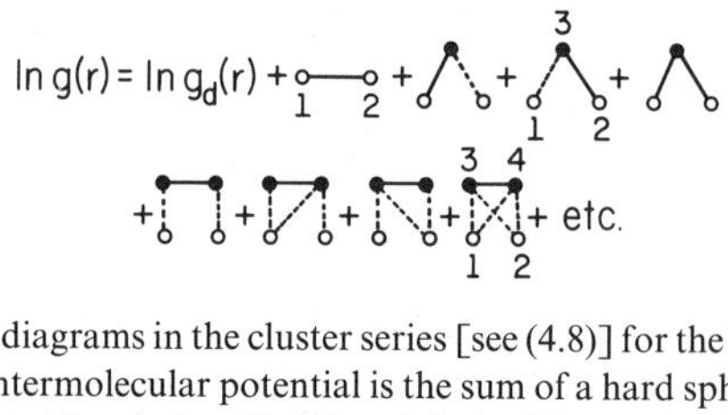

Fig. 5. The first few diagrams in the cluster series [see (4.8)] for the pair correlation function of a model fluid whose intermolecular potential is the sum of a hard sphere part plus a perturbation. Solid lines represent ϕ bonds [see (4.6)], and dashed lines represent h_d bonds [see (4.7)].

Suppose a diagram contains m field points, which are labeled $3, 4, m + 2$. Then the value of the diagram is

$$\frac{1}{\nu} \rho^m \int dr_3 \cdots dr_{m+2} \left[\prod h_d(r_{ij}) \right] \left[\prod \phi(r_{kl}) \right] \tag{4.9}$$

where each h_d or ϕ factor in the integral corresponds to an h_d or ϕ bond in the diagram. Also, ν is a numerical factor whose value depends on the topological structure of the diagram. For example the values of the first, third and eighth diagrams in Fig. 5 are, respectively,

$$\phi(r_{12}), \qquad \rho \int dr_3 \, h_d(r_{13}) \phi(r_{23}),$$

and

$$\tfrac{1}{2} \rho^2 \int dr_3 \, dr_4 \, h_d(r_{13}) h_d(r_{14}) h_d(r_{23}) h_d(r_{24}) \phi(r_{34}),$$

where $r_{ij} = |\mathbf{r}_i - \mathbf{r}_j|$. For a more detailed discussion of the meaning of the diagrams and for the derivation of (4.8), the reader is referred to Ref. 38.

B. Cluster Theory of Ionic Solutions

Mayer used his cluster theory in a calculation of the properties of a model ionic solution in which the effective interionic interactions are of the type given in (4.1). (An ionic solution should actually be regarded as a two-component fluid since at least two different species of ions must be present to satisfy electroneutrality. This adds some extra complications to the statement of (4.8) and to the definition of the cluster integrals. For simplicity we will ignore these (nonessential) complications in the following discussion of Mayer's derivation of a renormalized potential for ionic solutions.)

The first term in (4.8) for the cluster series for $\ln g - \ln g_d$ represents the contribution from direct Coulombic interaction between the particles. Its value is

$$\phi(r) = \frac{-Z_i Z_j \beta e^2}{\epsilon r} \tag{4.10}$$

and hence it represents a very long-ranged correlation. In the fluid however, the actual correlations are of much shorter range due to a cooperative screening effect. Hence, the sum of the subsequent diagrams in (4.8) must cancel this first diagram for large r. One of the essential features of Mayer's ionic solution theory is that a small subset of these diagrams, rather than the entire sum, can be regarded as responsible for this cancellation. In particular, the diagrams which are simple chains of Coulomb bonds are the

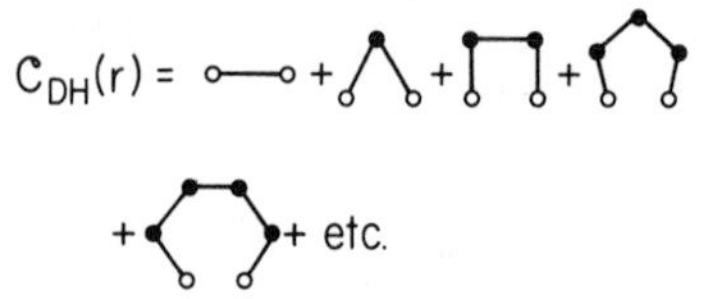

$$C_{DH}(r) = \text{○——○} + \bigwedge + \prod + \{\}$$

$$+ \langle\rangle + \text{etc.}$$

Fig. 6.　The first few diagrams in the series for the renormalized potential in the Mayer ionic solution theory. The solid lines are ϕ bonds.

important ones. See Fig. 6, which shows the diagram represented by (4.10) and the chains which cancel it.

　　The mathematical reason for the cancellation is easy to see if we consider the Fourier transform of the diagrams. The chain diagrams are convolution integrals. For example, the chain of two ϕ bonds is $\rho \int dr_3\, \phi(r_{13})\phi(r_{23})$. The Fourier transform of a convolution is simply a product of the Fourier transforms of the two functions in the integral; hence, the Fourier transform of the diagram is $\rho[\hat{\phi}(k)]^2$, where

$$\hat{\phi}(k) = \int \mathbf{dr}\ e^{i\mathbf{k}\cdot\mathbf{r}}\phi(r) \tag{4.11}$$

Similarly the Fourier transform of the chain with n ϕ bonds is

$$\rho^{n-1}[\hat{\phi}(k)]^n \tag{4.12}$$

The sum of the chains of two or more ϕ bonds is a geometric series whose sum is

$$\hat{\phi}(k)\left[\frac{\rho\hat{\phi}(k)}{1 - \rho\hat{\phi}(k)}\right] \tag{4.13}$$

The long-ranged nature of the Coulomb potential is displayed in the Fourier transform of $\phi(r)$ as a divergence of $\hat{\phi}(k)$ for small k, that is,

$$\hat{\phi}(k) \sim \frac{-1}{k^2} \tag{4.14}$$

for small k. It is easily seen from (4.13) that as $k \to 0$ the factor in square brackets approaches -1, and thus the sum of chains of two or more ϕ bonds approaches $-\hat{\phi}(k)$ as $k \to 0$. This exactly cancels the divergence of the Fourier transform of the first diagram in the cluster series and hence exactly cancels the long-range nature of the diagram. $\hat{C}_{DH}(k)$, the sum of all the chains including the first diagram in the series, is given by

$$\hat{C}_{DH}(k) = \hat{\phi}(k)[1 - \rho\hat{\phi}(k)]^{-1} \tag{4.15}$$

which leads to the Debye–Hückel potential in (4.2) upon Fourier inversion. (Here the subscript DH denotes Debye–Hückel). An extra advantage of adding these terms together is that it cancels the divergence in each of the chain diagrams [see (4.12) and (4.14)]. In fact, the chains with two or more links are represented by divergent integrals, but the sum of chains is a finite quantity. Thus far, we have

$$\ln g(r) = \ln g_d(r) + C_{DH}(r) + \text{ sum of all diagrams in (4.8)}$$
$$\text{which are not chains of } \phi \text{ bonds} \qquad (4.16)$$

At this stage, even though the cancellation of the long-range part of ϕ and the cancellation of the divergences in the chain diagrams are explicitly contained in the results, (4.16) is unsatisfactory because there remain many divergent integrals among the diagrams left. The next important feature of Mayer's theory is that these remaining divergences also cancel each other. This can be shown by expressing the sum of all the remaining diagrams in terms of C_{DH} rather than ϕ. The result can be expressed as follows:

$\ln g(r) = \ln g_d(r) +$ sum of all connected diagrams with two root points (labeled 1 and 2 and separated by a distance r), any number of field points, at most one h_d bond and any number of C_{DH} bonds connecting any two points, at least one C_{DH} bond, no articulation points, no reference pair of articulation points, such that the diagram does not become disconnected if the two roots are removed, and such that no field point has two C_{DH} bonds and no h_d bonds attached to it. \qquad (4.17)

The first few diagrams in this series are shown in Fig. 7. Just as the first diagram in Fig. 7 is the sum of chains of one or more ϕ bonds, in the same way every other diagram in Fig. 7 or (4.17) is the sum of an infinite number of diagrams in Fig. 5 or (4.8); for example, see Fig. 8. The fact that each diagram in (4.17) is expressed in terms of C_{DH} bonds rather than ϕ bonds shows that

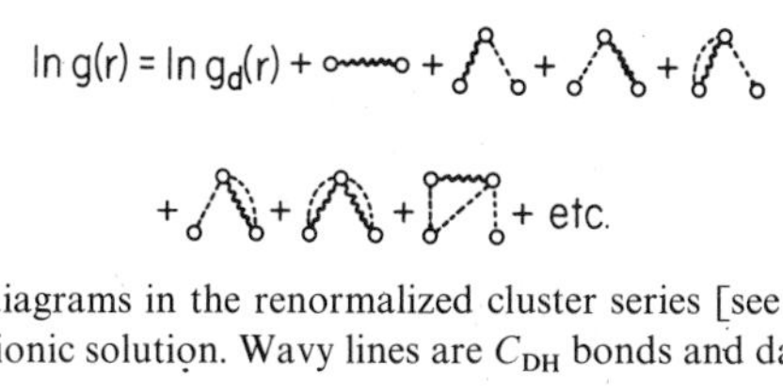

Fig. 7. The first few diagrams in the renormalized cluster series [see (4.17)] for the pair correlation function of an ionic solution. Wavy lines are C_{DH} bonds and dashed lines are h_d bonds.

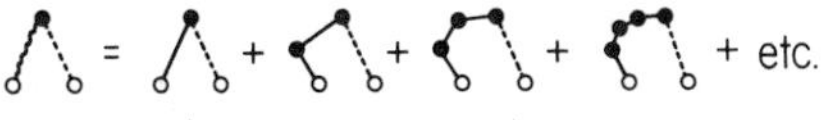

Fig. 8. The second diagram in Fig. 7 is the sum of an infinite number of diagrams in Fig. 5.

the long-ranged divergences of the diagram in (4.8) completely cancel one another. $C_{DH}(r)$ is sometimes called a "renormalized potential" since it enters the series in (4.17) in much the same way as ϕ enters (4.8).

Equation (4.17) is a good starting point for numerical calculations of the properties of ionic solutions, particularly for low concentrations of singly charged ions. Under these conditions

$$g_d(r) \cong e^{-\beta u_d(r)} \qquad (4.18)$$

and to lowest order in the density only the $C_{DH}(r)$ diagram in (4.17) needs to be retained. The result is

$$g(r) = e^{-\beta u_d(r)} e^{C_{DH}(r)} \qquad (4.19)$$

which is a nonlinear Debye–Hückel result.

We can summarize the major ideas of the Mayer ionic cluster theory in the following way: There is a cooperative screening effect in ionic solutions in which those correlations between two ions which are induced by their direct interaction is screened by the presence of other ions. A mathematical representation of the screening shows that the cluster diagram involving only the direct interaction of two ions is largely canceled by a set of other diagrams which have the topological structure of chains. The sum of the direct interaction and these chains is defined as the renormalized potential. The fact that the renormalized potential is of shorter range and is weaker than the direct interaction is a manifestation of this cancellation or screening. Furthermore the renormalized potential can be used to eliminate the original potential from the cluster series, thus showing that the cancellation and the elimination of divergences extends to all terms in the cluster series and is not confined just to the specific diagrams in the renormalized potential. It is this feature of the renormalization method that distinguishes it from other *ad hoc* truncations of cluster expansions, such as are suggested by various integral equations for the pair correlation function. The resulting series is an expansion in powers of the renormalized potential, rather than the Coulomb potential, and is much more suitable for numerical evaluation.

C. Comments about the Renormalization Technique

There are several important points about Mayer's renormalization techniques which merit further discussion. We do not know the relationship between the physical nature of the Coulomb screening effect and the topological structure (chains of ϕ bonds) of the diagrams chosen to be in the renormalized potential. This is unfortunate since a proper coupling of physical intuition with the mathematics of cluster expansions would be of enormous help in extending the theory of different types of fluids. Since Mayer's choice cannot be justified on physical grounds we must be satisfied with the mathematical justification discussed above; with this choice all the divergences of the diagrams due to the long-range nature of the potential are systematically eliminated. Moreover the resulting series is useful for numerical calculations and leads to the experimentally verified Debye–Hückel limiting laws for the thermodynamic properties of dilute ionic solutions. However, Mayer's is by no means the only choice yielding these desirable properties and in the next section we will consider other possible choices which are more appropriate for dense fluids.

D. Cluster Theory of Dense Fluids

In this section we will discuss a way of generalizing Mayer's renormalized potential to the case of dense liquids. For dense liquids of molecules interacting with slowly varying long-ranged forces, for example, the Kac model potential, one form of renormalized potential that has been suggested is a sum of chains of not only ϕ bonds but also h_s bonds, where h_s is the "short-ranged part" of h.[39–41] The simplest choice for h_s is h_d.[42] Using it, we define

$C_K(r) =$ sum of all diagrams in (4.8) for $\ln g(r)$ which are
    chains of ϕ bonds and h_d bonds.   (4.20)

Here K stands for Kac. Each member of the series contains at least one ϕ bond, and, as a result of one of the topological restrictions in (4.8), no two adjacent members of the chain can both be h_d bonds. See Fig. 9 for the first few diagrams in the series. These diagrams include all those in Mayer's definition, plus additional ones which describe the effect of the hard sphere forces.

These diagrams are all convolution integrals and their sum can be readily evaluated using Fourier transform techniques. The result is

$$\hat{C}_K(k) = \frac{\hat{S}_d(k)\hat{\phi}(k)\hat{S}_d(k)}{[1 - \rho\hat{S}_d(k)\hat{\phi}(k)]} \qquad (4.21)$$

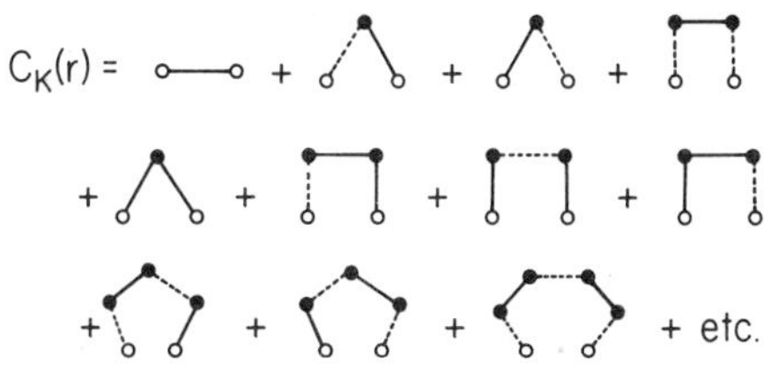

Fig. 9. The first few diagrams in the definition [see (4.20)] of the renormalized potential for the Kac model fluid. Solid lines are ϕ bonds and dashed lines are h_d bonds.

where

$$\hat{S}_d(k) = 1 + \rho \int d\mathbf{r} \, e^{i\mathbf{k}\cdot\mathbf{r}}(g_d(r) - 1) \qquad (4.22)$$

$\hat{S}_d(k)$ is called the structure factor of the hard sphere fluid. It is real and positive for all values of $\mathbf{k}$ and depends only on the magnitude of $\mathbf{k}$. This renormalized potential, like Mayer's, has the property that even if $\hat{\phi}(k)$ diverges as $-1/k^2$ for small k, the renormalized $\hat{C}_K(k)$ approaches a finite value, $-\hat{S}_d(0)$, as $k \to 0$. Thus in Fig. 9, the first diagram is to some extent cancelled by the others.

When this new renormalized potential is used for high-density fluids, the possibility of a new type of divergence arises. Namely, if $\hat{\phi}(k)$ is positive for some value of k and if ρ is large enough, we might have

$$\rho\hat{S}_d(k)\hat{\phi}(k) = 1 \qquad (4.23)$$

for some particular k, which leads to a divergence in $\hat{C}_K(k)$. This breakdown of the renormalized potential is sometimes misinterpreted as a breakdown of the fluid, that is, the onset of instability or a phase transition. In fact it is just an artifact of this particular choice of diagrams in (4.20). This divergence must be eliminated in order for the renormalized potential to be useful. One method of doing this is to make a better choice of h_s than h_d in the chain sum (4.20).[43-45] Another method is discussed by Andersen and Chandler[5] and leads to the theory called the optimized cluster theory. Here we will discuss a third method which involves finding those diagrams in (4.8) that cancel the divergence and including these additional diagrams in the definition of the renormalized potential. The results obtained are equivalent to those developed in the original formulation of the optimized cluster theory.

To find a set of diagrams that cancel the divergence in C_K, consider those graphs that can be described in the following way (see Fig. 10): Imagine the two root points and a set of one or more field points located on the circumference of a circle, so that by passing from root 1 and going clockwise around

Fig. 10. Illustration for the definition of the diagrams in $C_L(r)$, the renormalized potential appropriate for liquids. The points are on a circle, with a chain of $\phi + f_d\phi = \Phi$ bonds (represented by solid lines) and h_d bonds (represented by dashed lines) leading from one root to another. There are f_d bonds (represented by dotted lines) connecting some pairs of points.

the circle one passes through all the field points and then arrives at 2. Now imagine connecting each pair of adjacent points on this circle with either an h_d bond or a Φ bond, where

$$\Phi(r) = [1 + f_d(r)]\phi(r) \tag{4.24}$$

except that no such bond is to connect the roots directly. This generates a chain similar to, and in many cases identical with, the chains in C_K. However, now we are using Φ bonds rather than ϕ bonds. The former are "decorations" of the latter. Every diagram that is expressed with n Φ bonds is equal to a sum of 2^n diagrams expressed with ϕ and $f_d\phi$ bonds. These decorations prohibit the overlap of two particles which are directly connected by a ϕ bond since $1 + f_d(r)$ is zero for $r < d$. [Outside the hard core, of course, $\Phi(r) = \phi(r)$].

Overlap of nonadjacent particles in the diagrams should also be inhibited (if not prohibited). Thus further decorations are also required. The ones we consider are reminiscent of the cluster diagrams contributing to the Percus–Yevick equation for $y_d(r)$.[46] Imagine adding some or no f_d bonds between nonadjacent points on the circle in such a way that no two f_d bonds cross each other and so that no field point is left with only h_d bonds attached to it. An f_d bond may or may not be drawn between the roots.

Our new definition of the renormalized potential, which we call $C_L(r)$, is that it includes the diagram with just one Φ bond between the roots plus all the diagrams that can be drawn in the way just described. Some examples of the diagrams in C_L are shown in Fig. 11. This renormalized potential is given the subscript L because it is a very useful one for liquids. Some examples of diagrams not in C_L are shown in Fig. 12. Each diagram in C_L contains at least one Φ bond and has no reference pair of articulation points. Also note

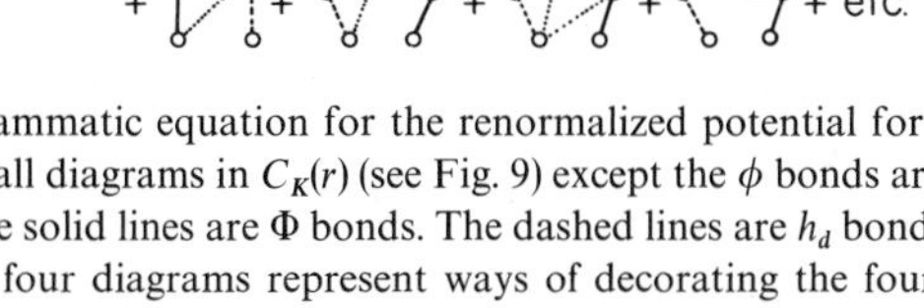

Fig. 11. Diagrammatic equation for the renormalized potential for liquids. The quantity $C'_K(r)$ is the sum of all diagrams in $C_K(r)$ (see Fig. 9) except the ϕ bonds are replaced by Φ bonds ($\Phi = \phi + f_d\phi$). The solid lines are Φ bonds. The dashed lines are h_d bonds. The dotted lines are f_d bonds. The first four diagrams represent ways of decorating the fourth diagram in Fig. 9. The next six represent ways of decorating the sixth, seventh, and eighth diagrams of Fig. 9. The next two are included in $C_L(r)$ even though they are not decorated versions of a diagram in $C_K(r)$. The last three are some of the ways of decorating the ninth diagram in C_K.

Fig. 12.   Diagrams which are not included in the definition of $C_L(r)$. The first diagram is excluded because it has crossing f_d bonds. If one or both of these bonds are removed, the diagram becomes acceptable. The next two are excluded because each has one point with only h_d bonds attached.

that for each diagram in C_L with one or more field points which does not have an f_d bond between the roots, there is another diagram in C_L which is identical except for the presence of an f_d bond between the roots, and vice versa. It follows that

$$C_L(r) = 0 \qquad \text{for } r < d \qquad (4.25)$$

since

$$f_d(r) = -1 \qquad \text{for } r < d \qquad (4.26)$$

and hence each such pair of diagrams is equal in magnitude but opposite in sign for $r < d$.

This choice of renormalized potential includes all the diagrams in $C_K(r)$. Moreover, we will show that it includes diagrams which cancel the divergence in $C_K(r)$. (These extra diagrams, containing f_d bonds, are obtained by breaking up the corresponding diagrams in (4.8).) Hence it is a more suitable choice of renormalized potential for dense liquids at low temperatures. Not only is it suitable, it is also very useful and leads to an accurate theory of the structure of simple liquids, as we shall see below.

We must now find a way of adding these diagrams. Let us note that some of the diagrams in C_L (i.e., all except for the first diagram $\Phi(r)$ and the diagrams with an f_d between the roots) have nodal points, that is, field points which if removed would disconnect the diagram into two parts, with one root point in each part. For example, in the first diagram in Fig. 11, the field point on the right is a nodal point. Let $\psi(r)$ be the sum of the diagrams in $C_L(r)$ with nodal points and let $\phi'(r)$ be the sum of all the nodeless diagrams. Then, obviously

$$C_L(r) = \phi'(r) + \psi(r)$$

Moreover, from the remarks above we have the following relationship:

$$\phi'(r) = \Phi(r) + f_d \psi(r) \qquad (4.27)$$

because every nodeless diagram with one or more field points has the structure of a nodal diagram with an f_d bond added between the roots. Next, note that all the diagrams in C_L may be regarded as chains, with the links in the chains being joined at nodal points. There are three types of links: h_d, $(1 + f_d)\phi = \Phi$, and f_d times a member of the series for ψ. For example, the last diagram shown in Fig. 11 has two nodal points, 3 and 5, and may be regarded as a chain of three links. The first link, connecting 1 and 3, is h_d. The third link, connecting 5 and 2, is Φ. The second link, connecting 3 and 5, is a product $f_d(r_{35})$ and a nodal diagram. This nodal diagram is included in $C_L(r)$ and is shown explicitly as the third diagram on the right side of the equation in Fig. 9. The only restriction on these chains is that no h_d links can be attached together. Then the general diagram in $C_L(r)$ is a chain whose links are h_d bonds and members of the series defining $\phi'(r)$ with the additional restriction mentioned above. Topologically, the series for $C_L(r)$ is similar in structure to $C_K(r)$, and we can use (4.21) to obtain the result

$$\hat{C}_L(k) = \frac{\hat{S}_d(k)\hat{\phi}'(k)\hat{S}_d(k)}{[1 - \rho\hat{S}_d(k)\hat{\phi}'(k)]} \tag{4.28}$$

Note that this is not a solution for $\hat{C}_L(k)$ but actually an integral equation, since $\phi'(k)$ on the right side is defined in terms of the diagrammatic structure of $C_L(r)$ in (4.27).

To solve this integral equation we will use a variational method. Let us define the following functional of ϕ':

$$F(\phi') = \frac{1}{(2\pi)^3\rho} \int d\mathbf{k}\{\rho\hat{S}_d(k)\hat{\phi}'(k) + \ln [1 - \rho\hat{S}(k)\hat{\phi}'(k)]\} \tag{4.29}$$

If we take the functional derivative of $F(\phi')$ with respect to $\hat{\phi}'(k)$ we obtain

$$(2\pi)^3 \frac{\delta F(\phi')}{\delta\hat{\phi}'(k)} = -\frac{\rho\hat{S}_d(k)\hat{\phi}'(k)\hat{S}_d(k)}{[1 - \rho\hat{S}_d(k)\hat{\phi}'(k)]} = -\rho\hat{C}_L(k) \tag{4.30}$$

from (4.28). This leads to

$$\frac{\delta F(\phi')}{\delta\phi'(r)} = -\rho C_L(r) \tag{4.31}$$

To use these results, note that from (4.27) we have

$$\phi'(r) = -\beta u(r) \qquad r \geq d$$

and

$$\phi'(r) = -\psi(r) \qquad r < d$$

and from (4.25) that

$$C_L(r) = 0 \qquad r < d$$

If we regard $\phi'(r)$ for $r < d$ as the quantity to be calculated we see from (4.31) and (4.25) that the behavior of $\phi'(r)$ for $r < d$ is such as to make $F(\phi')$ stationary with respect to changes in $\phi'(r)$ for $r < d$. Moreover, since

$$x + \ln (1 - x) \geq 0 \qquad \text{for all } x < 1 \tag{4.32}$$

we have

$$F(\phi') > 0 \tag{4.33}$$

for all ϕ' for which the integral exists. Hence if we imagine choosing $\phi'(r)$ for $r > d$ according to (4.27) and then varying $\phi'(r)$ for $r < d$ so as to minimize $F(\phi')$, at the minimum F will be stationary. Hence to sum the diagrams in $C_L(r)$ we must find the function $\phi'(r)$ for $r < d$ which minimizes $F(\phi')$. When $\phi'(r)$ is substituted into (4.28), we can obtain $\hat{C}_L(k)$ and by Fourier inversion obtain $C_L(r)$.

The process of minimizing $F(\phi')$ is a straightforward one to perform numerically. One way this can be done is to use a trial solution of the form

$$\phi'(r) = \sum_{n=0}^{m} a_n \left[1 - \left(\frac{r}{d} \right) \right]^n \qquad \text{for } r < d \tag{4.34}$$

$$= -\beta u(r) \qquad \text{for } r > d$$

and minimize $F(\phi')$ with respect to the coefficients $a_0, \ldots, a_m$ by a Newton–Raphson method.[47]

Now let us use this variational principle to discuss the nature of the renormalized potential, $C_L(r)$. We will show that $C_L(r)$ has no divergence of the type which occurs in $C_K(r)$. Moreover, we will see that at high densities, the first term in C_L, namely $\Phi(r)$, is largely canceled by the sum of the remaining terms. This is a mathematical representation of repulsive force screening, that is, of the fact that at high density the structure of a simple liquid is not greatly affected by the attractive forces.

To discuss this divergence, let us consider the integral in (4.29) for F that attains its minimum value when evaluated using the actual behavior of $\phi'(r)$ for $r < d$. The integrand is positive definite and is small only if $\rho \hat{S}_d(k)\hat{\phi}'(k)$ is small. In the variational calculation, $\phi'(r)$ will assume the functional form necessary to make $\rho \hat{S}_d(k)\hat{\phi}'(k)$ as small as possible for all values of k. Moreover, it will especially try to avoid the singularity associated with having $\rho \hat{S}_d(k)\hat{\phi}'(k) = 1$ for any value of k. We might expect, therefore, that

$$\rho \hat{S}_d(k)\hat{\phi}'(k) < 1 \tag{4.35}$$

for all values of k, as has been checked in several numerical tests of the minimization procedure for various model fluids. This means that the renormalized potential $C_L(r)$ contains no divergences, and so the singularities in $C_K(r)$ are canceled by the additional diagrams included in the definition of $C_L(r)$.

From the diagramatic definition of C_L (see Fig. 11), it is seen that at low density

$$C_L(r) = -\beta u(r) + 0(\rho) \qquad r > d \qquad (4.36)$$

that is, at low density the renormalized potential is the same as the actual perturbation potential.* [This result can also be easily verified using the low-density limit of the variational procedure given in (4.34)]. At higher densities, however, the minimization procedure tends to make $\rho \hat{S}_d(k)\hat{\phi}'(k)$ as small as possible. From (4.28) it can be seen that this makes $\hat{C}_L(k)$ small. The actual $\hat{\phi}(k)$, therefore, tends to be small for values of k where $\hat{S}_d(k)$ is large, and if necessary to achieve this, it is allowed to become large where $\hat{S}_d(k)$ is small. At low densities, $S_d(k)$ is approximately unity for all k and we obtain (4.36). At high density, $S_d(k)$ is very small for small k, that is, for $k \lesssim \pi/d$. This gives ϕ' some added flexibility in its attempt to minimize $F(\phi')$. The net result is that the renormalized potential is smaller at high density than at low density and hence actually decreases as ρ increases. This is the repulsive force screening effect mentioned above: for example, see Fig. 13, which shows $C_L(r)$ for various densities at the same temperature. The potential used in the calculations is a hard sphere potential plus an attractive perturbation of the Lennard–Jones type. It can be seen that as the density is increased the renormalized potential decreases in magnitude.

We have defined the renormalized potential as the sum of a subset of the diagrams in the original cluster series (4.8) for $\ln g$. An important consequence of this particular definition of $C_L(r)$ is that the entire series for $\ln g$ can be expressed in terms of C_L. The details of this topological reduction are straightforward and will be omitted. The result is

$$\ln g(r) = \ln g_d(r) + C_L(r) + \text{sum of all connected diagrams}$$
with two root points (labeled 1 and 2 and separated by a distance r), at least one field point, at most one h_d bond and any number of C_L bonds connecting any two points, no articulation points, no reference pair of articulation points, no field point with only a C_L bond and an h_d bond or only two C_L bonds attached to it, such that the diagram does not become disconnected if the two roots are removed.   (4.37)

---

* For ionic solutions at low density we find that

$$C_L(r) = C_{DH}(r) + O(\rho) \qquad r > d$$
$$= 0 \qquad r < d$$

so this method reduces to Mayer's under these conditions.

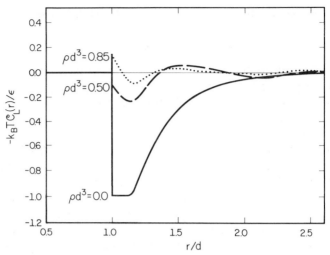

Fig. 13. Renormalized potential at three representative states for the system for which $w(r) = u_d(r) + u(r)$. Here, $u(r)$ is the Lennard–Jones attractive perturbation potential [see eq. (3.4)]. For all three states the temperature is $k_B T/\varepsilon = 1.15$. The densities are $\rho d^3 = 0$, 0.5, and 0.85.

This is analogous to Mayer's result, except that we have used a different form of the renormalized potential. In this renormalized potential, the direct interaction between two molecules is canceled to some extent by the addition of other diagrams, this cancellation being more effective at high density than at low density. Moreover, the fact that the entire cluster series can be expressed in terms of $C_L(r)$ indicates that this cancellation occurs systematically throughout the cluster series. Thus, this new series is in effect an expansion in powers of the renormalized potential and of the density, and for models of simple liquids the expansion converges quickly if either the density or the renormalized potential (or both) is small. The arguments given above and the numerical results presented below indicate that the renormalized potential is small if either the temperature is high or the density is large. Thus the series converges well at high density, at low density, and at high temperature. For any particular potential, however, it is necessary to perform explicit calculations to determine more precisely the conditions under which the renormalized potential is small and the series converges quickly.

If the terms in (4.37) decrease in magnitude quickly enough, we can discard all but the first term to get

$$\ln g(r) = \ln g_d(r) + C_L(r) \qquad (4.38)$$

or

$$g(r) = g_d(r)e^{C_L(r)} \qquad (4.39)$$

This is called the exponential (EXP) approximation and its accuracy for various fluids has been tested. Some of the results will be reviewed below.

E. Comments on the New Renormalization Technique

This new renormalization technique has many of the same features and justifications as did Mayer's. Here again we do not know how to relate the physical nature of the (repulsive force) screening effect to the topological structure of the diagrams included in C_L. However, the structure of the C_L diagrams, namely chains of h_d and ϕ bonds "decorated" with additional f_d bonds, suggest the following interpretation. Screening of a perturbation potential occurs even for non-Coulombic potentials and is accounted for approximately by summing chains of interactions. In dense liquids, the screening represents a competition between attractive and repulsive interactions, and so h_d bonds, representing the repulsions, must be included as links in the chains, as well as the ϕ bonds. For short-ranged perturbations, the field points in the chains are constrained to be close to one another—so close that nonadjacent members may come within a hard core distance of one another. This is physically impossible but is mathematically allowable in cluster integrals. To correct for the hard core interactions between points in the chain, we should include some diagrams in which these points are connected by f_d bonds.

As in Mayer's theory, our choice of renormalized potential can be justified only with mathematical and experimental reasons. We have seen that this choice eliminates the divergences in the chain sum and that the renormalized potential can be used to eliminate the original potential from the cluster series. We also will see below that the simplest approximation suggested by the theory, the exponential approximation, is in very good agreement with computer experiments on simple liquids. It is also accurate for ionic solutions.[47,48] The present choice of renormalized potential is certainly not unique. It is possible to make at least one other choice (see Appendix B) which retains desirable features of the present choice (but which does not lead to significantly different results), and there may be other choices which are better for a wider variety of liquids. The choice we have made, however, is certainly adequate for a treatment of simple liquids and ionic solutions.

F. Principal Results of the Renormalized Cluster Theory

The most important results of the preceding discussion are (4.37) and (4.39) for the pair correlation function of a fluid of attracting hard spheres. The former is an exact formal result; the latter is a truncation of the infinite series and represents a useful and tractable numerical approximation. The same type of renormalization methods can be applied to an analysis of the cluster series for the excess free energy density of such a fluid. An infinite series can be obtained in which all the diagrams are expressed in terms of the renormalized potential C_L. The reader is referred to Ref. 5 for the details. When the series is truncated the result is

$$\mathscr{A} = \mathscr{A}_d + a_{HTA} + a_{ORPA} + B_2 \qquad (4.40)$$

where $\mathscr{A}_d$ is the hard sphere excess free energy density,

$$a_{HTA} = -\frac{\beta}{2}\rho^2 \int d\mathbf{r}\, g_d(r)u(r) \qquad (4.41)$$

$$a_{ORPA} = -\frac{1}{2(2\pi)^3} \int d\mathbf{k}\{\rho\hat\phi(k)\hat S_d(k) + \ln\,[1 - \rho\hat\phi(k)\hat S_d(k)]\} \qquad (4.42)$$

$$B_2 = \frac{1}{2}\rho^2 \int d\mathbf{r}\, h_d(r)\frac{1}{2}[C_L(r)]^2 + \frac{1}{2}\rho^2 \int d\mathbf{r}\, g_d(r) \sum_{n=3}^{\infty} (n!)^{-1}[C_L(r)]^n \qquad (4.43)$$

Equation (4.40) is called the ORPA + B_2 approximation for historical reasons.

The conditions under which (4.40) is an accurate approximation for the entire cluster series are approximately the same as the conditions for the validity of (4.39). These results were originally derived in a somewhat different way[5] and are collectively referred to as the optimized cluster theory.

G. Tests of Optimized Cluster Theory

This theory for calculating the structural and thermodynamic properties of a fluid from its interatomic potential has been applied to a variety of potentials, such as the Lennard–Jones potential, primitive models of electrolyte solutions, and the square well potential. In this section we discuss tests of this theory for the Lennard–Jones potential described above in Section III.

The Lennard–Jones potential has a soft repulsive core rather than a hard sphere interaction. However, the optimized cluster theory results can easily be modified to take the softness into account, using the methods discussed in Section II. The details of the derivation will be omitted. The results are

$$g(r) = e^{-\beta[u_0(r) + u(r)]}y_d(r)e^{[C_L(r) - \Phi(r)]} \qquad (4.44)$$

which is analogous to (4.39) and (2.16), and

$$\mathscr{A} = \mathscr{A}_d + a_{HTA} + a_{ORPA} + B_2 \qquad (4.45)$$

In (4.44) and (4.45) the renormalized potential is the one calculated for a fluid whose potential is a hard core with diameter d plus the attractive part of the original potential. In (4.44), the function $C_L(r) - \Phi(r)$ for $r < d$ is to be interpreted as the smooth extrapolation of its functional behavior for $r > d$.

The hard sphere diameter should be chosen so that

$$\int \mathbf{dr}[e^{-\beta u_0(r)} - e^{-\beta u_d(r)}]y_d(r)e^{[C_L(r) - \Phi(r)]} = 0 \qquad (4.46)$$

However, in practice for potentials like the Lennard–Jones potential, $C_L(r) - \Phi(r)$ is slowly varying compared to $y_d(r)$ near $r = d$ and therefore it is adequate to use the diameter calculated from the simpler criterion in (2.11).

There are two types of unambiguous tests to which we can subject these results. First, we can compare the thermodynamic properties and the pair correlation function with molecular dynamics and Monte Carlo calculations for the Lennard–Jones fluid. Secondly, we can test the theory for internal consistency.

The first type of test is made in Table II and Figs. 14 and 15, which compares the OCT with computer simulation results. It can be seen that throughout the temperature–density plane the agreement is excellent and that the discrepancy between the OCT and computer simulation results are usually of the same magnitude as the statistical uncertainties of the latter.

The second type of test can be carried out by calculating the pressure (or the internal energy) in two ways. The first way is to calculate the excess free energy from (4.45) and then differentiate numerically with respect to density to obtain the pressure (or with respect to temperature to obtain the internal energy). The second way is to use the virial equation (or the

TABLE II

Pressure and Internal Energy for the Lennard–Jones Fluid Calculated Using the ORPA + B_2 Approximation of the Optimized Cluster Theory (OCT) Compared with the Results Computed from the High-Temperature Approximation (HTA) and from Monte Carlo Computer Simulations (MC)

		$\beta P/\rho$			$-\Delta E/N\epsilon$		
$\rho\sigma^3$	$k_B T/\epsilon$	HTA[a]	OCT[b]	MC[c]	HTA[a]	OCT[b]	MC[c]
0.1	0.75	0.42	0.23	0.23	0.56	1.15	1.15
0.1	1.35	0.77	0.72	0.72	0.55	0.78	0.78
0.2	1.35	0.53	0.50	0.51	1.16	1.51	1.50
0.5	1.35	0.18	0.30	0.30	3.24	3.36	3.37
0.65	1.15	0.13	0.22	0.31	4.41	4.47	4.45
0.85	1.15	2.85	2.84	2.86	5.66	5.69	5.67

[a] Obtained by numerical differentiation of (3.8).

[b] Obtained by numerical differentiation of (4.45).

[c] Machine calculation results taken from Verlet and Weis.[10] The uncertainty in these results at high densities is ± 0.05.

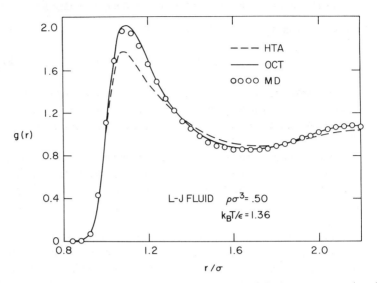

Fig. 14. Pair correlation function for the Lennard–Jones fluid for a state near the critical point. The open circles ar the molecular dynamics results of Verlet.[9] The dashed and solid curves are the results of the high-temperature approximation and the optimized cluster theory, respectively.

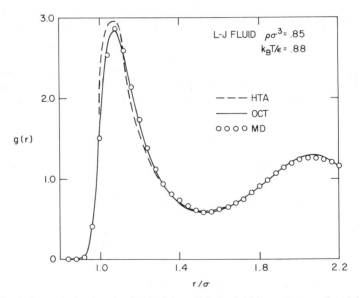

Fig. 15. Pair correlation function for the Lennard–Jones fluid for a state near the triple point. See Fig. 14 for additional information.

energy equation) and the exponential approximation for the pair correlation function to obtain the pressure (or the internal energy). Comparison of the two results gives a measure of the internal consistency and hence of the accuracy of the theory. There is no reason to expect that the two theoretical results for the pressure should contain errors that are consistently equal to each other, and this is why agreement between them is a measure of the accuracy of each. In fact there is good reason to expect that the errors of the two results are very different in magnitude. The expansion parameters in the cluster series for $\mathscr{A}$ and g are ρR^3 and $\bar{C}$, where R and $\bar{C}$ are the range and average magnitude of the renormalized potential, respectively. The cluster series corrections to the ORPA $+ B_2$ approximation for $\mathscr{A}$ are of order $\rho(\rho R^3)^2(\bar{C})^2$, whereas the corrections to the EXP approximation are of order $\rho R^3\bar{C}$. It follows that the compressibility factor calculated from the ORPA $+ B_2$ approximation contains errors of order $(\rho R^3)^2\bar{C}^2$, whereas the virial compressibility factor from EXP has errors of order $\rho R^3\bar{C}$. Thus the differences between them are of order $\rho R^3\bar{C}$. If the differences are found to be small, then the expansion parameters are small and the OCT approximations are accurate.

The entries in Table III show how well the OCT satisfies the consistency test for the Lennard–Jones fluid at several representative states. At high densities ($\rho\sigma^3 > 0.65$) the differences between the compressibility factors

TABLE III

Comparison of the Compressibility Factors of the Lennard–Jones Fluid as Calculated by Differentiating the ORPA $+ B_2$ Approximation ($\beta P_f/\rho$) and by Applying the Virial Equation to the EXP Approximation ($\beta P_v/\rho$)

$\rho\sigma^3$	$k_B T/\epsilon$	$\beta P_f/\rho$	$\beta P_v/\rho$
0.154	1.351 ·	0.60	0.60
0.107[b]	1.249[b]	0.65	0.66
0.318[a]	1.303[a]	0.30	0.28
0.558[b]	1.249[b]	0.17	0.18
0.748	1.351	1.67	1.70
0.757	1.249	1.52	1.54

[a] This state is the "experimental" critical point for the Lennard–Jones fluid.[49] At that state the experimental compressibility factor is 0.29.

[b] This state is in the critical region and directly adjacent to the liquid-gas coexistence curve.[49] See Fig. 1.

calculated by the two methods differ by no more than 0.03. A difference of this magnitude is expected because of the uncertainty in our knowledge of the hard sphere equation of state and of $g_d(r)$, both of which are used as input for the OCT. At lower densities, however, these quantities are known more accurately, and any differences larger than 0.01 are significant. At low and moderate densities, differences of this size are found only for states very close to the liquid-gas critical point. The failure of the OCT to satisfy the consistency test near the critical point is one manifestation of the fact that OCT provides a classical theory of phase transitions. A detailed study of the predictions of the OCT in the vicinity of the liquid-gas phase transition has been reported by Sung and Chandler.[49]

H. Concluding Remarks

The optimized cluster theory has been applied to the study of various types of liquids, namely the Lennard–Jones fluid,[47] ionic solutions,[47] and mixtures of hard spheres and square well molecules.[50] A generalization of the theory has been applied to water (D. Chandler and H. C. Andersen, unpublished) and to fused salts (S. Hudson and H. C. Andersen, unpublished). These calculations have met with varying degrees of success. For the first group of fluids, the accuracy of the theory is remarkably good. However for the last two types of fluids, the accuracy is very poor and the OCT is qualitatively incorrect as a description of these fluids.

From the derivation of the EXP and ORPA + B_2 approximations, one would expect them to be accurate whenever the product of ρR^3 and $\bar{C}$ is small. Here $\bar{C}$ and R are the strength and range of the renormalized potential. In all the cases tested, this was found to be a valid indicator of the accuracy of the theory. It follows that EXP and ORPA + B_2 are accurate whenever the perturbation is causing a relatively small change in the number of nearest neighbors of a molecule and in the overall distribution of molecules around a given molecule.* Moreover, the results of an OCT calculation (namely the strength and range of the calculated renormalized potential as well as the consistency check) provide a clear indication of how much to trust the ORPA + B_2 and EXP approximations for any particular fluid.

V. IMPLICATIONS

The results discussed in Sections III and IV for the structure of the Lennard–Jones liquid at high density are special cases of more general principles that may be applicable to liquids.

---

* An OCT calculation requires that the potential be separated into a reference part plus a perturbation. Only the perturbation is treated using the renormalization technique. For the fluids mentioned above for which OCT gave poor results, it might be true that a different separation of the potential would improve the accuracy of the results.

We have seen that the attractive part of the Lennard–Jones potential is not small compared with $k_B T$ at liquid temperatures, yet it has only a small effect on the structure of the liquid compared with the overwhelming effect of the short-ranged repulsions. We believe it is generally true that at high density the short-ranged repulsions are dominant and the effect of attractions is small; this principle applies to molecular as well as atomic liquids and to some nonequilibrium properties as well as to the equilibrium structure.

This idea means that at high densities the static and dynamic structures of a dense liquid are determined mainly by the shape of the molecules that comprise the fluid. We have shown that when the molecular shape is spherical (or nearly spherical) and if the hard sphere diameter is chosen properly, the static structure due to continuous short-ranged repulsions is related very simply to that of a hard sphere fluid at the same density. We believe that the dynamic as well as static properties of repulsive force systems are close to those in an appropriately chosen hard sphere system. Thus, for dense liquids composed of fairly spherical molecules, the motion of molecules in the fluid should be related simply to the particle motion in a model hard sphere fluid. Similarly, we expect that the static and dynamic properties of dense fluids of molecules that are approximately ellipsoidal in shape (or tetrahedral, or rod-like, etc.) should be simply related to those of model fluids of hard ellipsoids (or hard tetrahedra, or hard rods, etc).

In this section we will outline some of the available evidence for the validity of these ideas. Further, we will discuss reasons for possible exceptions to these principles.

A. Equilibrium Structure of Molecular Liquids

There is a growing amount of evidence to support the belief that the intermolecular correlations in dense molecular fluids are determined by the shape of the molecules in the fluid. Sung and Chandler[28] showed that repulsive forces dominate the center of mass radial distribution function determined by Berne and Harp from molecular dynamics computations on a model for liquids composed of diatomic molecules. Lowden and Chandler[51,52] have studied the intermolecular equilibrium correlations in molecular liquids for which the molecules are assumed to be composed of overlapping hard spheres which are fused together rigidly. The fluid structure for these nonspherical hard core models is obviously due to steric effects and nothing else. The calculations by Lowden and Chandler show that these steric effects produce the measured features of the intermolecular structures of the following liquids: N_2, CS_2, CSe_2, C_6H_6, and CCl_4.

There is no evidence that dipole–dipole (or quadrupole–quadrupole, etc.) terms in intermolecular potentials are significant in determining the microscopic intermolecular correlations in real one-component dense fluids.* These terms in the multipole expansion for interactions describe the slowly varying forces between molecules at large intermolecular separations. From the discussion presented in the Introduction, and from the calculations presented in Sections III and IV, one expects that slowly varying portions of intermolecular potentials play only a small role in determining the liquid structure.

B. Freezing

The concept that the harshly repulsive parts of the intermolecular forces determine the structure of a dense fluid leads naturally to the idea that excluded volume effects (and not attractive forces) are chiefly responsible for the liquid–solid freezing transition. This idea was used by Longuet-Higgins and Widom[6] when they developed a theory for the freezing of liquid argon. Their work, and that of others,[54,55] indicates that the freezing transition in simple fluids is intimately related to the fluid–solid transition that occurs in a hard sphere system.[56] Although the numerical values of the thermodynamic properties associated with the transition (e.g., heat of fusion, density discontinuities) do depend on attractive forces, the mechanism for freezing in simple fluids is the same instability in the fluid phase which causes the hard sphere to solidify at high densities.

C. Two Causes for Exceptions

When attractive forces do produce a significant structural effect in a dense fluid, the reason for it is easy to understand. The effect of hydrogen bonds in liquid water is one example already mentioned in the Introduction. The structure of some liquid mixtures provides another example. In this case, it is possible for the attractive interactions to produce structural effects which need not compete with the role of the repulsive forces. This point is illustrated most simply by the model system composed of hard spheres of diameter σ mixed with a square well species with the same hard core diameter and an attractive well of range 1.5 σ.[57] The attractions between the square wells tend to make the square well particles cluster together. This clustering can occur without changing the excluded volume correlations produced by the hard cores. As a result, the attractions are able to create important structural effects.

---

* However, the long-ranged nature of dipole–dipole interactions create a many-body cooperative phenomenon which significantly effects the long-range *asymptotic* decay of pair correlations. Indeed, for many-body systems containing dipoles, the asymptotic behavior of the pair correlations depend on the shape of the container of the macroscopic system. A review covering this subject is given by Deutch.[53]

Thus, there are two possible situations in which attractive interactions can play a significant role in determining the intermolecular correlations of dense fluids. In one, relevant to liquid water, the attractive forces are sufficiently large and quickly varying so that they can actually rupture the structure formed by the repulsive forces. When this situation occurs, the optimized cluster theory discussed in Section IV cannot describe the effects of the attractions.* In the other situation, which is frequently important for liquid mixtures, the attractive interactions can produce structural effects provided they do not compete with the correlations produced by the repulsions. For this latter case, it appears that the optimized cluster theory can successfully describe the effects of the attractions.[50,57]

D. Liquid Crystals

When molecules in a fluid are very long, the fluid can exist as a liquid crystal as well as an isotropic liquid. A rough estimate of the dimensions of a typical liquid crystal molecule [e.g., MBBA (p-methyloxybenzylidene-n-butylaniline) is approximately a spherocylinder 20 Å in length and 5 to 6 Å in width] shows that at liquid crystal densities (for MBBA at 1 atm pressure, $\rho \approx 2.3 \times 10^{-3}$ Å^{-3}), the molecules are indeed crushed extremely close to one another. As a result it is not likely that the dipole–dipole interactions, for example, are competitive with the harshly repulsive forces which define the shape of the molecules. The intermolecular structure of a liquid crystal is almost certainly determined by the long shape of the molecule. Indeed, this view is corroborated by recent light scattering measurements[59] of the Kirkwood g-factor as a function of density and temperature for the dense isotropic liquid phases of MBBA, MBA (p-methylbenzylidene-n-butylaniline) and EBA (ethylbenzylidene-n-butylaniline).

The expectation that the structure of a liquid crystal is determined mainly by the shape of the molecules in the fluid together with the current under-standing of the liquid–solid phase transition for simple systems leads us to expect that the mechanism for the phase transition between isotropic fluid and liquid crystal phases should be understood in terms of the packing of long hard particles. Onsager[60] used this physical picture when developing his theory of the isotropic–anisotropic fluid phase transition. However, his

---

* It is possible that one can devise a statistical mechanical theory for a repulsive force system which exhibits the tetrahedral intermolecular structure that occurs in liquid water. If this repulsive force system is used as the reference system for liquid water, then the hydrogen bonds will not cause appreciable changes in the fluid structure, and the optimized cluster theory will probably provide a useful way of describing the effects of hydrogen bonds in liquid water. However, if the chosen reference system for water does not exhibit tetrahedral ordering, the hydrogen bonds will cause too great an effect to be calculated from a simple application of optimized cluster theory (H. C. Andersen and D. Chandler, unpublished work). Instead a new renormalization of the hydrogen bond is required.[58]

theory also employed severe mathematical approximations which have been shown to be inaccurate.[61] It is our belief that the approximations, and not the physical picture, are the source of some of the incorrect predictions of the theory.

Our conjecture about the mechanism for the transition between liquid crystal and isotropic liquid phases contradicts the physical picture used in many of the theories of liquid crystals.[62,63] These theories attribute the properties of liquid crystals to dipole–dipole interactions. All of the arguments we have presented suggest that this physical picture is incorrect.

E. Molecular Motion in Liquids

The work of Kushick and Berne[64] provides direct evidence that the dynamical processes occuring in dense liquids are governed primarily by the short-ranged repulsive forces. In that work, computer simulations were performed for both the Lennard–Jones reference systems and the Lennard–Jones fluid itself. By comparing the results obtained for both systems, Kushick and Berne showed that at high density the velocity autocorrelation function for the reference system is similar to that for the total Lennard–Jones liquid. The attractive forces indeed play a minor role.

If the molecules in a liquid are fairly spherical in shape, it seems reasonable that the dynamics produced by the repulsive forces should be close to that occurring in a hard sphere fluid. This idea was probably first used by Enskog (see e.g., Chapman and Cowling[65]). The theory of transport developed by Enskog and the modern developments of present-day researchers provide evidence that molecular motion in dense fluids composed of fairly spherical molecules is related simply to particle motion in the hard sphere fluid.

Verlet and coworkers[13,66] have shown that the diffusion constant of the Lennard–Jones fluid is represented to within 10% by the diffusion constant of the hard sphere fluid. The diameter was, in effect, chosen to be the one associated with $u_0(r)$ according to (2.11). Protopapas et al.[67] have used the hard sphere model to calculate diffusion constants for a wide variety of liquid metals.

The diffusion constant is the zero-frequency Fourier component of the velocity autocorrelation function. Thus the work of Verlet and of Protopapas et al. indicates that the zero-frequency components for a fluid with a realistic interatomic potential and for its associated hard sphere system are, to a good approximation, the same. Kim and Chandler[68,69] have used a similar assumption in their development of a qualitatively accurate phenomenological theory of the velocity autocorrelation function for simple liquids. Chandler[70] has used the apparent connection between dynamics due to continuous repulsive forces and the dynamics of hard sphere particles to develop a simple theory for rotational and translational motion in molecular liquids. One of the principal results of this theory is a microscopic derivation of Gordon's J-diffusion model approximation.[71] The microscopic derivation allows one to arrive at these results without recourse to the unphysical assumptions usually attributed to Gordon's theory. Further, a useful expression is found for the relaxation time introduced phenomenologically by Gordon. Chandler[70] has shown that this expression predicts results that are in close agreement with those found from computer simulations on a model for liquid nitrogen.[72] The theory has also been used successfully to interpret high-pressure experiments which probe rotational and translational motions in liquids.[73–76]

F. Summary

There exists compelling evidence that at the high densities which characterize most of the liquid phase the dynamic and static structures of liquids are dominated by the short-ranged repulsive forces. It is our opinion that the phenomena that occur in most liquids are best understood by considering first the excluded volume effects produced by these forces. The attractive interactions and other slowly varying forces such as dipole–dipole interactions produce "second-order" effects which can usually be described by perturbation theory or ignored altogether. The application of this idea has produced the quantitative equilibrium theory of simple liquids discussed in this article. We believe subsequent applications will produce accurate theories for the dynamic and static correlations in complex molecular fluids.

APPENDIX A. SOME GRAPH THEORETIC TERMINOLOGY

In this appendix, we will give brief definitions of some of the graph theoretic terms used in the text. The graphs we are concerned with consist of points and bonds. The points represent particles are are of two types: root points (usually two of them are in a graph) and field points (any number of them may appear). There are various types of bonds, such as h_d, ϕ, C_L and f_d bonds. The bonds represent either interactions between the particles (e.g., ϕ or f_d) or correlations between the particles which are induced by the interactions (e.g., h_d or C_L).

A diagram is *connected* if it is possible to travel from any point to any other point along a path consisting of bonds and points. For example, in Fig. A-1, the first diagram is connected and the second is not.

Fig. A–1. Illustration of the definition of a connected diagram. The first diagram is connected and the second is not connected.

An *articulation point* is a root point or a field point which if removed would leave the diagram disconnected in such a way that at least one of the disconnected parts contains no root points. For example, the last two diagrams in Fig. A-2 have articulation points and the first one does not.

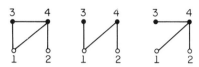

Fig. A–2. Illustration of the definition of articulation point. In the second diagram, point 1 is an articulation point. In the third, 4 is an articulation point. The first has no articulation points.

A *reference pair of articulation points* is a pair of points which if removed would disconnect the diagram in such a way that at least one of the disconnected parts contains no root point and one or more field points and only bonds that are related to the reference system (i.e., h_d and f_d bonds). For example, the last two diagrams in Fig. A-3 have a reference pair of articulation points.

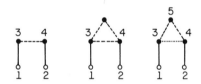

Fig. A–3. Illustration of the definition of a reference pair of articulation points. Here the dashed and dotted lines represent h_d and f_d bonds, respectively. These are reference system bonds. In the last two diagrams, points 3 and 4 are a reference pair of articulation points.

APPENDIX B. DIAGRAMATIC FORMULATION OF THE MEAN-SPHERICAL-MODEL INTEGRAL EQUATION

The mean-spherical-model (MSM) integral equation proposed by Lebowitz and Percus[77] has been the focus of great interest during the past few years. The solutions of this integral equation provide an approximation to $g(r)$ for classical fluids in which the pair interactions are of the form

$$w(1, 2) = \infty, \qquad r_{12} < d$$
$$= u(1, 2), \qquad r_{12} > d \qquad \text{(B.1)}$$

where $u(1, 2)$ is finite but otherwise arbitrary. It may, for example, depend on particle orientations as well as the separation between centers. The primary reason for the interest in the MSM equation is that it is exactly soluble for several nontrivial systems.

When the perturbation interaction, $u(1, 2)$, is exactly zero, the system is the hard sphere fluid, and the MSM equation becomes the Percus–Yevick equation for hard spheres. Wertheim[22] and Theile[21] have presented the exact solutions for this case. Wertheim[78] has also solved the MSM equation when the perturbation is the dipole–dipole interaction. For that case the particles in the fluid are hard spheres with permanent electric dipoles. Other one-component fluids for which exact solutions have been derived are hard spheres with Yukawa potentials as the perturbation,[79] polarizable hard spheres with permanent electric dipoles,[80] and charged hard spheres in a uniform neutralizing background.[81] The two-component fluids for which solutions are available are the neutral mixture of charged hard spheres in a dielectric continuum,[82] hard spheres mixed with hard spheres containing permanent point dipoles,[83] and hard spheres with point dipoles mixed with charged hard spheres.[84–86] More general multipole-electrostatic interactions have also been studied.[87] In addition to the analytic solutions, the MSM equation has been applied numerically to study the structure of real liquids with differing degrees of success.[88–90]

In this appendix we will discuss the cluster–diagramatic formulation of the MSM equation. This formulation shows the close connection of the MSM equation to the OCT. Further, it reveals the strengths and limitations of theories which are based on this equation. As a result, one is able to judge *a priori* whether a particular application of the MSM equation will be successful.

For notational simplicity, we consider only one-component systems with spherically symmetric perturbations. The generalizations to many components and to $u(1, 2)$ functions that depend on orientations are straightforward. The MSM equation is the Ornstein–Zernike equation for $g(r) - 1 = h(r)$ (see Ref. 38b)

$$h(r) = c(r) + \rho \int d\mathbf{r}' c(|\mathbf{r} - \mathbf{r}'|) h(r') \tag{B.2}$$

plus the closure relations

$$h(r) = -1, \qquad r < d \tag{B.3}$$

and

$$c(r) = -\beta u(r), \qquad r > d \tag{B.4}$$

Equation (B.3) is exact for particles with a hard core. Equation (B.4) is the only approximation. The quantity $c(r)$, which is defined by (B.2), is called the *direct correlation function*. By using Fourier transforms, (B.2) can be written as

$$\hat{h}(k) = \hat{c}(k)[1 - \rho\hat{c}(k)]^{-1} \tag{B.5}$$

Thus, one solves the MSM equation [(B.2), (B.3) and (B.4)] by finding the $c(r)$ for $r < d$ which makes $h(r) = -1$ inside the hard core. Then for $r > d$, $h(r)$ is determined by inverting (B.5).

The reader may notice the similarity between (B.5) and (4.15). The similarity implies that $h(r)$ is $c(r)$ plus the sum of all singly connected chains of two or more c functions. (This can be checked by solving (B.2) iteratively with ρ as the ordering parameter.) As a result we have the exact result that[38b,c]

$$c(r) = \text{sum of all nodeless diagrams in } h(r) \tag{B.6}$$

It is possible to find a subset of these nodeless diagrams which can be summed to form $c_{MSM}(r)$, the direct correlation function in the MSM approximation. To do this it is convenient to consider a class of diagrams which are similar to those used in Section IV to construct $C_L(r)$ (see Fig. 10). Imagine a polygon with three or more vertices. Make two adjacent vertices the root points; the remaining ones are the field points. Connect each pair of adjacent points with either a f_d bond or a Φ bond (but not both), except that no such bond is to connect the roots directly. (Here, $f_d(r)$ is the hard sphere Mayer cluster function, and $\Phi(r) = (1 + f_d)(-\beta u)$. Thus, $\Phi(r)$ is $-\beta u(r)$ for $r > d$, and it is zero for $r < d$.) Next, add some or no f_d bonds between nonadjacent vertices in such a way that no two f_d bonds cross. There are no other restrictions. Let $D(r)$ denote the sum of all such diagrams. Then the following equalities can be proven by straightforward topological considerations:

$$h_{MSM}(r) = f_d(r) + \Phi(r) + [1 + f_d(r)]D(r) \tag{B.7}$$

and

$$c_{MSM}(r) = f_d(r) + \Phi(r) + f_d(r)D(r) \tag{B.8}$$

To verify these results one must do three things. First, note that the right-hand side of (B.7) satisfies (B.3). Second, note that the right-hand side of (B.8) satisfies (B.4). Third, show that (B.7) and (B.8) satisfy (B.2). This is done by using the definition of $D(r)$ to establish that $c_{MSM}(r)$ does contain all the nodeless diagrams in $h_{MSM}(r)$ and that all the diagrams (and no more) contained in $h_{MSM}(r)$ are generated by summing all the singly connected chains of c_{MSM} bonds.

One may extract from $h_{MSM}(r)$ all the diagrams containing no Φ bonds. The sum of these diagrams is the Percus–Yevick correlation function for hard spheres,[46] $h_d^{(PY)}(r)$. By a topological reduction,[38c] the remaining diagrams in

$h_{MSM}(r)$ (those involving one or more Φ bonds) may be expressed in terms of Φ, f_d, and $h_d^{(PY)}$ bonds. The result is

$$h_{MSM}(r) - h_d^{(PY)}(r) = \Phi(r) + [1 + f_d(r)] [\text{sum of all diagrams formed in the same way as those in } D(r) \text{ except now on the exterior of the polygon we have } h_d^{(PY)} \text{ instead of } f_d \text{ bonds, and no diagram is permitted if any field point in the diagram is intersected by only } h_d^{(PY)} \text{ bonds}] \tag{B.9}$$

By comparing (B.9) with the definition of $C_L(r)$ given in Section IV, we see that

$$h_{MSM}(r) = h_d^{(PY)}(r) + C_L^{(PY)}(r) \tag{B.10}$$

where $C_L^{(PY)}(r)$ is $C_L(r)$ evaluated using the Percus–Yevick approximation for h_d rather than the exact hard sphere correlation function. Equation (B.10) has been derived before using a very different method.[5]

There are two important points that can be drawn from this analysis. First, since the Percus–Yevick theory is fairly accurate for hard spheres, there is only a small difference between $C_L^{(PY)}(r)$ and $C_L(r)$. Therefore, one may use the analytic solution of the MSM equation (if it is obtainable) to calculate

$$C_L(r) \cong h_{MSM}(r) - h_d^{(PY)}(r) \tag{B.10}$$

One may then use this analytic expression to calculate $g(r)$ from the OCT (in particular, the EXP approximation):

$$g(r) \cong g_d(r) \exp [h_{MSM}(r) - h_d^{(PY)}(r)] \tag{B.11}$$

Equation (B.11) is inherently more accurate than $g_{MSM}(r) \cong g(r)$.[47] However, its accuracy is limited, and this brings us to the second point. As discussed in Section IV, the OCT is accurate only if $\rho C_L(r)$ is small. Thus, (B.11) and the MSM equation even more so are accurate and thus useful only when the quantity $\rho[h_{MSM}(r) - h_d^{(PY)}(r)]$ is small. The MSM equation is not reliable if the equation predicts that the fluid structure is significantly different from that of the hard sphere system. Thus, whenever the structural properties of a fluid are qualitatively different from those of the noble gas fluids the MSM cannot be expected to give even a qualitatively correct description. For example, the MSM will not provide a correct theory for liquid water or for fused salts. Further, the MSM equation will not be accurate if an unsatisfactory division is made of the pair potential (see Section III). The reference potential must be harsh enough that its effects are nearly those of a hard core with diameter d; and the reference potential must contain all the quickly varying repulsive interactions. If these conditions are not met, $\rho[h_{MSM}(r) - h_d^{(PY)}(r)]$ will not be small.

Acknowledgments

We are grateful to Stephen H. Sung for providing us with his unpublished numerical results for the Lennard–Jones fluid. The information depicted in Fig. 4, and some of the entries in Tables II and III were obtained from his unpublished calculations. We also want to thank George Stell for helpful comments on a previous version of this paper.

This work was supported in part by grants from the National Science Foundation (GP 20058A and MPS 73–08591 A02), the Petroleum Research Fund as administered by the American Chemical Society, and the Alfred P. Sloan Foundation.

References

1. D. Chandler and J. D. Weeks, *Phys. Rev. Lett.*, **25**, 149 (1970).
2. J. D. Weeks, D. Chandler, and H. C. Andersen, *J. Chem. Phys.*, **54**, 5237 (1971).
3. *Ibid.*, **55**, 5421 (1971).
4. H. C. Andersen, J. D. Weeks, and D. Chandler, *Phys. Rev.*, **A4**, 1597 (1971).
5. H. C. Andersen and D. Chandler, *J. Chem. Phys.*, **57**, 1918 (1972).
6. H. C. Longuet-Higgens and B. Widom, *Mol. Phys.*, **8**, 549 (1964).
7. H. Reiss, *Adv. Chem. Phys.*, **IX**, 1 (1965).
8. J. A. Barker and D. Henderson, *J. Chem. Phys.*, **47**, 4714 (1967).
9. L. Verlet, *Phys. Rev.*, **165**, 201 (1968).
10. L. Verlet and J. J. Weis, *Phys. Rev.*, **A5**, 939 (1972).
11. J. A. Barker and D. Henderson, *Ann. Rev. Phys. Chem.*, **23**, 439 (1972).
12. H. C. Andersen, *Ann. Rev. Phys. Chem.* **26**, 145 (1975).
13. L. L. Lee and D. Levesque, *Mol. Phys.*, **26**, 1351 (1973).
14. W. A. Steele and S. I. Sandler, *J. Chem. Phys.*, **61**, 1315 (1974).
15. S. I. Sandler, A. D. Gupta, and W. A. Steele, *J. Chem. Phys.*, **61**, 1326 (1974).
16. M. H. Kalos, D. Levesque, and L. Verlet, *Phys. Rev.*, **A9**, 2178 (1974).
17. D. Shiff, *Nature*, **243**, 130 (1973).
18. G. Stell, J. C. Rasaiah, and H. Narang, *Mol. Phys.*, **27**, 1393 (1974).
19. B. J. Alder and T. E. Wainwright, *J. Chem. Phys.*, **33**, 1439 (1960).
20. J. K. Percus and G. J. Yevick, *Phys. Rev.*, **110**, 1 (1958).
21. E. Thiele, *J. Chem. Phys.*, **39**, 474 (1963).
22. M. S. Wertheim, *Phys. Rev. Lett.*, **10**, 321 (1963).
23. L. Verlet and J. J. Weis, *Mol. Phys.*, **24**, 1013 (1972).
24. J. A. Barker, *Proc. Roy. Soc. London*, **A241**, 547 (1957).
25. K. E. Gubbins, W. R. Smith, M. K. Tham, and E. W. Tiepel, *Mol. Phys.*, **22**, 1089 (1971).
26. T. L. Hill, *Statistical Mechanics*, McGraw Hill, New York, 1956.
27. J. P. Hansen and J. J. Weis, *Mol. Phys.*, **23**, 853 (1972).
28. S. H. Sung and D. Chandler, *J. Chem. Phys.*, **56**, 4989 (1972).
29. L. Verlet, *Phys. Rev.*, **159**, 98 (1967).
30. D. Levesque and L. Verlet, *Phys. Rev.*, **182**, 307 (1969).
31. R. Zwanzig, *J. Chem. Phys.*, **22**, 1420 (1954).
32. L. D. Landau and E. M. Lifshitz, *Statistical Physics*, Pergamon Press, London, 1958.
33. D. A. McQuarrie and J. L. Katz, *J. Chem. Phys.*, **44**, 2393 (1966).
34. B. Widom, *Science*, **157**, 375 (1967).
35. J. L. Lebowitz and O. Penrose, *J. Math. Phys.*, **7**, 98 (1966).

36. P. Debye and E. Hückel, *Physik Z.*, **24**, 185 (1923).
37. J. E. Mayer, *J. Chem. Phys.*, **18**, 1426 (1950).
38. Useful references on cluster theory include (*a*) H. J. Friedman, *Ionic Solution Theory*, Interscience, New York, 1962; (*b*) G. Stell in *The Equilibrium Theory of Classical Fluids*, H. L. Frisch and J. L. Lebowitz, Eds., Benjamin, New York, 1964; and (*c*) T. Morita and K. Hiroika, *Progr. Theor. Phys.*, **25**, 537 (1961).
39. J. L. Lebowitz, G. Stell, and S. Baer, *J. Math. Phys.*, **6**, 1282 (1965).
40. G. Stell, J. L. Lebowitz, S. Baer, and W. Theumann, *J. Math. Phys.*, **7**, 1532 (1966).
41. G. Stell and J. L. Lebowitz, *J. Chem. Phys.*, **49**, 3706 (1968).
42. P. Hemmer, *J. Math. Phys.*, **5**, 75 (1964).
43. G. Stell and K. Theumann, *Phys. Rev.*, **186**, 581 (1969).
44. G. Stell, *Phys. Rev.*, **184**, 135 (1969).
45. G. Stell, in *Phase Transitions and Critical Phenomena*, V. 5, C. Domb and M. S. Green, Eds., Academic, London, 1974.
46. G. Stell, *Physica*, **29**, 517 (1963).
47. H. C. Andersen, D. Chandler, and J. D. Weeks, *J. Chem. Phys.*, **57**, 2626 (1972).
48. S. Hudson and H. S. Andersen, *J. Chem. Phys.*, **60**, 2188 (1974).
49. S. H. Sung and D. Chandler, *Phys. Rev.*, **A9**, 1688 (1974).
50. S. H. Sung, D. Chandler, and B. J. Alder, *J. Chem. Phys.*, **61**, 932 (1974).
51. L. J. Lowden and D.Chandler, *J. Chem. Phys.*, **59**, 6587 (1973).
52. *Ibid.*, **61**, 5228 (1974).
53. J. M. Deutch, *Ann. Rev. Phys. Chem.*, **24**, 301 (1973).
54. J. J. Weiss, *Mol. Phys.*, **28**, 187 (1974).
55. R. K. Crawford, *J. Chem. Phys.*, **60**, 2169 (1974).
56. W. G. Hoover and F. H. Ree, *J. Chem. Phys.*, **49**, 3609 (1968).
57. B. J. Alder, W. W. Alley, and M. Rigby, *Physica*, **73**, 143 (1974).
58. H. C. Andersen, *J. Chem. Phys.*, **59**, 4714 (1973); *ibid.*, **61**, 4985 (1975).
59. G. R. Alms, T. D. Gierke, and W. H. Flygare, *J. Chem. Phys.*, **61**, 4083 (1974).
60. L. Onsager, *Ann. N.Y. Acad. Sci.*, **51**, (1949).
61. J. Vieillard-Baron, *Mol. Phys.*, **28**, 809 (1974).
62. G. Meier and A. Saupe, in *Liquid Crystals*, G. H. Brown, G. L. Dienes, and M. M. Labes, Eds., Gordon & Breach, London, 1966.
63. W. L. McMillan, *Phys. Rev.*, **A4**, 1238 (1971).
64. J. Kushick and B. J. Berne, *J. Chem. Phys.*, **59**, 3732 (1973).
65. S. Chapman and F. G. Cowling, *The Mathematical Theory of Non-Uniform Gases*, 3rd ed., Cambridge University Press, Cambridge, 1970.
66. D. Levesque and L. Verlet, *Phys. Rev.*, **A2**, 2514 (1970).
67. P. Protopapas, H. C. Andersen, and N. A. D. Parlee, *J. Chem. Phys.*, **59**, 15 (1973).
68. K. Kim and D. Chandler, *J. Chem. Phys.*, **59**, 5215 (1973).
69. D. Chandler, *Acc. Chem. Res.*, **7**, 246 (1974).
70. D. Chandler, *J. Chem. Phys.*, **60**, 3500, 3508 (1974).
71. R. G. Gordon, *J. Chem. Phys.*, **44**, 1830 (1966).
72. J. Barojas, D. Levesque, and B. Quentrec, *Phys. Rev.*, **A7**, 1092 (1973).
73. J. DeZwaan, R. J. Finney, and J. Jonas, *J. Chem. Phys.*, **60**, 3223 (1974).
74. J. DeZwaan and J. Jonas, *J. Chem. Phys.*, **62**, 4036 (1975).
75. J. H. Campbell, J. F. Fisher, and J. Jonas, *J. Chem. Phys.*, **61**, 346 (1974).
76. D. Chandler, *J. Chem. Phys.*, **62**, 1350 (1975).
77. J. L. Lebowitz and J. K. Percus, *Phys. Rev.*, **144**, 251 (1966).
78. M. S. Wertheim, *J. Chem. Phys.*, **55**, 4291 (1971).
79. E. Waisman, *Mol. Phys.*, **25**, 45 (1973).
80. M. S. Wertheim, *Mol. Phys.*, **26**, 1425 (1973).

81. R. G. Palmer and J. D. Weeks, *J. Chem. Phys.*, **58**, 4171 (1973).
82. E. Waisman and J. L. Lebowitz, *J. Chem. Phys.*, **56**, 3086, 3094 (1972).
83. S. A. Adelman and J. M. Deutch, *J. Chem. Phys.*, **59**, 3971 (1973).
84. *Ibid.*, **60**, 3935 (1974).
85. L. Blum, *Chem. Phys. Lett.*, **26**, 200 (1974).
86. L. Blum, *J. Chem. Phys.*, **61**, 2129 (1974).
87. *Ibid.*, **57**, 1862 (1972); **58**, 3295 (1973).
88. L. Blum and A. H. Narten, *J. Chem. Phys.*, **56**, 5197 (1972).
89. R. O. Watts, D. Henderson, and J. A. Barker, *J. Chem. Phys.*, **57**, 5991 (1972).
90. A. H. Narten, L. Blum, and R. H. Fowler, *J. Chem. Phys.*, **60**, 3378 (1974).

RECENT ADVANCES IN THE STUDY OF LIQUIDS BY X-RAY DIFFRACTION*

J. F. KARNICKY AND C. J. PINGS

*Division of Chemistry and Chemical Engineering,
California Institute of Technology,
Pasadena, California*

CONTENTS

* Work supported in part by a grant from the National Science Foundation.

I. INTRODUCTION

A. Content of This Paper

This paper is intended to inform the reader of the current state of the art of studying fluid structure by X-ray diffraction. The early X-ray diffraction studies of liquids were reviewed by Gingrich[34] in 1943. In 1962 Kruh[62] and Furukawa[32] separately wrote review papers on the study of liquid structure by X-ray diffraction techniques (including X-ray, neutron, and electron diffraction). The studies of the structure of monatomic liquids were reviewed by Pings[95] in 1968.

The present paper is concerned primarily with the development of fluid X-ray diffraction structural determinations since 1962, and the reader is strongly urged to examine the review papers by Gingrich,[34] Kruh,[62] Furukawa,[32] and Pings[95] in conjuction with the present article. It should be noted that the Kruh paper (404 references) reports some experiments in 1960 and 1961 which are not referenced in the Furukawa paper, and the Pings article contains a very comprehensive discussion of X-ray scattering theory and describes in detail the X-ray studies of the inert gas fluids.

The present paper is a report of the current state of development of methods for obtaining by X-ray diffraction, the distribution functions which describe the structure of molecular fluids in the liquid and dense gas states. It does not attempt to include low-angle studies of molecular clustering, neutron diffraction studies, or X-ray studies of amorphous solids, liquid crystals, or liquid metals. The reader is directed to the following sources for reviews of subjects closely related to the content of the present article:

1. The theory of liquid structure has been reviewed by Rowlinson[104] in 1965 (equations of state for dense fluid systems) and Cole[15] in 1968 (classical fluids and the superposition approximation).

2. Safford and Leung[107] discussed the study of electrolytic solutions by X-ray diffraction and neutron inelastic scattering. Hendren[46] has compiled a bibliography of over 900 references related to X-ray and neutron diffraction from liquids. Powles[98] has reviewed the use of slow neutron scattering methods for the study of the structure of molecular liquids.

3. References to low-angle X-ray diffraction are found in a survey of the field by Brumberger[14] in 1967, a bibliography compiled by Renouprez[102] in 1970, and a discussion by Brady[12] in 1971.

4. X-ray studies of amorphous substances (in particular, amorphous catalysts) are discussed by Ratnasamy and Leonard.[101]

5. A recent source of information (1973) about liquid metals is found in the *Proceedings of the 2nd International Conference on the Properties of Liquid Metals.*[99]

B. Limitations of This Experimental Technique

Perhaps it seems rather negative to start a discussion of liquid structure determination by X-ray diffraction with the consideration of what cannot be done, but it appears that such a course is necessary. The history of X-ray studies of liquids contains many examples of results which do not agree with the results of other X-ray experiments, conclusions drawn from features which were later shown to be artifacts of the data reduction approximations, and structural descriptions of liquids which stretch three-dimensional inferences far beyond what could be reasonably inferred from the one-dimensional X-ray diffraction data. (Particularly suspect are three-dimensional models which describe liquids as if they were crystalline solids with fixed sites for each molecule.) .

It must be emphasized that there is a fundamental limitation on the information one can obtain from X-ray studies of liquids. This is due to the fact that the X-ray diffraction data represent a one-dimensional quantity which is the average of quantities which, in general, vary with the three dimensions of the system, relative orientations within the system, and time.

In addition to this intrinsic limitation there are many possible sources of systematic error in the quite intricate process of going from the experimental count rate to the distribution functions which describe the liquid structure.

Improvement in the usefulness of liquid X-ray data over recent years is generally due to diminishment of these systematic errors by means of better monochromated incident radiation, more accurate atomic scattering factors, more reliable absorption corrections and so forth. Because the data reduction process is so intricate, it is virtually impossible for someone who has not himself performed such an experiment and carried through the data analysis to evaluate the accuracy of published results and the credibility of published conclusions about the structure being considered. Moreover, even someone who is familiar with the field has to rely on the integrity of the experimenter in reporting his results. Most experimenters, however, still do not present any quantitative estimates of their reliability. As pointed out by Furukawa,[32] even the simple procedure of reporting the experimental radial distribution function for values of r less than an atomic diameter would give some clue as to the quality of the data, but in many published papers concerned with X-ray studies of diffraction from fluids this is still not done.

There is no doubt that the results of X-ray diffraction from fluids can be used, and have been used, to obtain valuable information about liquid structure. It is not even necessary to argue the relative merits of X-ray and neutron diffraction, as these techniques emphasize different aspects of fluid structure and should complement (rather than contest) each other. However, the results of any structure determination can be worse than useless unless they are reported objectively and with some evaluation of or clue to their accuracy and reliability. Certainly, there is a need for some of the inferences and speculations which are made about the detailed underlying structure of complex systems which produce the X-ray data, but it is absolutely imperative that, in reporting these results, a clear and objective assessment of the reliability of the various conclusions be made.

II. DETERMINATION OF THE LIQUID STRUCTURE DISTRIBUTION FUNCTIONS

A. Experimental Aspects

The experimental apparatus basically consists of an X-ray source whose monochromated output irradiates a fluid sample, whereupon the diffracted radiation is detected and recorded as a function of angle.

1. X-ray Source

X-ray Power. Improved technology has enabled recent experiments on fluids to be performed at electron powers higher than previously obtainable, using fixed anodes (Kirstein,[55] 48 kV and 35 mA electron current) or rotating anodes (Karnicky,[51] 60 kV and 100 mA electron current). Presently available rotating anode machines suitable for diffraction studies produce power up to 60 kV and 1 A electron current.[103]

Source Stability. The typical time variation of X-ray intensity produced by modern tubes is on the order of 1 or 2% over a few hours. (The Rigaku–Denki Company advertises a feedback voltage stabilizer option which is claimed to regulate the X-ray output within 0.2%.[103] Furumoto and Shaw[33] reported intensity stability of $\pm 0.1\%$ over 12 hr but did not specify details.)

Diffraction studies of fluids have reached a stage of precision where this 1 to 2% variation must be corrected for. One method used for removing the effects of source instability involves returning the counter periodically to a reference point and scaling the data by the relative intensity of this reference value. This method was used by Achter and Meyer[1] in their helium studies and by Mikolaj[74,76] and Smelser[110] in argon studies. A more common practice is to repeat the scan several times to cancel random errors and ensure reproducibility. Kirstein[55] and Karnicky[51] each used a multiple sweep quick scanning technique which served to minimize variation due to long-term drift and acted as a multichannel analyzer in averaging out short-term noise. Licheri[68] randomized his counter positioning sequence to eliminate effects due to systematic temporal drift.

Choice of Wavelength. The wavelength of X-ray used is chosen according to the absorption of the sample and the range of s desired; s is the scattering parameter, or momentum transfer, and is related to the scattering angle, 2θ, by the following equation:

$$s = 4\pi\lambda^{-1}\sin\theta \qquad (2.1)$$

where λ is the incident wavelength.

If high transmission of the incident and diffracted rays through the sample is desired (as in a Debye–Sherrer[22] geometry), short wavelength X-rays (usually Mo $K\bar{\alpha}$ Ag $K\bar{\alpha}$) are chosen. If low transmission is desired (as in a Bragg–Brentano[22] geometry), a longer wavelength would be preferred.

The other important consideration is whether the experimenter wishes to spread out low-angle data (for which a long wavelength is desired) or obtain as large an s value as possible for the accessible range of θ (for which a short wavelength is used).

Copper radiation was used by Achter and Meyer[1] and by Hallock[39,41–43] in their helium studies. Bochynski[11] and Stirpe and Tompson[115] combined low-angle counting using Cu radiation ($\lambda K\bar{\alpha} = 0.561$ Å) with higher-angle data using Mo radiation ($\lambda K\bar{\alpha} = 0.710$Å). In addition, Stirpe and Tompson[115] repeated their wide-angle data accumulation using Ag radiation ($\lambda K\bar{\alpha} = 0.561$ Å), which serves as an internal consistency check for the absence of wavelength-dependent errors. Most experiments are now performed using Mo radiation in order to reach a reasonably large maximum s value while still having sufficient absorption by the sample in the Bragg–Brentano geometry.

2. Monochromatization

The techniques now used to achieve monochromatic incident radiation are, in decreasing order of spectral purity: (1) crystal monochromator; (2) balanced dual filters; and (3) β filter and pulse height discrimination. The relative merits of these methods are discussed in the *International X-ray Tables*.[70]

Crystal Monochromators.   Crystal monochromators were used in most of the fluid studies since 1962. The placement of the monochromator (in the incident beam or in the diffracted beam) is influenced by the amount of fluorescence from the sample. A large amount of fluorescence occurs when the incident radiation comes from a target metal with an atomic number slightly higher than the atomic number of one of the elements in the specimen, but the continuum in the tube output X-ray distribution can cause any specimen to fluoresce. A monochromator between the sample and detector can prevent the fluorescent radiation from reaching the detector.[70]

Balanced Filters.   A description of the use of balanced filters is found in the article by Soules, Gordon, and Shaw.[111] Berman and Ergun[7] discuss the use of balanced filters to select only one member of the $K\alpha_1$, $K\alpha_2$ wavelength pair.

Balanced filters were used in a series of argon studies at the California Institute of Technology by Smelser,[110] Kirstein,[55] and Karnicky,[51] as well as in a study of $AsBr_3$ by Hoge and Trotter.[47] As with monochromators, the placement of the balanced filters is sometimes determined by questions of fluorescence. In the $AsBr_3$ study of Hoge and Trotter[47] the balanced filters for the $CuK\bar{\alpha}$ radiation were placed between the source and sample to minimize the fluorescence from the cobalt filter reaching the detector. Then, an additional β filter was placed between the source and sample to absorb any fluorescence from the specimen.

Karnicky[51] measured the actual distribution of wavelengths present in the incident beam after dual filter subtraction and corrected for this distribution in all subsequent data analysis, rewriting the algebraic expressions involving λ as integrals over the actual λ distribution, and solving numerically for the part of the coherent scattered intensity due to the monochromatic $AgK\bar{\alpha}$.

β Filter.   A β filter and pulse height discrimination were used by Mikolaj[74,76] in his argon experiments and by Achter and Meyer[1] and Hallock[39,41–43] in the small s region of their helium experiments.

3. Sample Confinement

The type of sample confinement chosen depends on whether a Bragg–Brentano[22] parafocusing geometry (reflection) or a Debye–Scherrer[22] geometry (transmission) is to be used.

Reflection Geometry. Reflection from the surface of a liquid has some advantages over transmission through a liquid. If the substance being studied sufficiently absorbs X-rays, no absorption correction is required. If this is not true, corrections can be made.[78] Also, in the reflection geometry the scattering from the sample cell generally contributes less to the total scattering detected.

In order for the use of a reflection geometry to be feasible, the scattering from the vapor above the surface of the liquid must be small, preferrably negligible. The X-ray group at Oak Ridge National Laboratory has used a cryostat of this type[67] to study carbon tetrachloride,[86] benzene,[82] water,[85,88] aqueous solutions of ammonia,[83] lithium chloride,[89] tetra-n-butyl ammonium fluoride,[87] and ammonium halides,[23,84] and molten beryllium fluoride-lithium fluoride mixtures.[118] At the University of Arkansas the X-ray workers in the Department of Chemistry and the Department of Physics have been using a reflection geometry cryostat[93,109] to study argon and xenon at their triple points,[45] bromine,[38] ammonia,[73] silicon tetrachloride,[105] carbon tetrafluoride,[44] carbon tetrachloride,[37] methane,[93] and aqueous solutions of alkalihalides,[65] cobalt(II)chloride,[125] zinc(II)chloride,[64] and zinc bromide.[126] For the experiments on the more volatile substances (CF_4, CCl_4, CH_4), corrections were made for the scattering from the vapor above the surface of the liquid.

Transmission Geometry. Transmission sample cells are suitable for fluids with large vapor pressures or low absorption coefficients, for gases, and for fluids at temperatures near and above their critical points. They have been used most frequently for low-temperature inert gas studies.

The choice of material and geometry for a transmission cell are dictated by considerations of strength, transmission of X-rays, scattering power of the sample, and the need to make corrections to the experimental data for absorption and double scattering.

Achter and Meyer[1] (1969) used a cylindrical mylar cell for their low-angle studies and a flat beryllium cell for their high-angle studies of liquid and gaseous helium. Their cryostat was previously built by Narahara.[81] Hallock[39,41-43] studied liquid and gaseous ^{3}He and ^{4}He using a cryostat built by Safrata[27] for neutron studies. Hallock used a variety of target cells, including cylindrical mylar,[40,41] cylindrical beryllium,[41] and flat beryllium.[42,43] Thompson used a flat beryllium transmission cell[115] in his low-temperature studies of neon[115] and argon.[35] At the California Institute of Technology, Mikolaj,[74,76] Smelser,[110] and Kirstein[55] have used a crysotat and cylindrical beryllium cell designed by Honeywell[48] in low-temperature studies of fluid argon. Karnicky,[51] in the same laboratory, designed and built a new sample-confining system to study fluid argon at $-110°C$. Furumoto and Shaw[33] used a cylindrical beryllium cell previously built for helium[36] in their study of liquid N_2, O_2 mixtures. Transmission cells in the Debye–Scherrer geometry were also used by Hoge and Trotter[47] (AsBr$_3$, glass capillary) and Bochynski[10,11] (benzene and nitrobenzene, flat bails of unspecified material). Barnett and Hall[4] have designed a transmission cell useful for the study of liquids at very high pressure (50,000 atm).

Unconfined Samples. Licheri[68] scattered X-rays from a flowing stream of aqueous HCl, thus eliminating the sample holder entirely. Nukui[90] also disposed of a sample holder by diffracting from a molten SiO_2 drop which adhered to a solid SiO_2 rod by surface tension and melt viscosity.

4. Detectors

The use of counting systems with scintillation crystals (Tl activated LiF or NaF) and pulse height discrimination has become almost universal. This system produces a distribution of output pulses with mean voltage proportional to the X-ray energy, and is linear to very high count rates. The only recent exceptions are that of Hallock[39] who used a xenon-methane proportional counter and Bochynski[10,11] who still uses the photographic technique.

B. Correcting the Experimental Data

The data obtained from the experimental count rate contain both coherent and incoherent scatter as well as multiply scattered X-rays. The scattering is modified by absorption in the sample and cell, polarization by the scattering processes in the sample and monochromiter, and, in some cases, variation in the number of irradiated scattering elements with angle. The experimental data must be carefully analyzed and corrected for these effects to extract $I(s)$, the singly scattered, coherent, structure dependent X-ray intensity.

1. Polarization

The originally unpolarized incident beam is polarized by reflection within the monochromiter, if one is used, and by diffraction in the sample. The equations giving the modification of $I(s)$ by polarization[22,32,70] are for no monochromiter

$$I_{obs}(s) = \left(\frac{1 + \cos^2 2\theta}{2}\right)I(s) \tag{2.2}$$

where $I_{obs}(s)$ is the modified intensity and 2θ is the scattering angle.

If a monochromiter is used, the equation is

$$I_{obs}(s) = \left(\frac{1 + \cos^2 2\theta_m \cos^2 2\theta}{1 + \cos^2 2\theta_m}\right)I(s) \tag{2.3}$$

where $2\theta_m$ is the scattering angle of the monochromiter.

2. Absorption

Reflection Geometry. If the sample is strongly absorbing, no correction for absorption is required in the Bragg–Brentano geometry. Millberg[78] has derived corrections for weakly absorbing samples.

Transmission Geometry. If the cross-section of the incident X-ray beam is small compared to the cell dimensions, the correction for absorption by the cell and sample becomes a trivial exercise in one-dimensional integration. For flat cells, the absorption is not a function of the position of the beam on the cell face, and hence the absorption corrections remain simple, even for beams of extended cross-section or nonuniform intensity.

For an X-ray beam of extended cross-section irradiating a curved or cylindrical cell, the calculation of absorption coefficients can become quite difficult. Pings and co-workers have calculated the absorption factors for a beam with extended rectangular cross-section and uniform intensity incident on a cylindrical cell both numerically[91] and analytically.[53] The calculation for a cylindrical cell and an incident beam of rectangular cross-section and non-uniform intensity has been done numerically by Kirstein[55] and Smelser.[110]

Smelser[110] has examined the effect of uncertainties in the mass absorption coefficient, μ, on the data reduction, and found that for a transmission geometry the error incurred in $I(s)$ is about an order of magnitude less than the error in μ.

3. Incoherent Scatter

The incoherent scatter, $I_{inc}(s)$, is calculated theoretically as a function of angle and subtracted from the experimental data. Errors in this calculation can be very serious when scattering X-rays from elements with low atomic weights or when scattering at high angles. Under these conditions the incoherent scattering contributes the largest fraction of the total scatter.

Since the approximations involved in calculating the incoherent atomic scatter are the same as those used in calculating the coherent atomic scattering, $f^2(s)$, the theoretical calculations of $I_{inc}(s)$ and $f^2(s)$ will be considered together in the section on obtaining the radial distribution function (Section C).

Some experimenters adopt the technique of placing the monochrometer between the sample and the detector, which has the effect of experimentally eliminating much of the incoherent scatter.

4. Multiple Scattering

In recent years calculation of the amount of multiple X-ray scattering for various geometries has resulted in the knowledge that (1) it is sufficient to consider only double X-ray scattering; and (2) although the problem of multiple scattering of X-rays is not so severe as that for neutron scattering, it is sufficiently large to require a correction when accurately evaluating experimental X-ray data.

Calculation of the amount of double scattering involves the integration of the expression for $dI_2(2\theta)$ over the volume of the sample irradiated by the incident beam (v_1) and seen by the detector system (v_2)

$$dI_2(2\theta) = I_0 K^2 \frac{I_1(2\theta_1)I_2(2\theta_2)}{r_{12}^2 R^2} (2\theta_1, 2\theta_2, 2\theta)e^{-\mu l_1 - \mu l_2 - \mu l_3} \, dv_1 \, dv_2 \quad (2.4)$$

I_0 is the incident intensity, $I_2(2\theta)$ is the intensity of twice-scattered X-rays, I_1 and I_2 are the intensities at scattering angles $2\theta_1$ and $2\theta_2$ for each scattering event, $e^{-\mu l_1 - \mu l_2 - \mu l_3}$ corrects for absorption before, between, and after the two scattering events, $P(2\theta_1, 2\theta_2, 2\theta)$ corrects for polarization, r_1 is the distance between the two scattering sites, R is the distance from the sample to the detector, and K is $\rho e^4/m_0^2 c^4$ where ρ is the density. The value of this integral will depend strongly on the scattering geometry used.

Strong and Kaplow[116] and Warren and Mozzi[121] evaluated this integral for various conditions for a reflection geometry, and presented their results in a form which can be adopted to most reflection experiments. Typically, for reflection geometries, the double scattering represents a few percent of the total scattered X-rays. For example, from the paper by Warren and Mozzi,[121] when scattering RhK$\bar{\alpha}$ radiation from vitreous SiO_2 in the range $2\theta = 90$–$180°$, the ratio of double to single scattering is ~ 0.08.

Karnicky[51] and Wignall[127] have calculated the double scattering in transmission experiments. Wignall shows that for a cylindrical sample cell the double scattering is about 2% of the total scattering, but for a flat transmission cell in the $\theta, 2\theta$ geometry, the double scatter can be as high as 15% for liquid argon at high angles. Karnicky[51] found that using a stationary transmission cell rather than a θ–2θ transmission cell does not significantly change the results. He also evaluated the double scattering from dense gaseous argon ($\rho = 0.08$–0.31 g/cm^3) including sample-sample, cell-cell, cell-sample, and sample-cell double scattering and found that the double scattering can be as much as 8% of the argon single scatter at $s = 5$ Å^{-1}. Hallock[42] was able to obtain estimates for the multiple scattering in his helium transmission experiments by adapting the neutron-scattering calculations of Blech and Auerbach[9] for a cylindrical cell and those of Vineyard[119] for a plane slab transmission geometry.

In recent papers some X-ray diffraction experimenters report that they corrected for multiple scattering, but do not give an indication of how this correction was actually made.

5. Variation in the Number of Irradiated Atoms

At any angle, 2θ, the intensity of diffracted radiation is proportional to the number of scattering elements which are simultaneously irradiated by the incident beam and observed by the receiving slit system. Most experiments are designed to have this observational volume constant for all angles. For that case, the number of scattering elements is a constant for all values of 2θ and this constant can be evaluated by one of several available normalizing techniques. These techniques are intimately involved with the Fourier transform process and will be discussed in the next part of this paper.

In some experiments the observational volume changes with angle and the scattered intensity must be corrected for the variable number of scattering elements. This can be done by geometrical optics calculations, but it is more reliably accomplished by experimental calibration of the scattering volume. Achter and Meyer[1] and Hallock[39,41–43] calibrated their helium test cells by scattering from the same cells filled with neon under conditions which allow the structure factor of the neon to be calculated from the virial expansion. They were then able to back-calculate the observational volume from the neon scatter, and hence the variation of helium scattering elements with angle. Furumoto and Shaw[33] determined the variation in observational volume in their N_2, O_2 experiments by a combination of geometrical optics calculations and an experimental calibration using argon gas.

C. Fourier Transformations

The coherent, singly scattered X-ray diffraction pattern from liquids, $I(s)$, may be transformed to yield distribution functions describing the spatial correlations of the electron density, atomic density, or molecular density in the fluid being studied. The usefulness of the results of X-ray diffraction are, in a large measure, dependent on the reliability of these distribution functions. Accordingly, it is necessary to examine in detail the procedures used and approximations made in transforming from the experimentally derived $I(s)$ to these distribution functions.

1. Monatomic Fluids

The coherently scattered radiation from a monatomic fluid is given by the following equation:[95]

$$\frac{I(s) - Nf^2(s)}{NZ} = \int_0^\infty 4\pi r^2 (\rho(r) - \bar\rho - \rho_0(r)) \frac{\sin sr}{sr} dr \qquad (2.5)$$

$I(s)$ is the coherently scattered intensity corrected for the (experimentally inaccessible) zero-angle scattering; N is a scale factor which converts the experimental intensities to electron units; Z is the atomic number of the element studied; and $f^2(s)$ is the atomic scattering factor, that is, the coherent scattering from one atom in the fluid. $\rho(r)$ is the density of electron pairs separated by a distance r; $\bar{\rho}$ is the average electron density in the fluid; and $\rho_0(r)$ is the density of electron pairs within the same atom. Equation (2.5) may be Fourier transformed to give

$$4\pi r^2 (\rho(r) - \bar{\rho} - \rho_0(r)) = \frac{2r}{\pi} \int_0^\infty s \frac{I(s) - N f^2(s)}{NZ} \sin sr \, ds \qquad (2.6)$$

The quantity $4\pi r^2 (\rho(r) - \rho_0(r)) \, dr$ is the number of electrons in a spherical shell of thickness dr at a distance r from the origin averaged over all possible electron densities at the origin, exclusive of contributions from electron pairs in the same atom. $4\pi r^2 (\rho(r) - \rho_0(r))$ is called the electron radial distribution function (ERDF) and (2.6) can be written

$$\text{ERDF} = 4\pi r^2 \bar{\rho} + \frac{2r}{\pi} \int s \frac{I(s) - N f^2(s)}{NZ} \sin sr \, ds \qquad (2.7)$$

The units of the ERDF are electrons^2/Å.

For monatomic fluids it is possible to write the intensity of coherently scattered radiation as a function of the density of atomic centers.[95] Further manipulation of (2.5) produces the following equation:

$$\frac{I(s) - N f^2(s)}{N f^2(s)} = \int_0^\infty 4\pi r^2 (\rho_a(r) - \bar{\rho}_a) \frac{\sin sr}{sr} \, dr \qquad (2.8)$$

In (2.8) $\rho_a(r)$ is the density of atomic centers separated by a distance r, corrected for the singularity at $r = 0$, and $\bar{\rho}_a$ is the average atomic density. Equation (2.8) may be Fourier transformed to give

$$4\pi r^2 (\rho_a(r) - \bar{\rho}_a) = \frac{2r}{\pi} \int_0^\infty s \frac{I(s) - N f^2(s)}{N f^2(s)} \sin sr \, ds \qquad (2.9)$$

The quantity $4\pi r^2 \rho_a(r) \, dr$ is the average number of atoms in a spherical shell of thickness dr at a distance r from a chosen atom, averaged over all locations of the origin atom. $4\pi r^2 \rho_a(r)$ is called the atomic radial distribution function (ARDF) and (2.9) becomes

$$\text{ARDF} = 4\pi r^2 \bar{\rho}_a + \frac{2r}{\pi} \int_0^\infty s \frac{I(s) - N f^2(s)}{N f^2(s)} \sin sr \, ds \qquad (2.10)$$

The ARDF is related to $g(r)$, a quantity usually called the radial distribution function, by

$$g(r) = \frac{\rho_a(r)}{\bar{\rho}_a} = \frac{\text{ARDF}}{4\pi r^2 \bar{\rho}_a} \qquad (2.11)$$

or combining (2.11) and (2.10)

$$g(r) = 1 + \frac{1}{2\pi^2 r \bar{\rho}_a} \int_0^\infty s \frac{I(s) - N f^2(s)}{N f^2(s)} \sin sr \, ds \qquad (2.12)$$

The use of (2.7) or (2.10) with experimental data for $I(s)$ involves many difficulties and approximations even for the simplest systems. These will be discussed in the order in which they appear when treating experimental data.

Free Atom Scattering Factors.   The coherent free atom scattering factors, $f^2(s)$, are used to calculate the kernel of the Fourier integrals in (2.7), (2.9), or (2.10). Before this is done, the incoherent atomic scattering, $I_{inc}(s)$, must be calculated and subtracted from the experimental intensity to obtain $I(s)$. The accuracy of the calculated values for $f^2(s)$ and $I_{inc}(s)$ depends on the accuracy of the electronic wave functions used to calculate the scattering factor. An exact representation of an atomic wavefunction would include electron correlations. Such a representation is available only for the lightest elements. The best uncorrelated wave functions are, in decreasing order of accuracy,

1. Relativistic Hartree–Fock wave functions (RHF); self-consistent field approximation, relativistic interaction terms, exchange terms calculated exactly.

2. Hartree–Fock wave functions (HF); self-consistent field approximation, nonrelativistic interaction terms, exchange terms calculated exactly.

3. Dirac–Slater wave functions (DS); self-consistent field approximation, relativistic interaction terms, exchange terms approximately calculated.

4. Hartree–Fock–Slater wave functions (HFS); self-consistent field approximation, nonrelativistic interaction terms, exchange terms approximately calculated.

5. Hartree wave functions (H); self-consistent field approximation, non-relativistic interaction terms, no exchange terms.

6. Thomas–Ferri–Dirac (TFD); statistical model, nonrelativistic interaction terms, no exchange terms.

Since the relativistic effects are completely negligible for atoms with atomic number $Z \leq 40$,[16] the RHF wave functions and HF wave functions are equivalent for these lighter elements. The DS and HFS wave functions are equivalent to each other and slightly inferior to the RHF and HF wavefunctions. The H and TFD wave functions are significantly inferior for calculating $f^2(s)$ and $I_{inc}(s)$. Hartree–Fock scattering factors have been available for the lighter elements (up to $Z = 20$) since about 1958[6,29–31] with a complete tabulation of all elements being available in 1969.[17,19–21]

Therefore, improvement in the wave functions used would not have any effect on the analysis of the monatomic scattering experiments discussed in this paper, with the possible exception of the xenon ($Z = 54$) study by Harris and Clayton[45] which made use of wave functions tabulated in 1958.[108]

Atomic scattering factors for He[54] and Li[5] have been calculated from correlated wave functions (99.90% of the correlation energy included). The differences between the correlated (Hyleraas) and uncorrelated (HF) wave functions are on the order of 1% for He and 1/2% for Li, and are expected to be completely insignificant for larger atoms.[21] Achter and Meyer[1] and Hallock[39,41–43] used the scattering factors calculated from these correlated wave functions to reduce their experimental data on helium.

Karnicky[51] was able to measure experimentally his atomic scattering factor for argon by extrapolating his low-density scattering estimates of $I(s)$ to zero density. He showed that using scattering factors measured in the same experimental system as used for the liquid structure determinations has the effect of canceling a certain class of systematic errors.

For a complete set of Hartree–Fock scattering factors, refer to Refs. 19 or 21 for the coherent factors, Ref. 20 for spherically symmetric incoherent factors, Ref. 17 for aspherical incoherent scattering factors, and Ref. 18 for the anomalous scattering factor corrections. Reference 25 tabulates coherent scattering factors for selected heavy atoms using the relativistic Hartree–Fock wave functions.

Adequacy of the Free Atom Approximation.   All the researchers using the calculated $f^2(s)$ have, of necessity, made the assumption that the free atom $f^2(s)$ could be used in (2.10) to determine the ARDF. This assumption is inaccurate if the atomic scattering factor in the fluid under consideration is different from the free atom atomic scattering factor because of instantaneous, time-varying distortion of the electronic wave functions. Piliavin[94] has recently calculated the effect of these distorted scattering factors for liquid helium and liquid argon and found that the inclusion of these distortions lowered the maximum in the radial distribution function [$g(r)$ from (2.11)] by about 1%.

Normalization.   The value for N, which scales the experimental data to electron units in (2.9), (2.10), and (2.12) may be determined in two ways:

1. Integral method[61]—In the limit as r goes to zero, $g(r)$ goes to 0, and (2.12) becomes

$$\int_0^\infty s^2 \frac{I(s) - N f^2(s)}{N f^2(s)} \, ds = -2\pi^2 \bar{\rho}_a \qquad (2.13)$$

Equation (2.13) can be used with the calculated $f^2(s)$ and the experimental $I(s)$ to determine N.

Rahman[100] has developed a similar integral method which makes use of the fact that $g(r)$ is zero for a finite range about $r = 0$.

2. High s method—At large values of s, the coherent intensity approaches the structureless free atom intensity, so

$$I(s) = N f^2(s)|_{\text{at high } s} \qquad (2.14)$$

N can be determined by matching the experimental $I(s)$ at high s to the calculated $f^2(s)$.

These two methods should, in theory, result in the same value for N, and this is a good test of the internal consistency of the data. However, in a real experiment the two normalization techniques emphasize different regions of the data and are subject to different types of errors. Therefore, it is misleading and ambiguous for a researcher to report, as some do, that N was determined by both the fit to high s and the integral method, unless they specify the method by which these two techniques were combined. Of the two, the integral method is probably more accurate because it uses the entire set of data to determine N.

Truncation of the Fourier Integral. In (2.10) or (2.7) the integral over the scattering data is taken from $s = 0$ to $s = \infty$. In practice, the data for $I(s)$ is available out to some value $s_{\max}$ determined by the maximum angle obtainable in the experiment and the incident wavelength.

From (2.1),

$$s_{\max} = 4\pi\lambda^{-1} \sin \theta_{\max} \qquad (2.15)$$

The integrals in (2.7) or (2.10) must be truncated, therefore, and the equations actually used to determine the distribution functions are

$$\text{ERDF} = 4\pi r^2 \bar\rho + \frac{2r}{\pi} \int_0^{s_{\max}} s \, \frac{I(s) - N f^2(s)}{NZ} \sin sr \, ds \qquad (2.16)$$

and

$$\text{ARDF} = 4\pi r^2 \bar\rho_a + \frac{2r}{\pi} \int_0^{s_{\max}} s \, \frac{I(s) - N f^2(s)}{N f^2(s)} \sin sr \, ds \qquad (2.17)$$

The effect of this truncation has been much examined[32,62,95,92,122] and is known to produce spurious ripples in the radial distribution functions, and, in particular, nonzero values for $g(r)$ at low r. These spurious ripples due to termination error may be identified in the ERDF or ARDF in three ways:

1. The effect of varying $s_{\max}$ on the resultant distribution functions from (2.16) or (2.17) can be examined. This was done by Pings and Paalman[96] in a

nitrogen study, by Stirpe and Tompson[115] (1962) in a neon study, and by Gingrich and Tompson[35] (1962) in an argon study.

2. The kernel of the Fourier integral in (2.16) or (2.17) may be multiplied by a "modification function"[122] which broadens the features of the radial distribution function and suppresses the spurious structure. The effect of this modification on the physically meaningful part of the transform can be calculated by the methods of Waser and Schomaker.[122] Stirpe and Tompson[115] and Harris and Clayton[45] used modification functions of the form e^{-bs^2}. Harris and Clayton[45] optimized their value of b by analyzing the resultant radial distribution function in terms of ideal peaks from single interaction pairs.[122]

3. A method derived by Waser and Schomaker[122] gives the spurious structure in the Fourier transform as the transform of the data terminating step function convoluted with the real (unterminated) radial distribution. Harris and Clayton[45] (1967) used this method to identify spurious structure in their argon and xenon radial distribution functions.

It must not be thought that all low r oscillations or subsidiary peak phenomena are the effect of termination error.[8] Kahn[52] showed that these spurious details may arise from erroneous $I(s)$ values in the experimentally accessible region. Rahman[100] and Paalman and Pings[92] showed that spurious oscillations could arise from incorrect normalization. Fehder[26] found a subsidiary peak in his radial distribution function, which was determined from a molecular dynamics study and did not involve any Fourier transformation. Kirstein[55] conducted numerical experiments to determine the effect on $g(r)$ and $c(r)$, the direct correlation function, of truncation, normalization errors, uncertainty in low-angle data, and distortion of the $I(s)$ data. Karnicky[51] performed a detailed error analysis for his argon experiments by expressing the various possible errors as perturbations on the experimental data, and calculating the effect of these perturbations on the potential function obtained from $c(r)$ and $g(r)$ in the Percus–Yevick equation. He found that errors in termination, normalization, and perturbation of the peaks and cross-overs in $I(s)$ all produced subsidiary peak phenomena and low r ripple.

Kaplow, Strong, and Auerbach[50] and, later, Konnert and Karle[59] proposed methods for improving the experimental data by adjusting the normalization parameter, N, background, and the data termination, in such a way that the low r oscillations disappear.

Nomenclature. In many papers there are differences in nomenclature. For example, the term $\rho(r) - \rho_0(r)$ in (2.5) and (2.6) is written as $\rho(r)$, and ρ_0 is used for the term $\bar{\rho}$. In fact, throughout the literature there is an inconsistency in usage of such terms as radial distribution function, distribution function, or electron radial distribution function. In general, it is necessary to examine the Fourier transformation equations used by the author to determine what is being calculated.

2. Homonuclear Molecular Fluids

For a molecular system composed of only one type of atom, (2.7) and (2.10) can be used to determine the ERDF and ARDF. (We will continue to write the Fourier integral as an infinite integral, as in (2.7) and (2.10), but is it to be understood that the actual experimental data to be transformed is only available to some value, s_{max}.)

Now, however, the assumption that the free atom scattering factors can be used for $f^2(s)$ is likely to be worse than before, because the bonded atom's wave functions are considerably more distorted than the wave functions of a monatomic species, and will no longer be spherically symmetric. Calculations for diatomic molecules by McWeeney,[71] Stewart,[113,114] and Morrison and Pings[79] show that the distortion of the free atom $f^2(s)$ by bonding is largest for the smaller molecules ($\approx 12\%$ for H_2) and becomes smaller when a smaller fraction of the electrons in the atom is involved in the bonding. For N_2 the distortion is about 1% so that the reduction of N_2 data [Furumoto and Shaw[33] and Pings and Paalman[96]] and O_2 data [Furumoto and Shaw[33]] by the free atom $f^2(s)$ is acceptable. Gruebel and Clayton[38] compared their X-ray intensities with neutron scattering data to find an experimental value for $f^2(s)$ of Br in the Br_2 molecule. Surprisingly, the experimental curve agrees best with scattering factors calculated from Hartree wave functions[49] which, since they do not include exchange terms, are decidely inferior to later wave functions. The difference is too large to be accounted for by distortion of the free atom $f^2(s)$ by bonding.

Molecular Radial Distribution Function. An equation analogous to (2.10) can be written for molecules:

$$4\pi r^2(\rho_m(r) - \bar{\rho}_m)) = \frac{2r}{\pi} \int_0^\infty s \frac{I(s) - NF^2(s)}{NF^2(s)} \sin sr \, ds \qquad (2.18)$$

$4\pi r^2 \rho_m(r) \, dr$ is the average number of molecules in a spherical shell dr thick at a distance r from a molecule chosen as the origin. $4\pi r^2 \rho_m(r)$ is called the molecular radial distribution function (MRDF), and (2.18) can be written

$$\text{MRDF} = 4\pi r^2 \bar{\rho}_m + \frac{2r}{\pi} \int_0^\infty s \frac{I(s) - NF^2(s)}{NF^2(s)} \sin sr \, ds \qquad (2.19)$$

The units of MRDF are molecules/Å. $F^2(s)$ is the scattering from one molecule, and is assumed to be spherically symmetric. Equation (2.19) then, ignores angular correlations in both the molecular scattering factor, $F^2(s)$ and the distribution function for the molecules, $\rho_m(r)$.

Morrison and Pings[80] have shown that the neglect of these angular correlations produces errors in the scattering intensity of liquid Cl_2 which are larger than presently obtainable experimental accuracy limits. Steele and Pecora[112] have developed a method for taking angular correlations into account in determining $I(s)$ from the angular dependent molecular distribution function $\rho_m(\mathbf{r}_1, \mathbf{r}_2)$, but the inversion of their work to obtain an improved form of (2.19) has not been accomplished.

As is the case in determining the ARDF and ERDF, the kernel of the Fourier integral which gives the MRDF (2.19) may be multiplied by a modification factor and the effect of the modification factor on the resultant MRDF can be analyzed according to the formalism of Waser and Schomaker.[122]

3. Heteronuclear Molecular Fluids

For a heteronuclear molecular fluid, (2.7) becomes

$$\text{ERDF} = 4\pi r^2 \bar{\rho} + \frac{2r}{\pi} \int_0^\infty s \, \frac{I(s) - N \sum_i f_i^2(s)}{N \sum_i Z_i} \sin sr \, ds \qquad (2.20)$$

where the sums are taken over the atoms in each molecule. However, the Fourier transformation is seldom used in this form. Usually, the kernel of the Fourier integral is divided by a "sharpening function" of the form $\left[\sum_i f_i(s) \right]^2$ to obtain a sharpened ERDF which shall be called ERDF':

$$\text{ERDF}' = 4\pi r^2 \bar{\rho} + \frac{2r}{\pi} \int_0^\infty s \, \frac{I(s) - N \sum_{ii} f_i^2(s)}{\left(N \sum_i Z_i \right) \cdot \left[\sum_i f_i(s) \right]^2} \sin sr \, ds \qquad (2.21)$$

By comparing (2.21) with (2.10) one can see that the sharpened ERDF' looks very much like the ARDF for monatomic species. In fact, ERDF' represents the superposition of the distribution functions for the various atomic pairs in the fluid. Because each distribution is weighted by the product of the number of electrons on each of the two atoms in the pair, the units of the ERDF' are still electrons^2/Å, as for the ERDF. Note that the ERDF' contains contributions from the intramolecular as well as the intermolecular structure.

The part of the Fourier integral kernel which is multiplied by $s \sin sr \, ds$ is called $i(s)$. The actual form of $i(s)$ depends on how the integrand is defined. For example, in (2.21),

$$i(s) = \frac{I(s) - N \sum_i f_i^2(s)}{\left(N \sum Z_i\right)\left[\sum_i f_i(s)\right]^2}$$ (2.22)

The $i(s)$ used to calculate the ERDF' is expressed in several forms in the literature, such as

$$i(s) = \frac{I(s) - N \sum_i f_i^2(s)}{N\left[\sum_i f_i(s)\right]^2}$$ (2.23)

which sums over the atoms in the molecule, or

$$i(s) = \frac{I(s) - N \sum_i X_i f_i^2(s)}{N\left[\sum_i X_i f_i(s)\right]^2}$$ (2.24)

which sums over the mole fractions of the elements in the fluid, or

$$i(s) = \frac{I(s) - \sum_i N_i f_i^2}{f_e^2(s)}$$ (2.25)

where

$$f_e^2(s) = \frac{\left[\sum_i N_i f_i(s)\right]^2}{\left[\sum_i N_i Z_i\right]^2}$$ (2.26)

which uses an effective electron notation. These forms all give the same treatment of the data, with various constants being incorporated into the normalization of the experimental $I(s)$. All the studies of molecular systems reviewed in this paper make use of one form or another of (2.21).

Interpretation of the ERDF' in terms of the individual atomic pair distributions has been discussed by Warren,[120] Mendel,[72] Waser and Schomaker,[122] and Pings and Waser.[97] As is the case with the previously discussed Fourier transforms, the kernel of (2.21) is sometimes modified by an artificial temperature factor of the form[122] e^{-bs^2} (see Refs. 37, 44, 105); and the ERDF' can be interpreted in terms of ideal peaks[122] (see also Refs. 37, 44).

Since the ERDF' contains information about the intramolecular as well as the intermolecular structure, it is possible to infer molecular parameters from the ERDF'. This can be done by either assuming the molecular model and obtaining the parameters by fitting to the ERDF' (see Refs. 37, 44, 82, 85, 105) or by determining the molecular model itself by fitting to the ERDF' (see Refs. 60, 123). Once the molecular structure is known, it is possible to subtract the scattering due to the intramolecular structure from $I(s)$ and infer properties of the intermolecular structure from the ERDF' (see Ref. 82)

Molecular Radial Distribution Functions.   Equation (2.19) has been used by many researchers to calculate the radial distribution of heteronuclear molecules in a fluid. The principal difference in the various methods lies in the calculation of the scattering factor for the molecule, $F^2(s)$.

The most frequently used approximation to $F^2(s)$ is the spherically averaged value for the sums of the atomic scattering factors in the molecule:[24,73]

$$F^2(s) = \sum_{i,j} \frac{f_i(s)f_j(s) \sin r_{ij}s}{r_{ij}s} \tag{2.27}$$

where r_{ij} is the distance between atoms i and j in the molecule. A factor $e^{b_{ij}s^2}$ is sometimes included in (2.27) to allow for vibrations of the atoms about their average separations. Harris and Clayton,[44] Gruebel and Clayton,[37] Rutledge and Clayton,[105] Weidner,[124] Kratochwill,[60] and Narten, Danford and Levy[85] calculated the MRDF from (2.19) and (2.27), with molecular parameters, r_{ij}, determined from the experimental ERDF' as previously discussed. Bochynski[10,11] calculated the MRDF for benzene and nitrobenzene using measurements of the molecular parameters obtained from the crystalline state. Kruh and Petz[63] calculated the MRDF from their experimental data on NH_3 using (2.19), but approximated the molecular scattering factor $F^2_{NH_3}(s)$ by the atomic scattering factor for nitrogen, $f^2_N(s)$. In a similar manner, Kiselev and co-workers[56,57] approximated the molecular scattering factor $F^2_{H_2O}(s)$ by the atomic scattering factor $f_O{}^2(s)$ to calculate a MRDF for their studies of water in pelliculer films. Petz[93] calculated a MRDF for methane using a $F^2_{CH_4}(s)$ which he determined experimentally by X-ray scattering from methane gas. Narten and Levy[88] calculated a MRDF from their X-ray diffraction data on water using a $F^2_{H_2O}(s)$ based on a self-consistent-field, molecular orbital calculation. They applied the method of Steele and Pecora[112] and found that the only significant term in the basis set expansion was the first, spherically symmetrical term. They were thus able to invert the Fourier integral by retaining only this term and hence obtained a spherically symmetrical $F^2_{H_2O}(s)$ and $\rho_m(r)$. They concluded that angular correlation effects for H_2O are on the order of 1 % and negligible by comparison with other sources of error.

Of course, in addition to the new approximations incurred in treating data for heteronuclear systems these results are subject to all the sources of error discussed for monatomic fluids and polyatomic homonuclear fluids.

4. *Aqueous Solutions*

Aqueous solutions are usually treated as mixtures of solvent (H_2O) molecules and solute species (usually ions). Equation (2.21) is used to evaluate an ERDF', but there is no unambiguous meaning to the calculation of a MRDF.

There are three levels of information available from the ERDF'. As pointed out by Licheri et al.,[69] the only direct structural information obtained from such an experiment consists of some characteristic distances corresponding to the principal peaks in the ERDF', and a measure of the extent of meaningful order phenomena. However, much more useful information may be extrated from these studies by inference. This information, such as coordination numbers of ions, is determined indirectly, for example, by studying the effect of a change in concentration on the ERDF', or by comparing the ERDF' for related solutes. A third level of "information" consists of the author modeling a variable parameter crystalline description of the solution, calculating the X-ray diffraction curves or ERDF' curves from the model, and comparing these calculated results with experiments. With regard to the usefulness of such a procedure, one may refer to one of the early articles by Warren:[115] "That there is a close relationship in properties such as these between the liquid and crystalline states is obvious from elementary considerations of density. The whole point to the study of the structure of liquids is *to find out precisely what are the differences between the liquid and the crystal* [Warren's italics]."

III. EXPERIMENTAL REVIEW

This section summarizes the recent X-ray diffraction studies of fluids. The conclusions drawn from the experimental data are those of the authors, and are presented here without comment. In general, however, the reader is advised to keep in mind the very large number of systematic errors which are incurred in going from experimental data to a final radial distribution function.

The figures are from the original papers. These present either the ARDF or its reduced form $g(r)[g(r) = ARDF/4\pi r^2 \bar{\rho}_a]$, the ERDF', and the MRDF in its reduced form $g(R)[g_m(r) = MRDF/4\pi r^2 \bar{\rho}_m]$. The figures are chosen to show representative experimental results.

A. Hard Spheres

Brady and Gravatt[13] obtained diffraction patterns from a fluid distribution of polystyrene spheres and were thus able to make a direct comparison between experiment and theory for the hard-spheres equation of state. The experimental data at liquid densities is best described by the hypernetted chain equation.

B. Monatomic Fluids

1. Helium

Achter and Meyer[1] scattered X-rays from liquid ^{4}He and ^{3}He and from ^{4}He vapor. Their $I(s)$ values are compared to the Feynman formula for the structure factor of liquid helium. The distribution functions are in generally good agreement with variational calculations. The peak in $g(r)$ is at 3.60 Å for the liquid ^{3}He and at 3.47 Å for ^{4}He. Figure 1 shows the atomic $g(r)$ for two of the liquid He states. At 4.2°K, 0.98 atm, the ^{4}He vapor is distinctly nonideal, having a well-defined nearest-neighbor shell.

Hallock[39,41–43] measured the structure factors of liquid and gaseous ^{3}He and ^{4}He in the momentum transfer range 0.15 Å^{-1} < s < 2.00 Å^{-1} at a variety of temperatures. His liquid results confirm the Feynman relationship and disagree by about 10% with the results of Achter and Meyer.[1] He does not calculate complete radial distribution functions but evaluates the first moment of the pair correlation function.

2. Neon

Stirpe and Tompson[115] measured the ARDF for six states of liquid neon on the coexistence curve from the critical point to the triple point. These

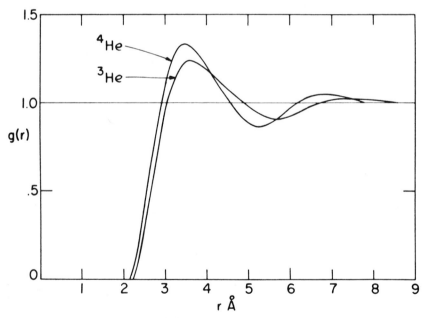

Fig. 1. Reduced atomic radial distribution functions for He^3 at $T = 0.56$°K and He^4 at $T = 0.79$°K. From Achter and Meyer.[1]

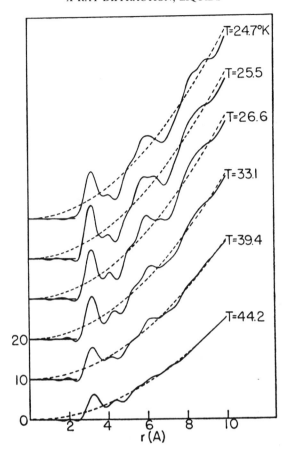

Fig. 2. Atomic radial distribution functions for liquid neon at its vapor pressure. From Stirpe and Tompson.[115]

ARDF functions are shown in Fig. 2. Areas of the first coordination shell varied from 8.4 atoms centered at 3.18 Å for the triple point to 4.3 atoms at 3.26 Å in the critical region. They also measured the low-angle scattering from 15 states within 2°K of the the critical point and calculated radii of gyration for these points.

3. Argon

Gingrich and Tompson[35] measured the ARDF for argon in the vicinity of the triple point. They observed five peaks in $i(s)$ and calculated a maximum of 10.5 nearest-neighbor atoms at 3.78 Å.

Harris and Clayton[45] also measured the argon ARDF at the triple point. They calculated 8.5 nearest neighbors at 3.79 Å, but the difference between this value and the 10.5 nearest neighbors found by Gingrich and Tompson[35] is largely a matter of using different techniques for defining "nearest neighbors." The positions of the maxima in the ARDF are seen to agree well with the predictions of either random-packed spheres or the convoluted hypernetted chain equation used with a Lennard–Jones potential.

A series of experiments at the California Institute of Technology have resulted in the determination of argon distribution functions over a wide range of temperatures and pressures.

Mikolaj[74, 76] studied 13 states in the general critical region ($0.280 \text{ g/cm}^3 \leq \bar{\rho}_a \leq 0.982 \text{ g/cm}^3$ and $-130°C \leq T \leq -110°C$). He studied the variation of the peaks in the ARDF with temperature and density, and subsequently calculated atomic $g(r)$'s, direct correlation functions,[77] and estimates of the pair potential.[75]

Smelser[110] studied six high-density states in the liquid region ($0.910 \text{ g/cm}^3 < \bar{\rho}_a \leq 1.261 \text{ g/cm}^3$, and $-165°C \leq T \leq -130°C$) and calculated radial distribution functions, direct correlation functions, and Percus–Yevick effective potentials.

Kirstein[55] carried out studies of five states of liquid argon ($0.91 \text{ g/cm}^3 \leq \bar{\rho}_a \leq 1.135 \text{ g/cm}^3$ and $-146°C \leq T \leq -130°C$) with experimental and data analysis techniques which were considerably improved over the work of Mikolaj and Smelser. He calculated radial distribution functions, direct correlation functions, and effective potentials. The $g(r)$ curves are shown in Fig. 3. He also carried out numerical studies of various sources of error in the data analysis.

Karnicky[51] studied the X-ray diffraction of argon at four densities along the $-100°C$ isotherm in the dense gas region ($0.0827 \text{ g/cm}^3 \leq \bar{\rho}_a \leq 0.3111 \text{ g/cm}^3$ in an experiment designed to determine the argon pair potential. The resultant potential is described by the parameters $\sigma = 3.389 \pm 0.015$ Å, $\varepsilon = 146.3 \pm 4.9°K, r_{min} = 3.86 \pm 0.05$ Å, with error limits calculated by an extensive error propagation perturbation analysis. Comparison of this pair potential well depth with the effective potentials from Kirstein's[55] data gives a variation in the well depth of the effective potential which agrees very well with the value calculated by Rowlinson for a triple–dipole nonadditive force.

4. Xenon

Harris and Clayton[45] (1966) measured the xenon ARDF at the triple point. They found 8.94 nearest neighbors at 4.38 Å, as was the case for their argon results, the position of the peaks in the ARDF agreed well with the calculations for close-packed spheres or the convoluted hypernetted chain equation with a Lennard–Jones potential. The ARDF is shown in Fig. 4.

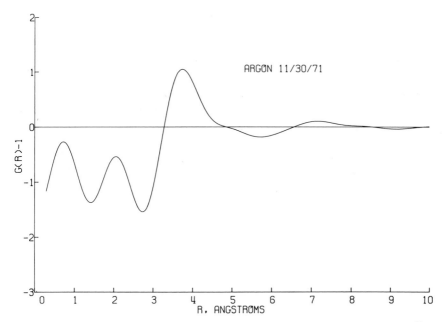

Fig. 3. Reduced atomic radial distribution functions for liquid argon. From Kirstein.[55]

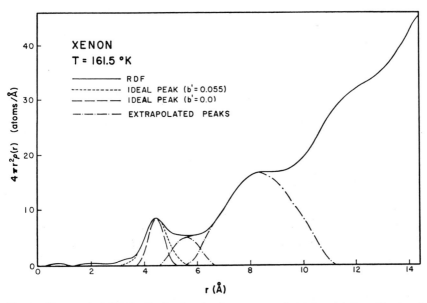

Fig. 4. The atomic radial distribution function for xenon near its triple point. From Harris and Clayton.[45]

C. Homonuclear Molecular Fluids

1. Nitrogen

Pings and Paalman[96] measured diffraction from N_2 at 7 atm and $-193°C$ and calculated the ERDF. The electronic $g(r)$ is shown in Fig. 5. They concluded that a nearest-neighbor shell centered at 4.0 Å contained 15 molecules, and that a previously reported subsidiary peak in the N_2 ERDF was spurious.

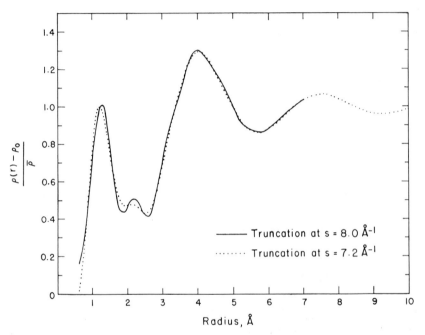

Fig. 5. Reduced electronic distribution function for liquid nitrogen. From Pings and Paalman.[96]

Furumoto and Shaw[33] studied liquid N_2 at its saturated vapor pressure at 64 and 77°K, O_2 at 64 and 77°K, and mixtures of N_2 and O_2 at 77°K (composition 20/80, 50/50, 80/20). For the mixtures, only the ERDF was calculated. The ARDF for N_2 showed a well-defined bonded atom peak at 1.12 Å and a nearest-neighbor coordination shell at 3.71 Å corresponding to 20.6 atoms. The first-neighbor shells show a separation into two peaks.

2. Oxygen

In the above-mentioned experiment Furumoto and Shaw[33] also measured the ARDF for O_2 at 64 and 77°K under its own vapor pressure. The results were very similar to those for N_2: a bonded atom peak at 1.24 Å, and a first-neighbor shell at about 3.34 Å corresponding to a total of 21 atoms separated into two peaks. The ARDF functions for N_2 and O_2 are shown in Fig. 6.

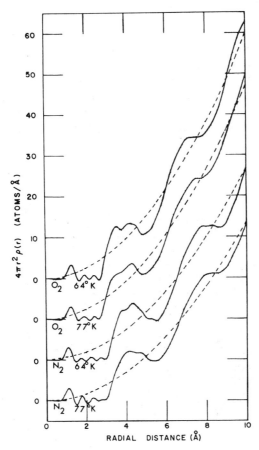

Fig. 6. The atomic radial distribution functions of nitrogen and oxygen. From Furumoto and Shaw.[33]

3. Bromine

Gruebel and Clayton[38] calculated the ARDF and MRDF from their X-ray study of Br_2 just above its freezing point. The first shell in the MRDF contains a total of 24 atoms, and is split into two peaks which the authors interpret as being different distances of closest approach for two different orientations of the molecules. Beyond the first shell, the MRDF is very similar to that of an inert gas.

D. Heteronuclear Molecular Fluids

1. Water

Narten, Danford, and Levy[85] scattered X-rays from water (4 to 200°C) and from D_2O at 4°C. Their results are interpreted in terms of an expanded ice-I lattice model for liquid water, with hexagonal symmetry and tetrahedral coordination of hydrogen-bonded water molecules. All lattice sites and about 50% of the interstitial cavities are occupied by water molecules, so that about 20% of the water molecules are in cavity positions (Fig. 7). In calculating intensity curves from the model, they retained discrete interactions to about 10 Å and used a continuum beyond this distance. Thermodynamic properties estimated from the model agree with the experimental values.

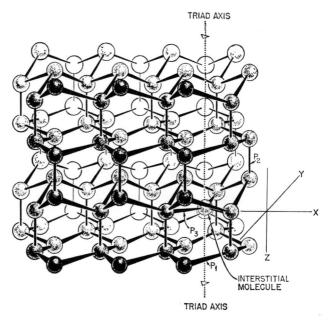

Fig. 7. The ice-I water (ball/stick ratio not to scale). From Narten, Danford, and Levy.[85]

Narten and Levy[88] analyzed a new set of scattering data for water in the same range of temperatures as above. They found a double maximum in the first peak of $i(s)$ at low temperatures, which had not been seen for any previous molecular structure. The molecular $g(r)$ functions are seen in Fig. 8.

Kiselev and co-workers[56,57] studied water as it exists in pellicular films on a mica surface by X-ray diffraction. They found that the local order persisted for longer distances than in free water, and the coordination numbers for the

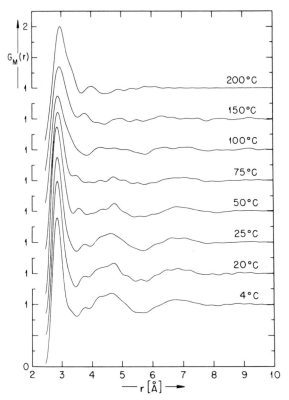

Fig. 8. Reduced molecular radial distribution functions for liquid water. From Narten and Levy.[88]

maxima in the MRDF were different from those of free water. (The first maximum in the free water MRDF represents 4.3 molecules and in the film water 6.2 molecules.)

2. Ammonia

Liquid NH_3 was studied at three temperatures by Kruh and Petz[63] and their results interpreted in relation to the crystal structure of solid ammonia. They conclude that hydrogen-bonded approaches occur at a mean distance of 3.56 Å and propose an explanation for the volume change of alkali metal ammonia solutions in terms of an approximate close-packed model. The ERDF' curves are shown in Fig. 9.

Narten[83] studied NH_3 as part of the water–ammonia system and decided his results for pure NH_3 were not explainable on the basis of the model of Kruh and Petz.[63] He found that his results fit the ice-I model[85] for water better than they fit the crystal structure of solid NH_3 or NH_3 clathrates.

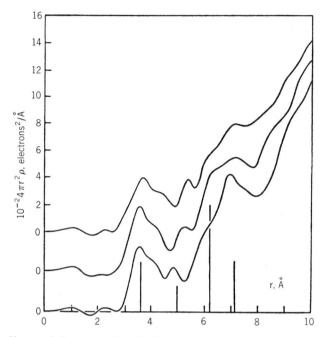

Fig. 9. Sharpened electronic radial distribution functions for NH_3 at 4, −45, and −74°C (top to bottom). Vertical lines show the separation and relative number of neighbors per molecule in CCP array of spheres of diameter 3.56 A. From Kruh and Petz.[63]

3. Silicon Tetrachloride

Klochkov and Skryshevskii[58] diffracted X-rays from liquid $SiCl_4$ to determine its inter- and intramolecular structure. They found tetrahedral molecules with a Si—Cl bond length of 2.1 Å and a first intermolecular coordination shell at 6.4 Å. They also studied a variety of silanes [organic silicon compounds such as $Si_2(CH_3)_6$] to determine the intramolecular Si—C and Si—Si bond lengths in these liquids.

Rutledge and Clayton[105] studied $SiCl_4$ at a variety of temperatures and obtained results that were significantly different from those of Klochkov and Skryshevskii.[58] They found the same Si—Cl bond length of 2.1 Å, but obtained a first coordination shell at 3.9 Å with about four molecules and another at 6.8 Å with about seven molecules. The MRDF functions at −65°C and 23°C are shown in Fig. 10. The authors conclude the presence of short-range intermolecular orientation effects.

4. Silicon Dioxide

Nukui and co-workers[90] (1972) studied a variety of molten silicas. They concluded that the higher-order bonds in the solid (noncrystalline) glass

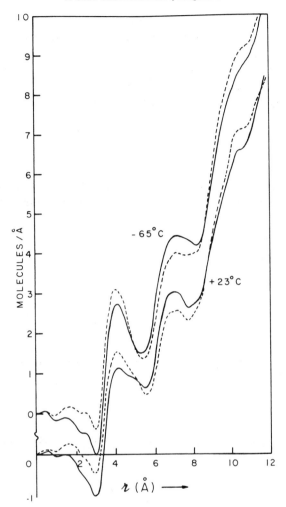

Fig. 10. Molecular radial distribution functions for $SiCl_4$ at $-65°C$ (top) and $+23°C$ (bottom). Dotted lines are the result of infinite sharpening for values of $s < 1$. From Rutledge and Clayton.[105]

were destroyed by melting and the liquid is characterized by a much more distorted conformation at far distances.

5. Arsenic Bromide

Hoge and Trotter[47] studied liquid $AsBr_3$ and determined that the intramolecular structure consists of discrete pyramidal molecules with dimensions essentially the same as in the solid and gas. Most of the intermolecular distances in the crystal are preserved in the liquid.

6. Sulfuric Acid

Weidner, Geisenfelder, and Zimmermann[123,124] scattered X-rays from 100% liquid H_2SO_4 at 20°C. They derived parameters for two molecular models by best fit to the ERDF'. The bond distances in both models agree with the bond distances in crystalline H_2SO_4. The intermolecular structure is described in terms of partial occupancy of unit cells. The molecules are described as existing in layers. The molecules within a layer are netted together by hydrogen bonds (bond length 2.55 Å) and hydrogen bonding also exists between the layers.

7. Methane

Petz[93] studied methane at four states ranging from the triple point to the critical point. He discovered that the structure and coordination shells vary

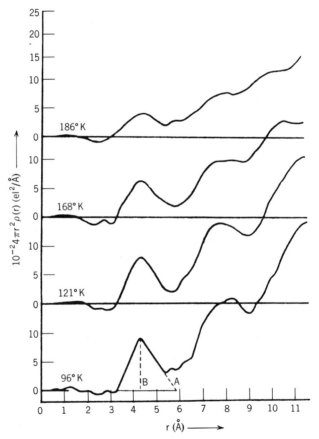

Fig. 11.   Sharpened electronic radial distribution functions for liquid methane. From Petz.[93]

in much the same manner as the inert gas structures and, like these, resemble the close packing of hard spheres. His ERDF' curves are shown in Fig. 11.

8. Carbon Tetrachloride

Gruebel and Clayton[37] studied liquid CCl_4 at 25, 52, and 115°C. From the ERDF' they obtained intramolecular bond distances somewhat larger than those found in the gas molecules. The MRDF curves (Fig. 12) show a close intermolecular structure consisting of a peak at 6.7 Å with a shoulder at 12 Å.

Beyond the first coordination shell the intermolecular structure is the same as that of a monatomic liquid.

Narten, Danford, and Levy[86] studied CCl_4 at 25°C. Their intramolecular bond distances were somewhat smaller than those measured by Gruebel

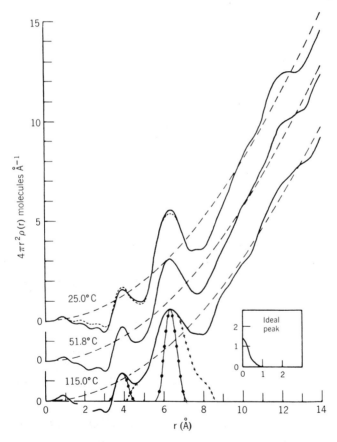

Fig. 12. Molecular radial distribution function for liquid CCl_4 with the inset ideal peak which represents one molecule pair. From Gruebel and Clayton.[37]

and Clayton,[37] but the intermolecular structure was in essential agreement. Narten and co-workers designed a model for CCl_4 consisting of a perturbed close packing of chlorine atoms surrounded by a continuous distribution of distances to exterior molecules. An intermolecular potential function derived from this model predicts thermodynamic properties in agreement with those found by experiment.

9. Carbon Tetrafluoride

Harris and Clayton[44] studied liquid CF_4 at -177, -153, and $-122°C$, with emphasis on determining the intermolecular structure and correlations. At all three temperatures they found peaks in the MRDF at 3.36, 5.20, and 9.00 Å. With increasing temperature, the height of the peaks decrease, the width increases, and the number of molecules in the coordination shells decreases.

10. Acetonitrile

Kratochwill, Weidner, and Zimmermann[60] studied liquid CH_3CN at $20°C$. They determined intramolecular bond distances by best fit procedures to the ERDF'. They interpreted the intermolecular structure in terms of various models, finding best agreement to the data for a model composed of correlation clusters, each cluster being about 11 Å in diameter and having eight molecules occupying lattice sites in an orthorhombic unit cell with 95 % site occupancy.

11. Benzene

Narten[82] investigated liquid C_6H_6 at $25°C$. He determined intramolecular bond distances that were somewhat larger than those in benzene vapor. The diffraction data agrees with the results calculated for a liquid model which consists of an origin molecule surrounded by 12 near neighbors in preferred orientations subject to the operations of the benzene crystal space group, with a random distribution of distances beyond. The ERDF' is shown in Fig. 13.

Bochynski[11] (1968) also studied liquid benzene. He used intramolecular parameters determined from the crystal to reduce his data to a MRDF and found preferred intermolecular distances at 4.11, 4.94, and 6.55 Å.

12. Nitrobenzene

Bochynski[10] studied the intermolecular distributions of liquid $C_6H_5NO_2$. He calculated a MRDF using bond lengths and angles for the $C_6H_5NO_2$ molecule determined from the crystal. He found preferred intermolecular distances at 4.10, 5.23, and 5.65 Å.

13. Molten Beryllium Fluoride, Lithium Fluoride Solutions

Vaslow and Narten[118] have obtained diffraction data on the liquid system BeF_2—LiF at a range of concentrations from pure BeF_2 to pure LiF. By

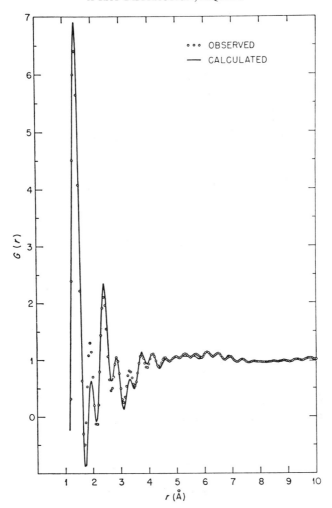

Fig. 13. Reduced sharpened electronic radial distribution function for liquid benzene at 25°C. From Narten.[82]

comparing their data with the diffraction patterns from solid crystalline and noncrystalline BeF_2, they conclude that liquid BeF_2 contains BeF_4 tetrahedra joined at the corners. With increasing LiF concentration, the tetrahedral network is distorted so that the Be^{2+} ions retain their tetrahedral environment, but the Li^+ ions' environment is strongly distorted from tetrahedrality. The ERDF' functions are shown in reduced ($g(r)$) form in Fig. 14.

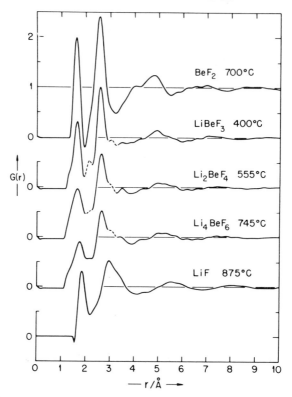

Fig. 14. Reduced sharpened electronic radial distribution functions for molten LiF–BeF$_2$ solutions. The dashed portions are artifacts caused by the cutoff. From Vaslow and Narten.[118]

E. Aqueous Solutions

1. Hydrochloric Acid

Terekhova[117] (1970) studied aqueous solutions of HCl having 30, 20, and 10 molecules of water per HCl molecule, and found ERDF' functions which did not greatly differ from pure water. Interpreting his scattering patterns in terms of an expanded ice-I structure, he concluded that the hydrogen exists as a H_3O^+ hydronium ion which is hydrogen bonded to three water molecules and spreads out the tetrahedral structure. The chloride ion occupies an interstitial void.

When Lee and Kaplow[66] studied HCl solutions of about the same molarity ($2N$, $4N$, $6N$), they found extreme departures from the pure water ERDF'. These changes involved the appearance of two new peaks which the authors interpret as being due to H—Cl pairs and O—Cl pairs.

Licheri, Piccaluga, and Pinna[68] (1971) studied $1N$ and $4.3N$ HCl and obtained results similar to those of Terekhova.[117] No new peaks were found and the ERDF' curves were qualitatively similar to those of pure water. The authors were unable to explain the discrepancy between Lee and Kaplow's[66] data and that of Terekhova[117] and their own.

2. Alkali Metal Halides

Lawrence and Kruh[65] (1967) studied eight alkali halides in varying aqueous concentrations: LiCl, LiBr, LiI, NaCl, NaI, CeCl, CeBr, CeI. They were able to decompose the peaks in the ERDF' curves into the component interactions and infer properties of the local structure of the solutions. Their principle conclusions were: the halide ions are surrounded by seven to nine water molecules in their first hydration sphere. Direct ionic contact exists in the cesium halides, but not in the sodium or lithium halides.

Licheri, Piccaluga, and Pinna[69] studied LiCl, LiBr, NaCl, and NaBr in varying aqueous concentrations. They obtained results for the first ionic hydration sphere which agreed with those of Lawrence and Kruh,[65] but concluded that the second and third hydration spheres seen by Lawrence and Kruh[65] were due to spurious oscillations. Their results indicate that the ions, in fact, produce a shorter range of order in the solutions than in the pure water.

Narten, Vaslov, and Levy[89] studied solutions of LiCl up to saturation concentration by both X-ray and neutron scattering. They interpret their results in terms of a model which describes octahedral coordination shells about the Cl^- ion and tetrahedral shells around the Li^+. Beyond these shells the water looks like the ice-I model of Narten, Danford, and Levy. [85] At concentrations of LiCl above a mole ratio of 1:10, the hydration spheres extend far enough to remove completely evidence of the ice-I tetrahedral water structure. Fig. 15 shows the X-ray ERDF' functions in reduced form.

3. Ammonia and Ammonium Halides

Narten[83] diffracted X-rays from liquid ammonia and aqueous ammonia solutions of 18.3 and 28.5% NH_3. He interprets the results for the solutions in terms of the ice-I model for water,[85] with NH_3 molecules replacing H_2O molecules at both framework and interstitial sites. The parameters for the model change, the nearest-neighbor distance being 2.91 Å for pure water, 2.95 Å for the 18.3% solution, 3.00 Å for the 28.5% solution, and 3.31 Å for pure ammonia. The $g(r)$ curves for the reduced ERDF' curves are shown in Fig. 16.

Narten and Lindenbaum[87] measured the structure of a concentrated solution of tetra-n-butylammonium fluoride in water at 25°C [$(C_3H_7)_4NF \cdot 41H_2O$]. their results indicate that the ice-I model of Narten, Danford, and

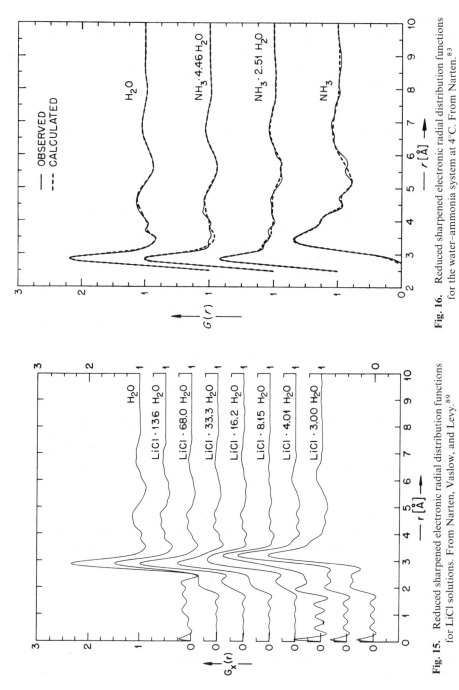

Fig. 16. Reduced sharpened electronic radial distribution functions for the water–ammonia system at 4°C. From Narten.[83]

Fig. 15. Reduced sharpened electronic radial distribution functions for LiCl solutions. From Narten, Vaslow, and Levy.[89]

Levy[85] adequately describes the solution. All water molecules, fluoride ions, and N^+ atoms are part of the network, with 11% of the network sites unoccupied. The butyl groups occupy interstitial positions. This model predicts nonunreasonable thermodynamic properties. Another model tested (the gas hydrate model) fails to explain either the X-ray data or the thermodynamic properties.

Danford[23] studied aqueous solutions of ammonium fluoride and concluded that the solution is essentially the same as pure water. That is, the solution is described by the ice-I model[85] with water molecules at lattice sites and interstitial positions replaced in a random manner by NH_4^+ and F^- ions. Differences between the ERDF' for water and the solutions are very slight.

Narten[84] diffracted X-rays from solutions of NH_4Cl, NH_4Br, and $NH_4I(NH_4X \cdot \sim 8H_2O)$. With increasing anion size, the first peak in the ERDF' resolves into two peaks, corresponding to resolution of the water-water and halide-water distributions. In terms of the ice-I model,[85] the NH_4^+ ions are treated as indistinguishable from the H_2O molecules. The anions occupy interstitial positions and cause the lattice to expand anisotropically. For the larger anions, more interstitial sites are occupied by anions or H_2O molecules until, for NH_4I solutions, all the interstitial sites are occupied.

4. Alkaline Earth Chloride Solutions

Albright[2] studied concentrated aqueous solutions of $MgCl_2$, $CaCl_2$, $SrCl_2$, and $BaCl_2$. The resultant ERDF' curves are consistent with the following description of the structure of the solution: The coordination number of the Cl^- ion is 8. The coordination numbers of the cations are between 6.5 and 8, with the hydration of the Mg^{2+} and Ca^{2+} ions insensitive to concentration. The Ca^{2+} ions markedly order the surrounding water structure, showing significant second nearest-neighbor coordination.

5. Transition Metal Halides

Kruh and Standley[64] studied very concentrated $ZnCl_2$ solutions (5 to 28 M) and determined that the Zn^{2+} ions were tetrahedrally coordinated At 28 M, the coordination consists of $3Cl^-$ ions and one water molecule, whereas at 5 M the coordination consists of about 2.4 Cl^- ions and 1.6 H_2O molecules. For the more concentrated solutions, the Zn^{2+} ions are coupled through the Cl^- ions, which explains the high viscosity and low electrical conductivity of concentrated Zn halide solutions.

Wertz, Lawrence, and Kruh[126] studied $ZnBr_2$ solutions, aqueous solutions of $ZnBr_2$ and HBr, and a solution of $ZnBr_2$ in acetone. The ERDF' curves for the $ZnBr_2$–HBr solutions are shown in Fig. 17. The results for the $ZnBr_2$ solutions were similar to the results of Kruh and Standley[64] for

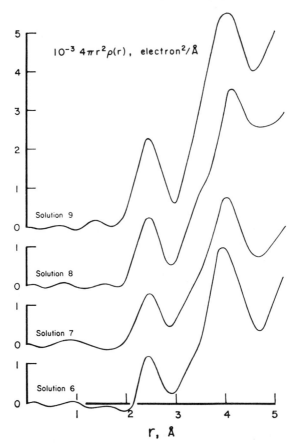

Fig. 17. Sharpened electronic radial distribution functions for ZnBr$_2$–HBr solutions: Solution 6 is 0.059 mole fraction Zn, 0.254 mole fraction Br; Solution 7 is 0.074 mole fraction Zn, 0.275 mole fraction Br; Solution 8 is 0.086 mole fraction Zn, 0.298 mole fraction Br; Solution 9 is 0.113 mole fraction Zn, 0.333 mole fraction Br; From Wertz, Lawrence, and Kruh.[126]

ZnCl$_2$, with Br$^-$ being less effective than Cl$^-$ in competing with H$_2$O for ligand sites around the Zn^{2+} ion. Addition of HBr replaces some or all of the H$_2$O ligands around the Zn^{2+} ion with Br$^-$ ions. The 4.2M solution of ZnBr$_2$ in acetone is described by Zn^{2+} tetrahedrally coordinated by two Br$^-$ ions and two acetone molecules.

Wertz and Kruh[125] examined solutions of concentrated CoCl$_2$ in water, methanol, and ethanol. Some of the curves for the ERDF′ are shown in Fig. 18. In water, the Co^{2+} is coordinated by six water molecules. In the alcohol solutions the cobalt ion is tetrahedrally coordinated, with four chloride nearest neighbors at the same distance as in the crystalline CoCl$_2$.

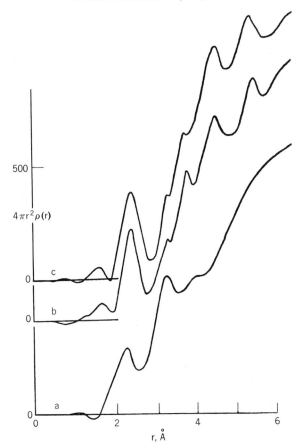

Fig. 18. Sharpened electronic radial distribution functions of $CoCl_2$ solutions: (*a*) in H_2O, (*b*) in CH_3OH, (*c*) in C_2H_5OH. From Wertz and Kruh.[125]

Molecular weight measurements of the solute in the alcohol solutions confirms the formation of $CoCl_4^{2-}$ ions at high concentrations.

6. Tetrafluoroboric Acid and Its Salts

Ryss and Radchenko[106] obtained ERDF' curves for aqueous solutions of HBF_4, $NaBF_4$, $LiBF_4$, and NH_4BF^4. The HBF_4 solutions have a structure similar to that of pure water, with a BF_4^- ion replacing a molecule of water. In concentrated $NaBF_4$ solutions the Na^+ ion is coordinated with six H_2O molecules. For $2.8 M$ solutions of $LiBF_4$ and NH_4BF_4, the cations are coordinated by four H_2O molecules. The concentrated $LiBF_4$ solutions have a structure similar to that of the crystallohydrate $LiBF_4 \cdot 3H_2O$.

7. Thorium Perchlorate and Thorium Oxyperchlorate

Dilute perchloric acid solutions of thorium perchlorate and aqueous solutions of thorium oxyperchlorate were studied by Bacon.[3] The most probable structural unit in the thorium(IV) perchlorate solution is a thorium(IV) ion surrounded by four perchlorate ions and three waters of hydration.

8. Trimethylamine Decahydrate

Folzer, Hendricks, and Narten[28] scattered X-rays from a solution of trimethylamine in water having the same composition as the solid hydrate, with s ranging from 0.02 to 16 Å^{-1}. They found no structure in the pair density beyond 8 Å. The X-ray data provide strong evidence for the existence of a three-dimensional network in the solution. Both the ice-I model[85] and the clathrate model can explain the observed short-range order in the solution. Figure 19 shows the reduced ERDF' $g(r)$'s for water and the trimethylamine decahydrate.

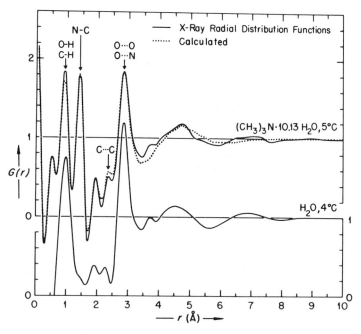

Fig. 19. Comparison of reduced sharpened electronic radial distribution functions for trimethylamine decahydrate and pure water. Calculated curve above ~2.5 Å based on simple tetrahedral network model. From Folzer, Hendricks, and Narten.[28]

References

1. E. K. Achter and L. Meyer, *Phys. Rev.*, **188**, 291 (1969).
2. J. N. Albright, *J. Chem. Phys.*, **56**, 3783 (1972).
3. W. E. Bacon, Ph.D. Thesis, Kent State University, 1968 (available from University Microfilms).
4. J. D. Barnett and H. T. Hall, *Rev. Sci. Instr.*, **35**, 175 (1964).
5. R. Benesch and V. H. Smith, Jr., *J. Chem. Phys.*, **53**, 1466 (1970).
6. J. Berghuis, I. Jbertha, H. Haanappel, and M. Potter, *Acta Cryst.*, **8**, 478 (1955).
7. M. Berman and S. Ergun, *Rev. Sci. Instr.*, **41**, 870 (1970).
8. A. Bienenstock, *J. Chem. Phys.*, **31**, 570 (1959).
9. I. A. Blech and B. L. Auerbach, *Phys. Rev.*, **A137**, A1113 (1965).
10. Z. Bochynski, *Acta Physica Polonica*, **34**, 185 (1968).
11. Z. Bochynski, *Acta Physica Polonica*, **34**, 557 (1968).
12. G. W. Brady, *Acc. Chem. Res.*, **Nov. 1971**, 367 (1971).
13. G. W. Brady and C. C. Gravatt, Jr., *J. Chem. Phys.*, **55**, 5095 (1971).
14. H. Brumberger, Ed., *Small Angle X-ray Scattering*, Gordon & Breach, New York, London, and Paris, 1967.
15. G. H. A. Cole, *Rept. Progr. Phys.*, **31**, 419 (1968).
16. D. T. Cromer, *Acta Cryst.*, **19**, 224 (1965).
17. D. T. Cromer, *J. Chem. Phys.*, **50**, 4857 (1969).
18. D. T. Cromer and D. Libermann, *J. Chem. Phys.*, **53**, 1891 (1970).
19. D. T. Cromer and J. B. Mann, *Los Alamos Scientific Laboratory Report LA-3816* (1967).
20. D. T. Cromer and J. B. Mann, *J. Chem. Phys.*, **47**, 1892 (1967).
21. D. T. Cromer and J. B. Mann, *Acta Cryst.*, **A24**, 321 (1968).
22. B. D. Cullity, *Elements of X-ray Diffraction*, Addison-Wesley, Reading, Mass., 1956.
23. M. D. Danford, "Diffraction Pattern and Structure of Aqueous Ammonium Fluoride Solutions," *ORNL Report No. 4244*, 1968.
24. P. Debye, *Ann. Phys.*, **46**, 809 (1915).
25. P. A. Doyle and P. S. Turner, *Acta Cryst.*, **A24**, 390 (1968).
26. P. L. Fehder, *J. Chem. Phys.*, **52**, 791 (1970).
27. T. R. Fisher, R. S. Safrata, E. G. Shelley, J. McCarthy, S. M. Clustin, and R. C. Barret, *Phys. Rev.*, **157**, 1149 (1967).
28. C. Folzer, R. W. Hendricks, and A. H. Narten, *J. Chem. Phys.*, **54**, 799 (1971).
29. A. J. Freeman, *Acta Cryst.*, **12**, 261 (1959).
30. *Ibid.*, 929 (1959).
31. *Ibid.*, **13**, 190 (1960).
32. K. Furukawa, *Rept. Progr. Phys.*, **25**, 395 (1962).
33. H. W. Furumoto and C. H. Shaw, *Phys. Fluids*, **7**, 1026 (1964).
34. N. S. Gingrich, *Rev. Mod. Phys.*, **15**, 90 (1943).
35. N. S. Gingrich and C. W. Tompson, *J. Chem. Phys.*, **36**, 2398 (1962).
36. W. L. Gordon, C. H. Shaw, and J. G. Daunt, *J. Phys. Chem. Solids*, **5**, 117 (1958).
37. R. W. Gruebel and G. T. Clayton, *J. Chem. Phys.*, **46**, 639 (1967).
38. *Ibid.*, **47**, 175 (1967).
39. R. B. Hallock, *Phys. Rev. Letters*, **23**, 830 (1969).
40. R. B. Hallock, *Rev. Sci. Instr.*, **41**, 1107 (1970).
41. R. B. Hallock, *J. Low Temp. Phys.*, **9**, 109 (1972).
42. R. B. Hallock, *Phys. Rev.*, **A5**, 320 (1972).
43. *Ibid.*, **A8**, 2143 (1973).
44. R. W. Harris and G. T. Clayton, *J. Chem. Phys.*, **45**, 2681 (1966).
45. R. W. Harris and G. T. Clayton, *Phys. Rev.*, **153**, 229 (1967).

46. Address requests to: Dr. James K. Hendren or Dr. Glen T. Clayton, University of Arkansas, Physics Department, Fayetteville, Arkansas 72701. Offered in *American J. Phys.*, **40**, 1343 (1972).
47. R. Hoge and J. Trotter, *Can. J. Chem.*, **43**, 2692 (1965).
48. W. I. Honeywell, C. M. Knobler, B. L. Smith, and C. J. Pings, *Rev. Sci. Instr.*, **35**, 1216 (1964).
49. R. W. James and G. W. Brindley, *Z. Krist.*, **78**, 470 (1931).
50. R. Kaplow, S. L. Strong, and B. L. Auerbach, *Phys. Rev.*, **138**, A1336 (1965).
51. J. F. Karnicky, Ph.D. Thesis, California Institute of Technology, 1974 (available from University Microfilms). J. F. Karnicky, H. H. Reamer, and C. J. Pings, submitted to *J. Chem. Phys.*
52. A. A. Khan, *Phys. Rev.*, **136**, A1260 (1964).
53. A. P. Kendig and C. J. Pings, *J. Appl. Phys.*, **36**, 1692 (1965).
54. Y. Kim and M. Inokuti, *Phys. Rev.*, **165**, 39 (1968).
55. B. E. Kirstein, Ph.D. Thesis, California Institute of Technology, 1972 (available from University Microfilms). B. E. Kirstein and C. J. Pings, submitted to *J. Chem. Phys.*
56. A. B. Kiselev, V. A. Liopo, and M. S. Metzik, *Izv. Vuz. Fiz.* (*USSR*), **6**, 158 (1971) (in Russian). Translated into English in *Sov. Phys. J.* (*U.S.A.*).
57. A. B. Kiselev, V. A. Liopo, M. S. Metzik, and A. A. Zhdanov, *Acta Cryst.*, **A28**, Supplement, p. 5126 (1972) (abstract only).
58. V. P. Klochkov and A. F. Skryshevskii, *Ukr. Fiz. Zh.* (*USSR*), **9**, 420 (1964) (in Russian) (English abstract).
59. J. H. Konnert and J. Karle, *Acta Cryst.*, **A29**, 702 (1973).
60. A. Kratochwill, J. U. Weidner, and H. Zimmermann, *Ber. Bunsenges Phys. Chem.* (*Germany*), **77**, 408 (1973) (in German) (English abstract).
61. J. Krogh-Moe, *Acta Cryst.*, **9**, 951 (1956).
62. R. F. Kruh, *Chem. Rev.*, **62**, 319 (1962).
63. R. F. Kruh and J. I. Petz, *J. Chem. Phys.*, **41**, 890 (1964).
64. R. F. Kruh and C. L. Standley, *Inorganic Chem.*, **1**, 941 (1962).
65. R. M. Lawrence and R. F. Kruh, *J. Chem. Phys.*, **47**, 4758 (1967).
66. S. C. Lee and R. Kaplow, *Science*, **169**, 477 (1970).
67. H. A. Levy, M. D. Danford, and A. H. Narten, "Data Collection and Evaluation with an X-ray Diffractometer Designed for the Study of Liquid Structure," *DRNL Report No. 3960*, 1966.
68. G. Licheri, G. Piccaluga, and G. Pinna, *Chem. Phys. Letters*, **12**, 425 (1971).
69. G. Licheri, G. Piccaluga, and G. Pinna, *J. Appl. Cryst.*, **6**, 392 (1973).
70. K. Lonsdale, Ed., *International Tables for X-ray Crystallography*, Volume III, The Knoch Press, Birmingham, England, 1962.
71. R. McWeeny, *Acta Cryst.*, **6**, 631 (1953).
72. H. Mendel, *Acta Cryst.*, **15**, 113 (1962).
73. H. Menke, *Z. Physik*, **33**, 593 (1932).
74. P. G. Mikolaj, Ph.D. Thesis, California Institute of Technology, 1965 (available from University Microfilms).
75. P. G. Mikolaj and C. J. Pings, *Phys. Rev. Letters*, **16**, 4 (1966).
76. P. G. Mikolaj and C. J. Pings, *J. Chem. Phys.*, **46**, 1401 (1967).
77. *Ibid.*, 1412 (1967).
78. M. E. Millberg, *J. Appl. Phys.*, **29**, 64 (1958).
79. P. F. Morrison and C. J. Pings, *J. Chem. Phys.*, **56**, 280 (1972).
80. P. F. Morrison and C. J. Pings, *J. Chem. Phys.*, **60**, 2323 (1974).
81. Y. Narahara, *J. Phys. Soc. Japan*, **24**, 169 (1968).

82. A. H. Narten, *J. Chem. Phys.*, **48**, 1630 (1968).

83. *Ibid.*, **49**, 1692 (1968).

84. A. H. Narten, *J. Phys. Chem.*, **74**, 765 (1970).

85. A. H. Narten, M. D. Danford, and H. A. Levy, *Disc. Faraday Soc.*, **43**, 97 (1967).

86. A. H. Narten, M. D. Danford, and H. A. Levy, *J. Chem. Phys.*, **46**, 4875 (1967).

87. A. H. Narten and S. Lindenbaum, *J. Chem. Phys.*, **51**, 1108 (1969).

88. A. H. Narten and H. A. Levy, *J. Chem. Phys.*, **55**, 2263 (1971).

89. A. H. Narten, F. Vaslow, and H. A. Levy, *J. Chem. Phys.*, **58**, 5017 (1973).

90. A. Nukui, H. Morikawa, S. Iwai, and H. Tagai, *Acta Cryst.*, **A28**, S5128 (1972) (abstract only).

91. H. H. Paalman and C. J. Pings, *J. Appl. Phys.*, **33**, 2635 (1962).

92. H. H. Paalman and C. J. Pings, *Rev. Mod. Phys.*, **35**, 389 (1963).

93. J. I. Petz, *J. Chem. Phys.*, **43**, 2238 (1965).

94. M. Piliavin, Ph.D. Thesis, California Institute of Technology, 1973 (available from University Microfilms).

95. C. J. Pings, in *Physics of Simple Liquids*, H. N. V. Temperley, J. S. Rowlinson, and G. S. Rushbrooke, Eds., North Holland Publishing Company, Amsterdam, 1968, Chapter 10.

96. C. J. Pings and H. H. Paalman, *Mol. Phys.*, **5**, 531 (1962).

97. C. J. Pings and J. Wasser, *J. Chem. Phys.*, **48**, 3016 (1968).

98. J. G. Powles, *Adv. Phys.*, **22**, 1 (1973).

99. *Proceedings of 2nd International Conference on the Properties of Liquid Metals, Tokyo, Japan, 3–8 September 1972*, Taylor and Francis, London, 1973. In particular the papers by J. E. Enderly and P. A. Egelstaff.

100. A. Rahman, *J. Chem. Phys.*, **42**, 3540 (1965).

101. P. Ratnasamy and A. J. Leonard, *Catalysis Rev.*, **6**(2), 293–322 (1972).

102. A. J. Renouprez, *Int. Union Crystallogr. Comm. Crystallogr. Appl. Bibliogr.*, *No. 4* (1970).

103. Rigaku Rota Unit RU-100, Rigaku Denki Co. Ltd., Cable Address: RIGAKUDENKI TOKYO.

104. J. S. Rowlinson, *Rept. Progr. Phys.*, **28**, 169 (1965).

105. C. T. Rutlege and G. T. Clayton, *J. Chem. Phys.*, **52**, 1927 (1970).

106. A. I. Ryss and I. V. Radchenko, *Ukr. Fiz. Zh.* (*USSR*), **9**, 416 (1964) (in Russian) (English abstract).

107. G. J. Safford and Pak. S. Leung, in *Techinques of Electrochemistry*, Vol. 2, E. Yaeger and A. J. Salkind, Eds., Wiley, New York, 1973, Chapter IV.

108. K. Sagel, *Tabellen zur Roentgensturktur-analyse*, Springer-Verlag, Berlin, 1958.

109. P. C. Sharrah, J. I. Petz, and R. F. Kruh, *J. Chem. Phys.*, **32**, 241 (1960).

110. S. C. Smelser, Ph.D. Thesis, California Institute of Technology, 1969 (available from University Microfilms).

111. J. A. Soules, W. L. Gordon, and C. H. Shaw, *Rev. Sci. Instr.*, **27** (1956).

112. W. Steele and R. Pecora, *J. Chem. Phys.*, **42**, 1863 (1965).

113. R. Stewart, *J. Chem. Phys.*, **42**, 3175 (1965).

114. *Ibid.*, **51**, 4569 (1969).

115. S. Stirpe and C. W. Tompson, *J. Chem. Phys.*, **36**, 392 (1962).

116. S. L. Strong and R. Kaplow, *Acta Cryst.*, **23**, 38 (1967).

117. D. S. Terekhova, *Zh. Strukt. Khim.* (*USSR*), **11**, 530 (1970) (in Russian) (translated into English).

118. F. Vaslow and A. H. Narten, *J. Chem. Phys.*, **59**, 4949 (1973).

119. G. H. Vineyard, *Phys. Rev.*, **96**, 93 (1954).

120. B. E. Warren, *J. Appl. Phys.*, **8**, 645 (1937).

121. B. E. Warren and R. L. Mozzi, *Acta Cryst.*, **21**, 459 (1966).

122. J. Waser and V. Schomaker, *Rev. Mod. Phys.*, **25**, 671 (1953).
123. J. U. Weidner, H. Geisenfelder, and H. Zimmermann, *Ber. Bunsenges Phys. Chem. (Germany)*, **75**, 800 (1971) (in German) (English abstract).
124. *Ibid.*, **76**, 628 (1972) (in German) (English abstract)
125. D. L. Wertz and R. F. Kruh, *J. Chem. Phys.*, **50**, 4313 (1969).
126. D. L. Wertz, R. M. Lawrence, and R. F. Kruh, *J. Chem. Phys.*, **43**, 2163 (1965).
127. G. D. Wignall, J. A. J. Jarvis, W. E. Munsil, and C. J. Pings, *J. Appl. Cryst.*, **7**, 366 (1974).

DIFFRACTION BY
MOLECULAR LIQUIDS*

L. BLUM

Department of Physics,
University of Puerto Rico,
Rio Piedras, Puerto Rico

AND

A. H. NARTEN

Chemistry Division,
Oak Ridge National Laboratory,
Oak Ridge, Tennessee

CONTENTS

* Research sponsored by the U.S. Atomic Energy Commission under contracts with the Union Carbide Corporation and with Oak Ridge Associated Universities.

I. INTRODUCTION

The statistical theory of the liquid state has the aim of calculating the properties of a liquid from knowledge of the intermolecular potential; this intermolecular potential is inserted into the partition function integral from which all the properties are calculated by standard formulas. Although this sounds very simple, it can seldom be done in practice with useful results.

One reason for this is that it is very difficult to measure the intermolecular potential and only very few systems have been studied by a direct technique such as atomic beam scattering, which could eventually give detailed information about the molecular forces. Therefore, the theoretical effort has shifted to the interpretation of the properties in terms of the structure function, or pair correlation function, which can be obtained from diffraction experiments, to a rather satisfactory degree of precision.

For monatomic fluids this has been done, and the agreement between experiment, molecular dynamics, diffraction studies, and the general theory is, to this date, quite satisfactory. For molecular fluids the situation is not so clear. In the first place, beam scattering experiments of the type needed to get the full angular dependence of the intermolecular pair potential (polarized molecules) have not been done, and therefore the proposed potentials are often inferred from theoretical calculations.

In the second place, the diffraction experiments (X-rays, neutrons, electrons) cannot give the full angular dependence of the structure function. Therefore, only a few properties, such as the compressibility, can be related to the structure function obtained from diffraction. The reason for this is that the structure function is now a function which depends on six variables: one distance between the molecules, two angles to point the direction of the second molecule, and three Euler angles that determine the orientation of the second molecule relative to the first one. The diffraction experiment, however, is one-dimensional and cannot possibly map a six-dimensional function. The situation is similar to the analysis of a powder-crystal pattern in which one sees an angular average of all different configurations. As we will see below, unpolarized radiation yields information about a cylindrically symmetric configuration in which all atoms are essentially smeared around the intermolecular axis.

There is some loose correspondence between this picture and the more traditional way of interpreting diffraction experiments on molecular fluids in terms of atom-atom pair correlation functions. For liquids with two atomic species, such as CO_2, H_2O, H_3N, CCl_4, C_6H_6, a minimum of three independent experiments is necessary to determine all possible atom-atom pair correlation functions. For water, for example, the three correlation functions are g_{OO}, g_{OH}, and g_{HH}. Unfortunately, only two satisfactory experiments (X-ray, and neutron on D_2O) have been done. Because the hydrogen has a large incoherent cross-section, the neutron scattering experiment on light water has not been done with sufficient precision. For homonuclear diatomic liquids, such as N_2, Cl_2, O_2, Br_2, only one pair correlation function $g_{AA}(R)$ is needed in this picture.

The atom-atom pair correlation functions do not tell us about the angular distribution of the molecules and are not related in any direct way to orientation-dependent properties such as the dielectric constant of the fluid. In the case of homonuclear diatomics, $g_{AA}(R)$ does not tell us if, for example, first nearest neighbors are aligned or perpendicular to each other. It seems therefore that a more systematic approach, such as the multipole expansion of molecular properties, is a reasonable alternative. Quite clearly, only simple and reasonably "spherical" molecules could be treated, but quite definitely all the simple liquids listed above fall into this category.

The central idea of this approach is that the expansion of the pair correlation function $g(\mathbf{R}, \mathbf{\Omega})$, where $\mathbf{R}$ is the separation vector of two molecules and $\mathbf{\Omega} \equiv \alpha, \beta, \gamma$ are the Euler angles for the relative orientation, is reasonably convergent. The physical requirement for this is that the angular interactions not be too strongly dependent on the orientations. We believe that even in a liquid such as water, where there are very strong orientational interactions (the hydrogen bonds), these are not like spikes pointing to the corners of a tetrahedron, but rather like bulges, which viewed from the center of the tetrahedron, have a sizable angular spread.

From group theoretical considerations we know that there are, for each molecular symmetry, a limited number of coefficients in the expansion of $g(\mathbf{R}, \mathbf{\Omega})$, and that the best set of coefficients are the ones of the so-called irreducible representations. These are described in the next section. The irreducible representations are labeled by their "chirality," or helicity, χ. The unpolarized radiation scattering only sees the $\chi = 0$ coefficients, which for cylindrical molecules correspond to arrangements of the molecules that are "smeared" about the intermolecular axis. The relation of the coefficients in this expansion to the thermodynamic functions is described at the end of this section.

Section III is devoted to the theory of the cross-sections for unpolarized radiation, X-rays, neutrons, and also electrons. We include also corrections for vibrations, and, in the case of neutrons, the recoil or dynamic corrections are calculated from a first-principles approach.

In Section IV we discuss experimental aspects of diffraction by molecular liquids, with emphasis on X-ray and thermal neutron scattering. Some recent results from diffraction studies are described in Section V.

II. MOLECULAR CORRELATIONS

A. Invariant and Irreducible Expansions

The molecular pair distribution function can be written as a series expansion in the complete set of orthogonal polynomials in the orientation variables, the coefficients of such series being a function of the intermolecular distance only.[1-4]

Consider a molecular fluid of number density $\rho = N/V$, where N is the number of molecules and V the volume. The pair correlation function can be written in a symmetric manner as

$$g(\mathbf{X}_1, \mathbf{X}_2) \tag{2.1}$$

where $\mathbf{X}_1 \equiv \mathbf{R}_1, \mathbf{\Omega}_1$; $\mathbf{R}_1$ is a vector giving the position of molecule 1; and $\mathbf{\Omega}_1 \equiv \alpha_1, \beta_1, \gamma_1$ are the three Euler angles that define its orientation in an arbitrary reference frame.

Comparing the definition of $g(\mathbf{R}, \mathbf{\Omega})$ of the last section, we notice that what we gained in symmetry is lost by the fact that we have a function of 12 variables, instead of six, and that we have to take out the three translations and three rigid rotations of the system of two molecules, which, in an isotropic fluid, should leave $g(\mathbf{X}_1, \mathbf{X}_2)$ invariant.

We can satisfy this invariance condition if we expand as

$$g(\mathbf{X}_1, \mathbf{X}_2) = \sum_{\substack{mnl \\ \mu\nu}} \hat{g}_{\mu\nu}^{mnl}(R_{12})\hat{\Phi}_{\mu\nu}^{mnl}(\mathbf{\Omega}_1, \mathbf{\Omega}_2, \hat{\mathbf{R}}_{12}) \tag{2.2}$$

where $\hat{g}_{\mu\nu}^{mnl}(R_{12})$ is a coefficient that depends only on the intermolecular distance, and

$$\hat{\Phi}_{\mu\nu}^{mnl}(\mathbf{\Omega}_1, \mathbf{\Omega}_2, \hat{\mathbf{R}}_{12})$$

$$= [(2m + 1)(2n + 1)]^{1/2} \sum_{\mu'\nu'\lambda'} \begin{pmatrix} m & n & l \\ \mu' & \nu' & \lambda' \end{pmatrix} D_{\mu\mu'}^m(\mathbf{\Omega}_1)D_{\nu\nu'}^n(\mathbf{\Omega}_2)D_{0\lambda'}^l(\hat{\mathbf{R}}_{12}) \tag{2.3}$$

Here $\hat{\mathbf{R}}_{12} = \theta_{12}, \phi_{12}$ is a compact notation for the two angles that define the orientation of the intermolecular distance vector $\mathbf{R}_{12}$, the functions $D_{\mu\nu'}^m(\mathbf{\Omega})$ are generalized Wigner spherical harmonics,[5] and we have used the standard notation for Wigner's 3-J symbols. $\hat{\Phi}_{\mu\nu}^{mnl}$ is invariant under rigid rotations of the reference frame, and therefore only five of the eight angles that appear in the argument are independent.

The invariants $\hat{\Phi}_{\mu\nu}^{mnl}$ form an orthonormal set since

$$\frac{1}{64\pi^4} \int d\mathbf{\Omega}_1 \, d\mathbf{\Omega}_2 \; \hat{\Phi}_{\mu\nu}^{mnl}[\hat{\Phi}_{\mu'\nu'}^{m'n'l'}]^* = \delta_{mm'}\delta_{nn'}\delta_{ll'}\delta_{\mu\mu'}\delta_{\nu\nu'} \qquad (2.4)$$

where $\delta_{mm'}$ is the Kronecker delta.

Since the reference frame is arbitrary, we may put the vector $\mathbf{R}_{12} = \mathbf{R}_1 - \mathbf{R}_2$ along the z-axis of our reference frame. Then, $\hat{\mathbf{R}}_{12} = (0, 0)$ and, using the fact that[5]

$$D_{0\lambda'}^l(\mathbf{0}) = \delta_{0\lambda'} \qquad (2.5)$$

and a property of the 3-J symbols, we can write

$$\hat{\Phi}_{\mu\nu}^{mnl} = [(2m + 1)(2n + 1)]^{1/2} \sum_\chi \begin{pmatrix} m & n & l \\ \chi & -\chi & 0 \end{pmatrix} D_{\mu\chi}^m(\mathbf{\Omega}_1) D_{\nu-\chi}^n(\mathbf{\Omega}_2) \qquad (2.6)$$

Substituting into (2.2) and rearranging we get

$$g(\mathbf{X}_1, \mathbf{X}_2) = \sum_{\substack{mn \\ \mu\nu, \chi}} g_{\mu\nu, \chi}^{mn}(R_{12})\Phi_{\mu\nu, \chi}^{mn} \qquad (2.7)$$

with

$$\Phi_{\mu\nu, \chi}^{mn} = [(2m + 1)(2n + 1)]^{1/2} D_{\mu\chi}^m(\mathbf{\Omega}_1) D_{\nu-\chi}^n(\mathbf{\Omega}_2) \qquad (2.8)$$

$$g_{\mu\nu, \chi}^{mn} = \sum_l \begin{pmatrix} m & n & l \\ \chi & -\chi & 0 \end{pmatrix} \hat{g}_{\mu\nu}^{mnl}(R_{12}) \qquad (2.9)$$

Equation (2.7) is an alternative way of expanding the pair correlation function and will be called the irreducible expansion. The coefficients of the irreducible expansion and the invariant expansion (2.2) are related by the simple orthogonal transformation (2.9). The inverse of (2.9) is

$$\hat{g}_{\mu\nu}^{mnl}(R_{12}) = (2l + 1) \sum_\chi \begin{pmatrix} m & n & l \\ \chi & -\chi & 0 \end{pmatrix} g_{\mu\nu, \chi}^{mn}(R_{12}) \qquad (2.10)$$

As far as the structural information is concerned, both expansions are equivalent and interchangeable. However, some properties are simpler in the invariant form (2.2), whereas, quite definitely, the irreducible expansion is mathematically more convenient.

Before going into the discussion of the geometrical meaning of these expansions, let us explore the consequences of the other symmetries that the two-molecule system has.

(a) Exchange symmetry

$$g(\mathbf{X}_1, \mathbf{X}_2) = g(\mathbf{X}_2, \mathbf{X}_1) \qquad (2.11)$$

means that the coefficients of the irreducible representation (2.7) satisfy the condition

$$g_{\mu v, \chi}^{mn}(R_{12}) = (-)^{n+m} g_{v\mu, -\chi}^{nm}(R_{12}) \tag{2.12}$$

This property can be easily deduced from (2.7). It means that for every value of χ, only about half of the coefficients are really independent. As we will see below, for most simple optically inactive fluids, the coefficients of χ are equal to those of $-\chi$, and then (2.12) becomes a symmetry condition for the matrix $\mathbf{g}_\chi$ [with elements defined by (2.12)].

(b) Molecular symmetry restricts the allowed values for the pairs (m, μ), (n, v). Since the expansion is invariant, we need to satisfy the symmetry requirements for each molecule individually. Therefore,[2] the allowed values of m, μ are such that $C_\mu^m(\hat{\mathbf{R}})$ (the Racah spherical harmonic, see Ref. 5) should belong to the totally symmetric representation of the molecular symmetry group.[6] Let us list a few physically interesting cases.

Group symmetry (Schoenflies notation):

C_{2V}: The water molecule has this symmetry

$$\mu, v = \text{even}$$

and

$$g_{\mu v, \chi}^{mn} = g_{\pm \mu \pm v, \chi}^{mn} \tag{2.13}$$

D_{6h}: Benzene is a typical example. The requirements are

$$\mu = \text{multiple of 6}, \qquad m = \text{even} \tag{2.14}$$

$C_{\infty v}$: Linear heteronuclear molecules such as HF, CO belong to this group. The requirement is

$$\mu = 0 \tag{2.15}$$

$D_{\infty h}$: This is the group of the homonuclear diatomics H_2, N_2, Cl_2, Br_2, and also linear molecules such as CO_2. We need

$$m = \text{even} \qquad \mu = 0 \tag{2.16}$$

T_d: Typical examples are CH_4, CCl_4, $GeBr_4$, and neopentane. Only certain linear combinations transform properly under the group operations.[7]

$$C_0^0, (C_2^3 - C_{-2}^3), [C_0^4 + 0.598(C_4^4 + C_{-4}^4)],$$
$$[C_0^6 - 1.871(C_4^6 + C_{-4}^6)] \tag{2.17}$$

The molecular symmetry requirements produce a considerable reduction in the number of independent coefficients of the expansions (2.2) and (2.7). Consider, for example, the case of a fluid with only dipole interactions.[8] Then, the only allowed coefficients in the invariant expansion are

$$\hat{g}_{00}^{110}; \qquad \hat{g}_{00}^{111}; \qquad \hat{g}_{00}^{112} \qquad\qquad (2.18)$$

A short calculation will show that the angular parts are proportional to

$$\hat{\Phi}_{00}^{110} \sim \mathbf{e}_1 \cdot \mathbf{e}_2 \qquad\qquad (2.19)$$

$$\hat{\Phi}_{00}^{111} \sim [\mathbf{e}_1 \times \mathbf{e}_2] \cdot (\mathbf{R}_{12}) \qquad\qquad (2.20)$$

$$\hat{\Phi}_{00}^{112} \sim 3(\mathbf{e}_1 \cdot \mathbf{R}_{12})(\mathbf{e}_2 \cdot \mathbf{R}_{12}) - (\mathbf{e}_1 \cdot \mathbf{e}_2) \qquad\qquad (2.21)$$

where $\mathbf{e}_1$ is the unit vector pointing in the direction of the dipole of molecule 1, and we have used the customary notation for vector and scalar products. The meaning of this is that the invariants are just all possible ways to form scalar products from the three vectors $\mathbf{e}_1$, $\mathbf{e}_2$, and $\mathbf{R}_{12}$. Φ_{00}^{111} (2.20) clearly corresponds to a helical interaction, which is absent in an optically inactive fluid.

From the irreducible expansion (2.8) we get the coefficients of

$$g_{00,0}^{11}; g_{00,1}^{11}; g_{00,-1}^{11} \qquad\qquad (2.22)$$

which are

$$\Phi_{00,0}^{11} = 3 \cos \beta_1 \cos \beta_2 \qquad\qquad (2.23)$$

$$\Phi_{00,\pm1}^{11} = -\tfrac{3}{2} \sin \beta_1 \sin \beta_2 \, e^{\pm i(\gamma_1 - \gamma_2)} \qquad\qquad (2.24)$$

These functions are similar to the LCAO (Linear Combination of Atomic Orbitals) molecular orbitals in quantum chemistry.[9] Equation (2.23) is just like a σ orbital formed by the linear combination of two p_z atomic orbitals on atoms 1 and 2. The two other ways of combining p atomic orbitals into degenerate π molecular orbitals are the real and imaginary parts of $\Phi_{00,\pm1}^{11}$, (2.24). For cylindrical molecules, the generalization to higher multipoles is simple, because $\chi = 0$ always corresponds to the cylindrical symmetric σ molecular orbitals, $\chi = \pm1$ to linear combinations of π orbitals, $\chi = \pm2$ to δ orbitals, and so on.

For molecules that are not cylindrical, the physical meaning of the irreducible coefficients is not so simple. The index χ is associated with the heliticity, or multiplicity of the translation–rotation axis, S_χ, in Schoenflies notation.[6] Clearly, the $\chi = 0$ coefficients correspond to configurations with no heliticity, or, in other words, configurations with a center of inversion located midway between molecules 1 and 2.

B. Thermodynamic Functions

As was stated above, the expansions of $g(X_1, X_2)$ are simply related to the bulk properties of the fluids.[6] The internal energy is[1]

$$E = E_{kin} + \frac{\rho^2}{128\pi^4 V} \int dX_1, dX_2\, U(X_1, X_2)g(X_1, X_2) \tag{2.25}$$

where the kinetic energy is $5/2\, kT$ for linear molecules and $3\, kT$ for nonlinear ones.

The pressure can be obtained in various different ways from the correlation function. From the virial equation

$$P = \rho kT - \frac{\rho^2}{384\pi^4 V} \int dX_1, dX_2\, g(X_1 X_2)\left\{\left[\frac{\partial U(X_1, X_2)}{\partial R_{12}}\right]\cdot R_{12}\right\} \tag{2.26}$$

or from the compressibility equation

$$\kappa kT = \frac{1}{\rho} + \frac{1}{64\pi^4 V} \int dX_1, dX_2\, h(X_1, X_2) \tag{2.27}$$

where the compressibility is defined by

$$\kappa = -\frac{1}{V}\left(\frac{\partial V}{\partial p}\right)_T \tag{2.28}$$

and we have used

$$h(X_1, X_2) = g(X_1, X_2) - 1 \tag{2.29}$$

There is still another way of getting the pressure, via the temperature integral of the internal energy (2.25), which gives the free energy[10] from which the pressure is obtained by a standard formula. This procedure does not give a closed formula like (2.26) and (2.27), because we must know the explicit dependence of $g(X_1, X_2)$ on the temperature.

Although all the different expressions for the pressure are, in principle, equivalent, the theoretical approximations and experimental limitations cause the various results obtained from different expressions. This disagreement is considered a measure of the goodness of the theory or the quality of the experiment.

Substituting the invariant expansions for $g(X_1, X_2)$, $h(X_1, X_2)$, and $U(X_1, X_2)$ into (2.25ff) we get

$$E = E_{kin} + 2\pi\rho^2 \int_0^\infty dR\, R^2 \sum_{\substack{mnl \\ \mu\nu}} (2l + 1)^{-1} \hat{g}_{\mu\nu}^{mnl}(R)[\hat{U}_{\mu\nu}^{mnl}(R)]^* \tag{2.30}$$

The irreducible coefficients give the alternative expression

$$E = E_{\text{kin}} + 2\pi\rho^2 \int_0^\infty dR\ R^2 \sum_{\substack{mn \\ \mu\nu\chi}} g^{mn}_{\mu\nu,\chi}[U^{mn}_{\mu\nu,\chi}]^* \tag{2.31}$$

The invariant pressure and compressibility equations are

$$P = \rho kT - \tfrac{2}{3}\pi\rho^2 \int_0^\infty dR\ R^3 \sum_{\substack{mnl \\ \mu\nu}} (2l+1)^{-1} g^{mnl}_{\mu\nu}\left[\frac{\partial U^{mnl}_{\mu\nu}}{\partial R}\right]^* \tag{2.32}$$

$$\kappa kT = \frac{1}{\rho} + 4\pi \int_0^\infty \hat{h}^{000}_{00}(R)R^2\ dR \tag{2.33}$$

Equation (2.33) is much simpler than (2.32), both from the theoretical and experimental points of view. Experimentally, however, it has to be determined through a special technique, small-angle scattering, which is very exacting and delicate. Therefore, in practice, the compressibility is often used to get the asymptotic behavior at small angles, rather than the opposite.

Another property of principal interest is the dielectric constant, ε. As has been shown recently by Deutch, Ramshaw, and Wertheim,[11] this property should be related to the coefficients of the direct correlation function, which is obtained by solving the Ornstein–Zernike equation,

$$h(\mathbf{X}_1, \mathbf{X}_2) - C(\mathbf{X}_1, \mathbf{X}_2) = \frac{1}{8\pi^2}\rho \int d\mathbf{X}_3\ h(\mathbf{X}_1, \mathbf{X}_3)C(\mathbf{X}_3, \mathbf{X}_2) \tag{2.34}$$

Again, the direct correlation function $C(\mathbf{X}_1, \mathbf{X}_2)$, defined by (2.34), can be expanded as

$$C(\mathbf{X}_1, \mathbf{X}_2) = \sum \hat{C}^{mnl}_{\mu\nu}(R_{12})\hat{\Phi}^{mnl}_{\mu\nu} \tag{2.35}$$

and the dielectric constant is given by the modified Clausius–Mossotti equation[12]

$$\frac{3}{4\pi}\left[\frac{\varepsilon - 1}{\varepsilon + 2}\right] = \frac{\rho\mu^2}{3kT}\left[1 - \frac{\rho}{\pi}A\right] \tag{2.36}$$

$$A = -\sqrt{3}\int_0^\infty dR\ R^2 \hat{C}^{110}_{00}(R) \tag{2.37}$$

The computation of the quantity A requires the solution of the Ornstein–Zernike equation for $\hat{C}^{110}_{00}$, but in order to do this, we need to know the pair correlation function irreducible coefficients for all values of the heliticity χ. As we shall see in the next section, only the coefficients with $\chi = 0$ are accessible from scattering experiments.

III. SCATTERING CROSS-SECTIONS

A. Basic Equations

The underlying assumption of the theory of radiation (X-rays, neutrons, electrons) scattering from isotropic fluids is that the first Born approximation[13] is good enough for the elementary scattering process.

For a weak coupling process, such as the electromagnetic interaction in the scattering of X-rays by the electrons of the fluids, this is an excellent approximation, and the scattering amplitude (see, for example, Ref. 13 for a detailed discussion) is given by the free-electron, Born approximation formula,

$$f_\alpha^x(Q) = \frac{e}{mc^2}(1 + \cos^2\theta)^{1/2} \int \rho(\mathbf{R}_\alpha)e^{i\mathbf{Q}\cdot\mathbf{R}_\alpha}\, d\mathbf{R}_\alpha \qquad (3.1)$$

where e is the electric charge, m the mass, c the velocity of light, θ the scattering angle, and $\rho(\mathbf{R}_\alpha)$ the electron density at $\mathbf{R}_\alpha$. $\mathbf{Q} = \mathbf{k}_0 - \mathbf{k}_s$, with $\mathbf{k}_0$ the incident wave vector and $\mathbf{k}_s$ the scattered wave vector.

For neutrons, Born's approximation can also be used, but for a different reason. In fact, the interaction of a neutron with a nucleus is very strong, and is also quite complex if we include the magnetic interactions of the neutron nuclear spins. But as was shown by Fermi,[14] it is still possible to use Born's approximation, if in the region of the effective interaction between the nucleus and the particle, the wave function of the particle changes only slightly in the scattering process. This is so far the neutron interaction since the range of the interaction, which is about 10^{-12} cm, is much smaller than the wavelength of the thermal neutron and interatomic distances (10^{-8} cm). Then we can use Fermi's pseudopotential,

$$V(\mathbf{R}) = \frac{2\pi\hbar^2}{m}\hat{b}_\alpha\delta(\mathbf{R} - \mathbf{R}_\alpha) \qquad (3.2)$$

In general, $^{14}\hat{b}_\alpha$ will be a spin-dependent operator, and the scattering amplitude for the elementary process of a neutron being scattered by a nucleus is, in the Born approximation,

$$f_\alpha^n(\lambda_0, \lambda_s) = -\frac{m}{2\pi\hbar^2}\int d\mathbf{R}\,\psi_{\lambda_s}^*(\mathbf{R})V(\mathbf{R})\psi_{\lambda_0}(\mathbf{R}) \qquad (3.3)$$

where $\psi_{\lambda_s}(\mathbf{R})$ is the wave function for the system in the final state, and $\psi_{\lambda_0}(R)$ the wave function in the initial state. For the neutrons, these will be normalized plane wave functions times spin eigenfunctions. In the "rigid molecules at rest" approximation, the nuclear wave function enters only through its spin and can be factored from the wave functions in (3.3). The scattering amplitude is therefore a function of the isotope, total spin (mass) and spin states of the nucleus in question. Different isotopes of the same element may have very different scattering amplitudes. This fact makes part of the scattering of the neutrons incoherent.

For the case of electron scattering, the validity of the Born approximation has not been fully established. If we assume that the higher corrections are unimportant, then the scattering amplitude[15] is

$$f_\alpha^{el}(Q) = \frac{1}{Q^2} [Z_\alpha - f_\alpha^x(Q)] \tag{3.4}$$

where $f_\alpha^x(Q)$ is the scattering amplitude for X-rays, and Z_α is the nuclear charge.

The most general theory of the scattering of radiation from fluids in the Born approximation is due to Van Hove.[16] The angle and frequency dependent cross-section is given by the equation[17]

$$\frac{d^2\sigma}{d\Omega\, d\omega} = \epsilon(k_s)(k_s/k_0)\mathscr{S}(\mathbf{Q}, \omega) \tag{3.5}$$

where $\epsilon(k_s)$ is the detector efficiency (which is included for reasons that will be apparent below), and the function $\mathscr{S}(Q, \omega)$ is defined by

$$\mathscr{S}(\mathbf{Q}, \omega) = \frac{1}{2\pi} \int_{-\infty}^{\infty} dt\, e^{-i\omega t} \sum_{\alpha,\beta} \langle f_\alpha^* f_\beta e^{-i\mathbf{Q}\cdot\mathbf{R}_\alpha(o)} e^{i\mathbf{Q}\cdot\mathbf{R}_\beta(t)} \rangle \tag{3.6}$$

where the brackets indicate quantum-mechanical ensemble average. We have used the generic notation f_α to indicate the Q-space, Fourier transforms of the Born approximations of the scattering amplitudes for X-rays, neutrons or electrons [(3.1), (3.3), and (3.4)].

The physical meaning of this expression is that since we do not detect the scattering amplitudes but the intensities, which are the amplitudes squared, we only see the result of the interference of two quanta of radiation from events that were not simultaneous when they occurred. The time t of (3.6) is a measure of this delay. Clearly, the phase incoherence that results is not random for short times and depends on the ratio of the speed of the radiation to the range of the molecular correlations. For X-rays, which travel at the speed of light, the effect is small and observable only if very coherent sources, such as Mösbauer sources of ultrasoft nuclear γ-radiations, or the yet-to-be-developed X-ray laser, are used. As we will see below, measurement of this effect provides a way to calculate the corrections to the neutron scattering experiments that will be discussed at the end of this section.

For neutrons, the situation is different, since the neutrons used in diffraction experiments are slow, and therefore the time-delay effects are not negligible.

Theoretically, however, the instantaneous configuration diffraction pattern, which is given by

$$S(\mathbf{Q}) = \left\langle \sum_{\alpha, \beta} f_\alpha^* f_\beta e^{-i\mathbf{Q}\cdot(\mathbf{R}_\alpha - \mathbf{R}_\beta)} \right\rangle \tag{3.7}$$

could be obtained by simply integrating (3.6) at constant $\mathbf{Q}$ since due to the fact that

$$\int_{-\infty}^{\infty} d\omega \, e^{-i\omega t} = 2\pi\delta(t) \tag{3.8}$$

we get

$$S(\mathbf{Q}) = \int_{-\infty}^{\infty} d\omega \, \mathscr{S}(\mathbf{Q}, \omega) \tag{3.9}$$

This means that, in order to get the scattering functions, we should in principle position a black detector at constant angle θ, which, for elastic (and only elastic) scattering, implies constant $\mathbf{Q}$, from the relation

$$Q = 2k_0 \sin \theta \tag{3.10}$$

In practice this is not so simple: just because in the laboratory frame no collision is perfectly elastic, constant angle does not imply constant $\mathbf{Q}$, and, therefore, a black detector at a fixed angle does not integrate the scattered intensity at constant $\mathbf{Q}$, as required by (3.9) (clearly inelastic collisions can occur in a wide variety of circumstances).

Furthermore, what is really measured is

$$\left(\frac{d\sigma}{d\Omega}\right)_\theta = \int_{-\infty}^{\omega_0} d\omega \cdot \frac{d^2\sigma}{d\Omega \, d\omega}\bigg|_{\text{constant }\theta} \tag{3.11}$$

where $\omega_0 = E_0/\hbar$ is the angular frequency of the incident neutron, and therefore, we are in fact, integrating (3.5) and not (3.6). In other words, the measured quantity is

$$\left(\frac{d\sigma}{d\Omega}\right)_\theta = \int_{-\infty}^{\omega_0} d\omega \, \epsilon(k_s/k_0)\mathscr{S}(\mathbf{Q}, \omega)\bigg|_\theta \tag{3.12}$$

which differs from (3.9) because the detector sensitivity depends on the scattered wave vector, the flux in an inelastic process is less due to the slowing down of the particles, and the frequency has an upper limit, which is the frequency of the incident neutrons ω_0. However, all these effects can be accounted for by a method introduced by Placzek[18] which will be discussed at the end of this section, and in which the leading term will turn out to be $S(Q)$

$$\left(\frac{d\sigma}{d\Omega}\right)_\theta = \varepsilon(k_0)S(Q) + \text{corrections} \tag{3.13}$$

The approximation that neglects these corrections is called the static or recoilless approximation. This is so because, for most fluids, the main contribution to the corrections comes from center of mass-motion terms, which are also recoil terms. For electron scattering the recoil corrections are negligible anyway, and the present status of the experiments does not warrant the discussion of these corrections.

B. Static Approximation

In the static approximation the scattering cross-section is proportional to the scattering function $S(Q)$ defined by (3.9). Let us now analyze the structure of this function for the scattering of molecular fluids. $S(Q)$ is a function of the molecular structure and the intermolecular correlation functions. The approximation of considering the molecular structure in the many liquids as being the same as in other phases of the substance is physically sound and is corroborated in some cases by direct experimental evidence,[19] such as the agreement between bond lengths obtained from electron diffraction in the gas phase and X-ray diffraction in the liquid phase. However, the molecules are not rigid, and there is evidence that the distortions, due to vibrational motion, are not quite the same in the gas phase as in the liquid phase, a fact which is probably due to the perturbation of the molecular fields caused by the interactions with the surrounding molecules. In the case of water, the structure of the gas molecule has a different H—O—H angle than in the ice-I crystal, and consequently, one should expect some random variations of this angle in the liquid.

If we consider that the molecular structure is constant and independent from changes in the surroundings, then we can separate the contributions from the molecules and from the intermolecular correlations. We write

$$S(\mathbf{Q}) = S_s(\mathbf{Q}) + S_m(\mathbf{Q}) + S_d(\mathbf{Q}) \tag{3.14}$$

with

$$S_s(\mathbf{Q}) = \sum (\overline{f_\alpha^2} - \overline{f_\alpha}^2) \tag{3.15}$$

the "self" contribution which is due to the interference of two quanta of radiation scattered from the same atom,

$$S_m(\mathbf{Q}) = \sum_{\substack{\alpha, \beta \\ \epsilon \text{ same} \\ \text{molecule}}} \langle f_\alpha^* f_\beta e^{-i\mathbf{Q}\cdot(\mathbf{R}_\alpha - \mathbf{R}_\beta)} \rangle \tag{3.16}$$

This is the molecular scattering function, which is due to the contributions from all pairs of atoms in the molecule, and

$$S_d(\mathbf{Q}) = \sum_{\alpha, \beta} \langle f_\alpha^* f_\beta e^{-i\mathbf{Q} \cdot (\mathbf{R}_\alpha - \mathbf{R}_\beta)} \rangle \tag{3.17}$$

is the "distinct" part, coming from atoms in different molecules. Now consider atom α in molecule i, calling $\mathbf{R}_i$ the position of the center of mass, then

$$\mathbf{R}_\alpha = \mathbf{R}_i + \mathbf{r}_{\alpha_i}(\mathbf{\Omega}_i) \tag{3.18}$$

where $\mathbf{r}_{\alpha_i}(\mathbf{\Omega}_i)$ is the position of α relative to the center $\mathbf{R}_i$, and depends on the orientation of the molecule i, which is given by the Euler angles $\mathbf{\Omega}_i$, Rearranging the sums we get

$$S_m(\mathbf{Q}) = \sum_{i=1}^{N} \left\{ \sum_{\alpha_i \neq \beta_i} \langle f_{\alpha_i}^* f_{\beta_i} e^{i\mathbf{Q} \cdot [\mathbf{r}_{\alpha_i}(\mathbf{\Omega}_i) - \mathbf{r}_{\beta_i}(\mathbf{\Omega}_i)]} \rangle \right\} \tag{3.19}$$

and

$$S_d(\mathbf{Q}) = \sum_{ij}^{N} \left\langle \left[\sum_{\alpha_i} f_{\alpha_i} e^{i\mathbf{r}_{\alpha_i}(\mathbf{\Omega}_i) \cdot \mathbf{Q}} \right]^* \left[\sum_{\beta_j} f_{\beta_j} e^{i\mathbf{r}_{\beta_j}(\mathbf{\Omega}_j) \cdot \mathbf{Q}} \right] e^{-i\mathbf{Q} \cdot (\mathbf{R}_i - \mathbf{R}_j)} \right\rangle \tag{3.20}$$

Using now the Rayleigh expansion formula[5] we obtain

$$e^{i\mathbf{Q} \cdot \mathbf{R}} = \sum_{l=0}^{\infty} i^l(2l + 1)j_l(QR) \sum_{\lambda=-l}^{l} D_{0\lambda}^{l}(\hat{\mathbf{Q}})[D_{0\lambda}^{l}(\hat{\mathbf{R}})]^* \tag{3.21}$$

where $\hat{\mathbf{Q}} \equiv (\theta, \phi)$ are the polar angles of the vector $\mathbf{Q}$, and a similar definition applies to $\hat{\mathbf{R}}$; $j_l(x)$ is the spherical Bessel function of order l. With these equations, the fact that the molecular averages are independent, in our approximation, of the intermolecular averages, and also, the expansions of the second section, we arrive at

$$S_m(\mathbf{Q}) = N \sum_{m, \mu} (2m + 1)|a_\mu^m|^2 \tag{3.22}$$

$$S_d(\mathbf{Q}) = N\rho \sum_{\substack{mnl \\ \mu\nu}} (-)^\nu (a_\mu^m)^* a_\nu^n h_{\mu\nu}^{mnl}(Q) \begin{pmatrix} mnl \\ 000 \end{pmatrix} \tag{3.23}$$

where we have used a slightly different definition for the molecular form factors of Steele and Pecora[20]

$$a_\mu^m = i^m[2m + 1]^{-1/2} \sum_{\alpha_i} \int d\mathbf{r}_{\alpha_i} f_{\alpha_i} j_m(Qr_{\alpha_i}) D_{\mu 0}^m(\hat{\mathbf{r}}_{\alpha_i}) \tag{3.24}$$

and also the Fourier-Bessel transform of the correlation function coefficient

$$\hat{h}_{\mu\nu}^{mnl}(Q) = 4\pi i^{-l} \int_0^\infty dr \, r^2 j_l(Qr) \hat{h}_{\mu\nu}^{mnl}(r) \tag{3.25}$$

where $\hat{h}_{\mu\nu}^{mnl}(r)$ is the invariant coefficient (2.2) of the indirect correlation function

$$h(\mathbf{X}_1, \mathbf{X}_2) = g(\mathbf{X}_1, \mathbf{X}_2) - 1 \tag{3.26}$$

The "distinct" or intermolecular part $S_d(Q)$ (3.23) can be written in a more compact way using the irreducible expansion (2.7)

$$S_d(Q) = N\rho \sum_{\substack{mn \\ \mu\nu}} (-)^\nu (a_\mu^{\ m})^* a_\nu^{\ n} \mathcal{H}_{\mu\nu,\,0}^{mn}(Q) \tag{3.27}$$

where

$$\mathcal{H}_{\mu\nu,\,0}^{mn}(Q) = \sum_l \binom{mnl}{000} h_{\mu\nu}^{mnl}(Q) \tag{3.28}$$

Notice that, as anticipated in Section II, the scattering cross-section has contributions from the $\chi = 0$ irreducible representation. Furthermore, notice that the contributions to $S_d(Q)$ are, for small Q, or scattering angle θ

$$(a_\mu^{\ m})^*(a_\nu^{\ n}) \hat{h}_{\mu\nu}^{mnl}(Q) \sim O(QL)^{m+n+l} \tag{3.29}$$

where we have used the asymptotic property of the spherical Bessel functions

$$j_l(x) \sim O(x)^l \qquad \text{for } x \to 0 \tag{3.30}$$

and where L is the range of the correlations.

We expect that the higher l coefficients, which correspond to the sharper angular-dependent correlations, to be of rather short range (perhaps the molecular size). Therefore, we expect that in most cases (3.27) will be a reasonably fast converging series.

Summarizing, from (3.15), (3.22), and (3.27) we get, in the static approximation,

$$\frac{1}{N} S(\mathbf{Q}) = \sum_{\alpha_i} (\overline{f_{\alpha_i}^{\ 2}} - \overline{f_{\alpha_i}}^2) + \sum_{m,\,\mu} (2m + 1)|a_\mu^{\ m}|^2 + \rho \sum_{\substack{mn \\ \mu\nu}} (-)^\nu (a_\mu^{\ m})^* a_\nu^{\ n} \mathcal{H}_{\mu\nu,\,0}^{mn}(Q) \tag{3.31}$$

The aim of the diffraction experiments is to find the coefficients $\mathcal{H}_{\mu\nu,\,0}^{mn}(Q)$ as a function of $S(\mathbf{Q})$. In theory, we need an infinite number of experiments (X-rays, neutrons) with different molecular form factors $a_\mu^{\ m}$ (which we know from the molecular structure) to solve (3.31) for the $\mathcal{H}_{\mu\nu,\,0}^{mn}(Q)$. This can never be done, and therefore, either we are dealing with a fairly spherical molecule, or we have to make use of theoretical models to interpret the data.

Let us now turn to the actual evaluation of the molecular form factors $a_\mu^{\ m}$. Here we have to make a distinction between the X-rays, neutrons, and electrons.

For X-rays and, also, for electrons, a rather simple explicit formula based on the independent atom approximation, can be derived. The approximation is based on the assumption that valence bonds do not change the electron density distribution of the electrons of the inner shells of the atoms. For heavy elements, the valence bond electron density distortion is small, and the approximation should be very good. For light elements, such as hydrogen, lithium, beryllium, boron, and perhaps carbon, the valence bond distortion has been investigated carefully by McWeeney, by Tavard, and by Stewart,[21] in connection with high precision solid state X-ray crystallography where these effects can be important. For liquid state X-ray diffraction[22] the independent atom approximation fails badly only for H_2 and LiH.

The atomic X-ray form factors in the Born approximation (3.1) for most atoms and ions[23] have been fitted as sums of Gaussians:

$$f_\alpha^x(Q) = \sum_{p=1} A_{\alpha p} e^{-B_{\alpha p} Q^2} + C_\alpha \tag{3.32}$$

where $A_{\alpha p}$, $B_{\alpha p}$, and C_α are tabulated constants. Substituting into the defining relation (3.24), we get after a short calculation

$$(a_\mu^m)_{\text{rigid}} = \sum_{\alpha_i} f_{\alpha_i}^x(Q)(i)^m (2m+1)^{-1/2} j_m(r_{\alpha_i} Q) D_{\mu 0}^m(\hat{r}_{\alpha_i}) \tag{3.33}$$

For electron scattering, we have to use the atomic electron form factor (3.4)

$$f_\alpha^{el}(Q) = \frac{1}{Q^2} \left[Z_\alpha - \left(\sum_{p=1} A_{\alpha p} e^{-B_{\alpha p} Q^2} + C_\alpha \right) \right] \tag{3.34}$$

For neutrons, the independent atom approximation is exact in the absence of spin-spin nuclear correlations. These effects are usually ignored, but have been computed for some isolated cases. For methane gas,[24] at room temperature, the incoherent, inelastic cross-sections are modified by less than 0.7%.

We believe that these effects may be larger in molecular fluids containing N, O^{17}, or C^{13} than in fluids like $O^{16}D_2$ or $C^{12}D_4$ where the central atom is spinless. A detailed calculation of the possible magnitude still remains to be done.

In the independent atom approximation, the neutrons sample uniformly the isotropic composition and spin orientations of every element of the fluid. Calling [see (3.3)]

$$\bar{f}_\alpha^n = \overline{f_\alpha^n(\lambda_0, \lambda_s)} \tag{3.35}$$

the average scattering amplitude for element α (the bar over $\bar{f}$ indicates the average of isotopic composition and spin orientation) we will also get the molecular form factor from (3.33) in which the nuclear scattering amplitude $\bar{f}_{\alpha_i}^n$ replaces the X-ray amplitude $f_{\alpha_i}^x(Q)$.

Consider now the vibrating molecule: the atoms in every molecule are subject to vibrational motion, which causes them to "smear" around their equilibrium positions. Assuming a three-dimensional Gaussian distribution, the probability density of atom α around its equilibrium position is[25]

$$\rho(\mathbf{r}_\alpha) = \frac{1}{\sqrt{8\pi^3\sigma_x{}^2\sigma_y{}^2\sigma_z{}^2}} \exp\left\{-\frac{1}{2}\left[\frac{x^2}{\sigma_x{}^2} + \frac{y^2}{\sigma_y{}^2} + \frac{z^2}{\sigma_z{}^2}\right]\right\} \tag{3.36}$$

where x, y, and z are the Cartesian components of $\mathbf{r}_\alpha - \mathbf{r}_{n_\alpha}$, and $\mathbf{r}_{n_\alpha}$ is the "equilibrium" or rest position of nucleus α. The mean-square displacements are given by

$$\sigma_x{}^2 = \frac{1}{2}\sum_\lambda C_{\lambda x}{}^2\left(\frac{\hbar}{\omega_\lambda}\right)\operatorname{ctnh}\left(\frac{\hbar\omega_\lambda}{2kT}\right) \tag{3.37}$$

where the sum is over all normal modes λ of the molecule, ω_λ is the angular frequency of normal mode λ, and $C_{\lambda x}$ is a matrix element of matrix $\mathbf{C}$ that transforms normal coordinates into Cartesian coordinates.

Two approximations can be made, which are consistent with the precision of the experiments.

(a) In most cases, the energy necessary to excite the first vibrational level is much larger than the thermal energy kT, and therefore, $\hbar\omega/\omega T \gg 1$ and since the function ctnh $(x) \sim 1$ for large x, this factor can be ignored. The matrix elements can be obtained from infrared data, and there are excellent sources of reference for them.[26] Then

$$\sigma_x{}^2 = \frac{1}{2}\sum_\lambda\left(\frac{\hbar}{\omega_\lambda}\right)C_{\lambda x}{}^2 \tag{3.38}$$

(b) It will be sufficient to approximate the ellipsoid with axes $\sigma_x, \sigma_y, \sigma_z$, by the average sphere, with radius σ_α

$$\sigma_\alpha{}^2 = \tfrac{1}{3}(\sigma_x{}^2 + \sigma_y{}^2 + \sigma_z{}^2) \tag{3.39}$$

This approximation is based on the empirical finding that deviations from sphericity are second-order effects, even in crystals.

Therefore for an atom α, the density function is approximately

$$\rho(\mathbf{r}_\alpha) = \frac{1}{(2\pi\sigma_\alpha{}^2)^{3/2}} e^{-(\mathbf{r}_\alpha - \mathbf{r}_{n_\alpha})^2/\sigma_\alpha{}^2} \tag{3.40}$$

Inserting this result into the calculation that led, from (3.24) to (3.33), we find that the only effect of the vibrational effect is to introduce a factor

$$e^{-\sigma_\alpha{}^2 Q^2} \tag{3.41}$$

for each atom. The final expression for the form factors is then

$$a_\mu{}^m = \sum_{\alpha_i} e^{-\sigma_{\alpha_i}{}^2 Q^2} f_\alpha(i)^m (2m+1)^{-1/2} j_m(r_{\alpha_i} Q) D_{\mu\nu}{}^m(\hat{\mathbf{r}}_{\alpha_i}) \tag{3.42}$$

where f_α is the scattering amplitude given by either (3.32) for X-rays, (3.34) for electrons, or (3.35) for neutrons.

Substitution of these corrected form factors into (3.31) and taking into account the fact that for the self part the vibrational corrections do not have to be included we get the final expression for the scattering function $S(Q)$ in the static approximation:

$$\frac{1}{N} S(Q) = \sum_{\alpha_i} [\overline{f_{\alpha_i}{}^2} - \bar{f}_{\alpha_i}{}^2 e^{-\sigma_\alpha{}^2 Q^2}] + \sum_{m\mu} (2m+1)|a_\mu{}^m|^2$$

$$+ \rho \sum (-)^\nu (a_\mu{}^m)^* a_\nu{}^n \mathscr{H}^{mn}_{\mu\nu,0}(Q) \tag{3.43}$$

In this equation we have made a distinction between $\overline{f_\alpha{}^2}$ and $\bar{f}_\alpha{}^2$, which is only significant for neutron scattering. The averages $\overline{f_\alpha{}^2}$ and $\bar{f}_\alpha{}^2$ are defined by (3.35) and subsequent discussion.

The above equation goes to the prescribed asymptotic limits for $|Q| \to 0$ (free molecules limit) and $Q \to \infty$ ("free atom" limit). In the case of neutrons, it also displays the correct Q-dependence of the scattering cross-sections.

C. Dynamic Corrections for Neutron Scattering

As was mentioned in Section A, the measured elastic cross-section is given by (3.12), namely

$$\left(\frac{d\sigma}{d\Omega}\right)_\theta = \int_{-\infty}^{\omega_0} d\omega \, \varepsilon(k_s)(k_s/k_0) \mathscr{S}(\mathbf{Q}, \omega) \bigg|_\theta \tag{3.44}$$

while what we really want is $S(\mathbf{Q})$, defined by

$$S(\mathbf{Q}) = \int_{-\infty}^{\infty} d\omega \, \mathscr{S}(\mathbf{Q}, \omega) \bigg|_{\mathbf{Q}} \tag{3.45}$$

We recall that $\varepsilon(k_s)$ is the detector efficiency for the scattered radiation of wave vector k_s and $\hbar\omega_0 = E_0$ is the maximum incident neutron energy. Clearly also

$$E_0 = \frac{\hbar^2 k_0{}^2}{2m} \tag{3.46}$$

with m the neutron mass. Now from the energy conservation law

$$\omega = \omega_0 - \omega_s = \frac{1}{\hbar}(E_0 - E_s) = \frac{\hbar}{2m}(k_0{}^2 - k_s{}^2) \tag{3.47}$$

where E_s is the energy and ω_s the angular frequency of the scattered neutron. But the vector wave change is

$$Q_\theta^2 = (\mathbf{k}_0 - \mathbf{k}_s)^2 = k_0^2 + k_s^2 - 2k_0 k_s \cos\theta \qquad (3.48)$$

where θ is the scattering angle. Therefore, if we use (3.46) and (3.47), we get

$$Q_\theta^2 = k_0^2 \left[2 - \frac{\omega}{\omega_0} - 2\left(1 - \frac{\omega}{\omega_0}\right)^{1/2} \cos\theta \right] \qquad (3.49)$$

This expression tells us that the modulus of the vector exchange $\mathbf{Q}$ at constant angle is a function of the relative energy loss:

$$x = \frac{\omega}{\omega_0} = \frac{\Delta E}{E_0} \qquad (3.50)$$

In an elastic collision $x = 0$, and

$$Q_\theta = 2k_0 \sin\frac{\theta}{2} \qquad (3.51)$$

as required. Also we see that the inelastic Q_θ is always equal to or less than the elastic Q, as it should be.

Notice, furthermore, that the static approximation is equivalent to taking $x = 0$, since then

$$\left(\frac{d\sigma}{d\Omega}\right)_\theta = \varepsilon(k_0)S(\mathbf{Q}) \qquad (3.13)$$

The logical approach, taken by Plazcek,[27] is to expand the integrand in (3.44) in a series in ω (or equivalently in x), and hope that the resulting series converges rapidly. The conditions for this are that the neutron energy be much larger (or much smaller) than the excitation energy of the molecular states, and also, that the ratio of (m/M), the neutron mass to the atomic mass, be small.

After some lengthy algebra, we get (up to order Q^2)

$$\frac{1}{\varepsilon(k_0)}\left(\frac{d\sigma}{d\Omega}\right)_\theta = S(\mathbf{Q}) - \frac{1}{2\omega_0}\left[(1 + k_0\epsilon_1) + Q^2\frac{\partial}{\partial Q^2}\right]\langle\omega\rangle$$

$$- \frac{1}{8\omega_0^2}\left[(1 - k_0\epsilon_1 - k_0^2\epsilon_2) - (Q^2 + 2k_0^2 + 2Q^2k_0\epsilon_1)\frac{\partial}{\partial Q^2}\right.$$

$$\left. - Q^4\left(\frac{\partial}{\partial Q^2}\right)^2\right]\langle\omega^2\rangle + \cdots \qquad (3.52)$$

where we have used the following definitions:

$$\varepsilon_n = \frac{1}{\varepsilon(k_0)} \left(\frac{\partial}{\partial k} \right)^n \varepsilon(k) \Bigg|_{k=k_0} \tag{3.53}$$

$$\langle \omega^n \rangle = \int_{-\infty}^{\infty} d\omega \, \omega^n \mathscr{S}(\mathbf{Q}, \omega) \tag{3.54}$$

The derivatives ε_n of the detector efficiency function can be found for each case rather easily, and depend explicitly on the particular type of detector $(1/k, 1 - e^{ak_0/k}, \text{etc.})$.

The moments $\langle \omega^n \rangle$ however, must be evaluated for every system individually. At first sight, it would seem that we need to have some knowledge of $\mathscr{S}(\mathbf{Q}, \omega)$. Fortunately, as has been shown by Plazcek,[27] De Gennes,[28] and others, the first few moments are due mostly to recoil of individual uncorrelated molecules, and can be computed without knowing $\mathscr{S}(\mathbf{Q}, \omega)$ in detail. Evaluation of the first two moments is outlined in the Appendix.

The computed final expression for the dynamic correction could be written as the sum of two terms:

$$\frac{1}{N} \left[\frac{1}{\varepsilon(k_0)} \left(\frac{d\sigma}{\partial\Omega} \right) - S(\underline{Q}) \right] = C_s(Q) + C_d(Q) \tag{3.55}$$

The first term $C_s(Q)$ is due to incoherent or self-scattering terms, and does not show features that depend on the molecular structure. It is of parabolic shape

$$C_s(Q) = C_0 + Q^2 C_2 \tag{3.56}$$

The other contribution $C_d(Q)$ is modulated by the molecular structure. The shape of the function is approximately sinusoidal, and it involves spherical Bessel functions of the zeroth and second order. The period of the oscillations is determined by the internuclear distances in the molecule. We would expect that this part of the corrections should have a significant effect on the structure functions inferred from neutron diffraction experiments.

IV. SCATTERING EXPERIMENTS

The diffraction pattern of a macroscopically isotropic liquid is symmetric about the incident beam, and the accessible range of scattering angles is $0° < \theta < 180°$. The aim of diffraction experiments is to obtain the scattering function $S(Q)$, defined in Section III, over a wide range of the momentum transfer coordinate:

$$Q = \frac{4\pi}{\lambda_0} \sin \frac{\theta}{2} \tag{4.1}$$

The accessible range of momentum transfer $0 < Q < (4\pi/\lambda_0)$ is limited only by the wavelength λ_0 of the incident radiation. Measurements on molecular liquids have been reported over a range $0.04 \text{ Å}^{-1} \lesssim Q \lesssim 36 \text{ Å}^{-1}$, the lower value for small-angle X-ray scattering[32] and the higher one for neutron scattering using a linear electron accelerator.[33] Conventional X-ray and neutron diffraction experiments are usually restricted to values of $Q \gtrsim 15 \text{ Å}^{-1}$. We will show that even this limited range of momentum transfer is adequate for the study of positional and orientational correlations in molecular liquids.

A. Conventional Diffraction Experiments

In conventional diffraction experiments a continuous source of radiation is used.[34] Commonly used sources are standard X-ray tubes and, in the case of neutrons, nuclear reactors. Two types of scattering geometries may be distinguished. In transmission mode, a parallel incident beam penetrates the sample and the scattered radiation leaves, ideally, without excessive absorption. This geometry is widely used in neutron scattering experiments[35] and, less frequently, in X-ray experiments with liquids containing light nuclides.[36] In reflection mode, the incident beam is scattered from the surface of the sample. This geometry, with a free horizontal sample surface, is widely used in X-ray scattering experiments[37] because it permits the use of a divergent beam and eliminates absorption and scattering from the sample container. In neutron and, less frequently, X-ray scattering experiments monochromatization is achieved by Bragg reflection from a crystal mounted in the incident beam. In most X-ray scattering experiments the mono-chromator crystal is mounted in the scattered beam because this arrangement largely eliminates the Compton scattering at large scattering angles.

In all conventional diffraction experiments a detector is used to count the scattered radiation at a preset angle θ. The theoretical expressions presented in Section III are based on the assumption that all regions of the fluid are exposed to the same incident radiation, that all radiation scattered from a given point at a given angle reaches the detector, and that the scattering process involves only a single event. For practical application of these expressions the measured intensity of scattered radiation must be corrected for deviations from the above conditions.

B. Data Reduction

1. Experimental Corrections

The measured intensities must be corrected for background radiation, for attenuation of both the incident and scattered radiation within the sample and also within the walls of any sample container, and for multiple scattering. All these corrections are highly dependent on the scattering geometry and they are the same for molecular as for monatomic liquids. For a detailed discussion of these corrections the reader is referred to recent review articles.[35,38] In the case of X-ray scattering, further corrections must be made for polarization and for Compton scattering, the latter being again geometry dependent.[36,37]

The absolute scale can be established[35,38] by normalizing the corrected intensities to the theoretical scattering from independent atoms at large values of the variable Q. For molecular liquids the oscillations of the scattering cross-sections about their asymptotic values are of sizeable amplitude even at the largest Q values accessible in scattering experiments. In practice, this presents no serious problem if the difference between theoretical and measured cross-sections is minimized by least squares, and if care is taken to assure that the resulting scale factor is independent of Q.

In the case of neutron scattering, the absolute scale is usually established by measuring and analyzing the scattering from a vanadium slab having the same dimensions as the sample.[35] Although this method also relies on the tabulated cross-sections of vanadium, it presents a distinct advantage of the neutron scattering technique.

2. Dynamic Corrections

The relationship between the measured cross-section and the scattering function $S(Q)$ is straightforward only in the static approximation (3.13), which assumes that the scattering particles are rigidly bound, so that all exchange of energy between radiation and sample can be neglected. In a liquid the particles may be considered as "bound" only if their mass M is very large compared to the mass m of the scattered photons or neutrons, and if the energy E_0 of incident radiation is much larger or smaller than the energy transfer in scattering processes. If these conditions are not met, departures from the static approximation must be considered, and lead to corrections of the order m/M to the cross-sections obtained from the scattering experiment. These dynamic corrections have been discussed in Section III.

For neutrons ($\lambda_0 = 1.1$ Å, $E_0 = 0.07$ eV) the incident energy is comparable to the energy transfer in molecular librations, and the dynamic corrections are significant. For X-rays ($\lambda_0 = 0.71$ Å, $E_0 = 1.7 \times 10^4$ eV) the dynamic corrections are negligible, and this is a distinct advantage of the X-ray scattering technique.

3. Refinement of the Scattering Function

The function $S(Q)$ oscillates around the asymptotic value $S_s(Q)$, defined in (3.15), with frequencies determined by the mean separation between pairs of atoms and molecules in the liquid. The scattering functions derived from diffraction data always contain additional error components, which can be eliminated if their frequency is sufficiently different from that due to the liquid structure.

Random errors arising predominantly from counting statistics can be minimized by choosing an appropriate[39] counting scheme for the scattering experiment. Remaining "noise" can be eliminated by standard smoothing procedures.[40]

Systematic errors in the function $S(Q)$ arise predominantly from imperfections in the correction procedures described in the preceding paragraphs. The main result of systematic errors is a smoothly varying, low-frequency "background." Consequently, a tolerable separation of this background can be accomplished by standard techniques used for many years in the reduction of gas diffraction data.[15]

It is almost always possible, through the use of these refining techniques, to obtain scattering functions which meet the available criteria[41,42] for the overall precision of diffraction data from liquids.

C. Data Analysis

The static scattering function, accessible from diffraction experiments, may be written as the sum of three terms (3.14), repeated here for convenience:

$$S(Q) = S_s(Q) + S_m(Q) + S_d(Q) \tag{4.2}$$

Only the distinct part $S_d(Q)$ of the scattering function contains information about the coefficients $\mathscr{H}^{mn}_{\mu\nu,\,0}(Q)$, defined in Section III, which are descriptive of the liquid structure. In order to obtain the function $S_d(Q)$, we must subtract from the measured $S(Q)$ the self term $S_s(Q)$ and the molecular term $S_m(Q)$. We describe in the following paragraphs a method of analysis that can always be applied to diffraction data from molecular liquids, and we will illustrate this procedure with neutron and X-ray diffraction results on carbon tetrachloride.[43]

1. Total Structure Function

The function $S(Q)$ for CCl_4, derived from X-ray diffraction, is shown in Fig. 1, together with the theoretical scattering from independent atoms, $S_s(Q)$. The corresponding curves for the case of neutron diffraction are shown in Fig. 2. The X-ray scattering curves attenuate very rapidly with increasing Q, whereas the neutron scattering function oscillates around the constant value of $S_s(Q)$ characteristic of point scatters. The first step in the data analysis is to construct a total structure function:

$$H(Q) \equiv [S(Q) - S_s(Q)]M(Q) \tag{4.3}$$

The factor $M(Q)$, arbitrary in principle, is chosen to change the scale to that characteristic of one molecule and, in the case of X-ray and electron scattering,

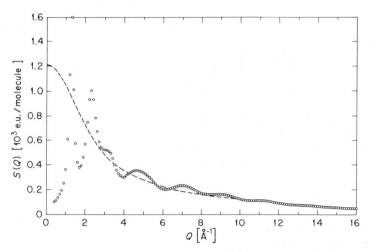

Fig. 1. Total scattering function, $S(Q)$, for liquid CCl_4 at 25°C from X-ray diffraction (points), scaled to the computed "self" scattering, $S_s(Q)$, from independent atoms in one CCl_4 molecule (dashed curve).

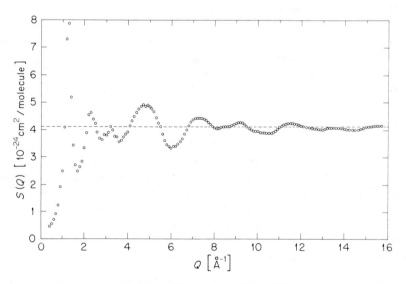

Fig. 2. Total scattering function, $S(Q)$, for liquid CCl_4 at 25°C from neutron diffraction (points), scaled to the computed "self" scattering, $S_s(Q)$, from independent atoms in one CCl_4 molecule (dashed curve).

to remove some of the breadth arising from the electron distribution. A convenient choice is

$$M(Q) = \left[\sum_\alpha | f_\alpha | \right]^{-2} \qquad (4.4)$$

with f_α the coherent scattering factor of atom α and summation over the atoms of one molecule. Alternate choices of the factor $M(Q)$ will be discussed below.

The total structure function may be written as

$$H(Q) = H_m(Q) + H_d(Q) \qquad (4.5)$$

with $H_m(Q) = S_m(Q)M(Q)$ and $H_d(Q) = S_d(Q)M(Q)$. The next step in the data analysis is the determination of the molecular structure function, namely (3.16) that corresponds to

$$H_m(Q) = M(Q) \sum_\alpha \sum_{\beta \neq \alpha} f_\alpha f_\beta \, e^{-b_{\alpha\beta}Q^2} j_0(QR_{\alpha\beta}) \qquad (4.6)$$

Equation (4.6) is equivalent to (3.22) in the SFA-approximation adopted throughout this discussion. This approximation, discussed in Section III, assumes that the molecular scattering factors can be computed, in good approximation, from the tabulated[23] scattering factors for spherical free atoms (SFA). To determine the mean interatomic distances $R_{\alpha\beta}$ and their mean-square variations $2b_{\alpha\beta}$ needed for the calculation of the molecular scattering factors (3.42), standard procedures for the analysis of gas diffraction data[15] may be used. An important assumption, verified by the available experimental data on liquids, is that the function $H_d(Q)$ decays to zero value much faster than the function $H_m(Q)$, so that $H(Q) \approx H_m(Q)$ at large values of Q. The parameters $b_{\alpha\beta}$ and $R_{\alpha\beta}$ descriptive of the molecular structure are then obtained by least-squares refinement of (4.6) against the high Q part of the structure function $H(Q)$ derived from experiment. Results for CCl_4 are shown in Fig. 3.

The computed curves $QH_m(Q)$ in Fig. 3 are in close agreement with the measured functions $QH(Q)$ beyond values of $Q \gtrsim 6$ Å^{-1}. The values of $b_{\alpha\beta}$ and $R_{\alpha\beta}$ derived from X-ray data[43] for liquid CCl_4 are, within experimental error, identical with values derived from gas diffraction data.[26]

We conclude that the molecular structure in liquids can be determined from X-ray diffraction data, and that the range of momentum transfer accessible in conventional diffraction experiments is adequate for this purpose. It is not clear whether the molecular structure can be determined from neutron data, because this information is needed to compute the dynamic corrections discussed in Section III.

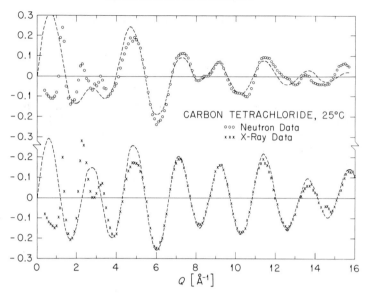

Fig. 3. Q-weighted total structure functions $QH(Q)$ derived from the data shown in Figs. 1 and 2 (points) and computed molecular structure functions $QH_m(Q)$ (dashed curves).

2. Distinct Structure Function

With the molecular structure thus determined, the function $H_m(Q)$ can be subtracted from the total structure function $H(Q)$ to yield the distinct structure function

$$H_d(Q) = S_d(Q)M(Q) \tag{4.7}$$

Results for our example, CCl_4, are shown in Fig. 4 and will be discussed in Section V.

The choice of the factor $M(Q)$ in (4.7) merits some further discussion. We recall that the distinct scattering function is related to the coefficients $\mathscr{H}^{mn}_{\mu\nu,0}(Q)$ by (3.27), which may be rewritten as

$$S_d(Q) = N\rho\left\{|a_0{}^0(Q)|^2 h_c(Q) + \sum_{\substack{mn \\ \mu\nu}} (-)^\nu (a_\mu{}^m)^* a_\nu{}^n \mathscr{H}^{mn}_{\mu\nu,0}(Q)\right\} \tag{4.8}$$

In (4.8) the leading term $\mathscr{H}^{00}_{00,0}(Q) = h^{000}_{00}(Q)$ in the expansion (3.27) has been designated $h_c(Q)$ because this coefficient is often descriptive of molecular centers.

For spherical molecules only the leading term in (4.8) needs to be retained, and the obvious choice is

$$M(Q) = |a_0{}^0(Q)|^{-2} \tag{4.9}$$

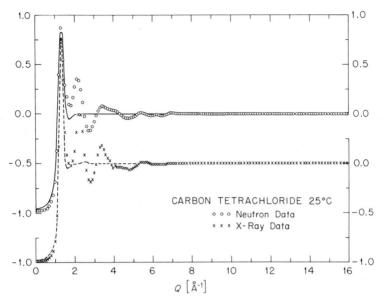

Fig. 4.   Distinct structure functions, $S_d(Q)$, derived from the curves shown in Fig. 3 (points). The dashed curves represent the spherical part, $|a_0^0|^2 h_c(Q)M(Q)$, with $h_c(Q)$ computed for a one-center hard sphere model of CCl_4, and $M(Q)$ from (4.10). (Reproduced from Ref. 47).

For nonspherical molecules the factor $a_0^0(Q)$ may assume zero value (3.42), and the use of (4.9) in (4.7) can lead to singularities in the function $H_d(Q)$. For nearly spherical molecules the spherical term $|a_0^0(Q)|^2$ in the scattering factor expansion does not differ greatly from the sum $\sum_{m\mu} |a_\mu^m(Q)|^2$ which is always nonzero. A useful choice is then

$$M(Q) = \left\{ \sum_{m\mu} |a_\mu^m(Q)|^2 \right\}^{-1} \qquad (4.10)$$

In general, the obvious choice of the factor $M(Q)$ is the molecular analog of (4.4), namely

$$M(Q) = \left\{ \sum_{m\mu} |a_\mu^m(Q)| \right\}^{-2} \qquad (4.11)$$

because this make the products $(a_\mu^m)^* a_\nu^n M(Q)$ nearly independent of the variable Q and hence assures minimum broadening and distortion of the correlation function obtained by Fourier inversion of the function $H_d(Q)$.

3. Fourier Inversion

It is always possible, and often useful, to construct a correlation function, namely

$$G(R) \equiv 1 + (2\pi^2 \rho R)^{-1} \int_0^{Q_{max}} QH_d(Q) \sin(QR) \, dR \qquad (4.12)$$

The function $G(R)$, accessible from a single diffraction experiment, is a combination of the coefficients $g_{\mu v, 0}^{mn}(R)$ which are convoluted with known functions[37] of the molecular scattering factors. The coefficients $g_{\mu v, 0}^{mn}(R)$ themselves cannot in general be extracted from the function $G(R)$.

V. RECENT STUDIES

Molecular liquids were among the first to which the method of X-ray diffraction was applied, and a number of systems have been studies over the years.[44] Neutron diffraction studies have been reported since 1953,[45] and the electron diffraction method has been successfully applied to only one molecular fluid,[46] namely water. The experimental data reported up to about 1960 must be regarded as crude by present-day standards, systematic errors of 10% and more being the rule rather than the exception. In the analysis of these data it was generally assumed that the scattering molecules have spherical symmetry; explicit treatments of nonspherical molecules were either qualitative in nature or based on specific models.

It has now become clear[47] that diffraction data of high precision are necessary to obtain information on the positional and orientational correlations in molecular liquids. We therefore restrict the following discussion to relatively recent studies involving diffraction data with an overall accuracy of 5% or better. Also, we shall consider only liquids that can be, or have been, interpreted in terms of models not specific to a particular system, but of a more general validity. These two criteria limit the following discussion to liquids with linear and with nearly spherical globular molecules.

A. Linear Molecules

Fluids with linear molecules have been attractive candidates for both theoretical and experimental studies. Molecular symmetry restricts the allowed values for the index pairs (m, μ), (n, v) in the expansions for the cross-sections and correlation functions, and this reduces the number of terms and simplifies their geometrical interpretation. For homonuclear molecules, the expansion coefficients for the molecular scattering factors for neutrons and X-rays (in the FSA approximation) differ only by a factor $f^X(Q)/f^N$, and this makes possible the direct comparison of experimental data obtained by the two methods.

Sandler, Das Gupta, and Steele[48] have reported extensive calculations for models of fluids with homonuclear diatomic molecules. The molecules are approximated by rigidly connected, overlapping (fused) spheres of dimensions similar to those of N_2, O_2, and the halogens. The fused spheres are interacting with two-center repulsive forces, intermolecular attractions being completely neglected. The model[49,50] permits the construction of scattering functions which can be compared with results from diffraction experiments. Neutron scattering functions for the fused sphere model of liquid nitrogen are compared with the experimental results of Henshaw, Hurst, and Pope[52] in Fig. 5. The agreement appears good, but it must be pointed out that the neutron data shown in Fig. 5 are in very poor agreement with results from various X-ray diffraction studies,[53] which in turn do not agree among themselves. In a similar manner, the test of the fused sphere model for liquid O_2, Cl_2, and Br_2 is not conclusive because of the low quality of the available diffraction data. Das Gupta, Sandler, and Steele[51] conclude that the large discrepancies between the experimental scattering functions of liquid Cl_2 and Br_2 and those computed for the model are probably due to the neglect by the model of attractive forces from quadrupolar interactions, which must be significant in the liquid halogens.

The only recent study of triatomic molecular liquids is that of Suzuki and Egelstaff[54] on CS_2 and CSe_2 using neutron diffraction. The structure

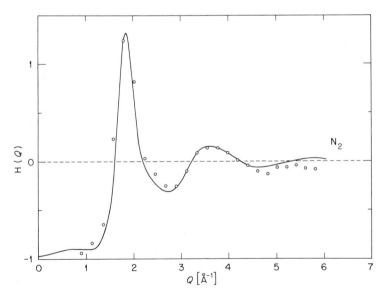

Fig. 5. Total structure function, $H(Q)$, derived from neutron diffraction data of Ref. 52 (points), and computed for fused sphere model of Ref. 51 (solid line).

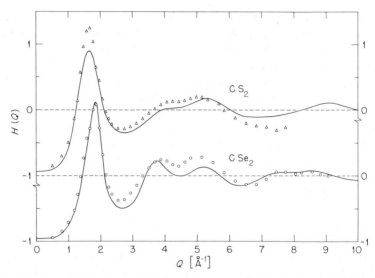

Fig. 6. Total structure functions, $H(Q)$, derived from neutron diffraction data of Ref. 54 (points), and computed for fused sphere models of Ref. 55 (solid lines).

functions, taken from Table 2 of their paper, are shown in Fig. 6. According to the authors[54] the curves may be in error by as much as $\pm 10\%$, probably because departures from the static approximation were not considered in the reduction of the data. Suzuki and Egelstaff interpret their data in terms of simple models, all of which are based on the assumption that the orientational correlations are independent of the radial correlations. This assumption may be reasonable for liquids with nearly spherical molecules, but it fails badly[48] for linear molecules.

The neutron data of Fig. 6 have also been analyzed by Lowden and Chandler[55] in terms of a fused sphere model with only repulsive interactions. Structure functions calculated for this model are compared with the curves from neutron diffraction in Fig. 6. In view of the quality of the neutron data, Lowden and Chandler consider only the discrepancies in the region of the first maximum of $H(Q)$ significant. They ascribe this discrepancy to the neglect of attractive forces by the model.

B. Globular Molecules

In liquids with globular molecules of the AB_4-type the positional correlation between molecular centers is much more pronounced than the orientational correlations between pairs of molecules. Even for molecules interacting with long-range and strongly directional forces, for example, H_2O, the orientational correlations are of rather short range. Hence, we expect in most cases the expansion (3.27) for the distinct scattering function to be reasonably fast converging, and the spherical term to be much larger than the nonspherical contributions. There is then hope that the structure function $h_c(Q)$ descriptive of molecular centers (4.8) and perhaps one or two of the nonspherical coefficients $\mathscr{H}_{\mu\nu,\,0}^{mn}(Q)$ can be determined from diffraction data.

Among the tetrahalides of group IV elements, liquid CCl_4 has been studied with both X-rays and neutrons. The X-ray data of Narten, Danford, and Levy[56] (NDL) are in good agreement with results from more recent X-ray studies.[43,57] We note that NDL were the first to report a quantitative determination of the molecular structure function $H_m(Q)$. The authors then analyzed their data in terms of a model based on a perturbed crystal structure. This model, though specific of CCl_4, has been confirmed as essentially correct by later investigators.[55,57,58] Liquid CCl_4 has been studied with thermal neutrons by Rao,[59] by Egelstaff, Page, and Powles,[60] and by Narten.[43,47] Agreement among the three data sets is not good, which is not surprising because Rao's data were not corrected for multiple scattering, and neither Rao nor Egelstaff et al., considered departures from the static approximation. Egelstaff, Page, and Powles analyzed their CCl_4 data (in addition to $GeBr_4$) in terms of the models mentioned in connection with our discussion of linear molecules. The molecular center structure functions and the arrangement of nearest neighbors in CCl_4 proposed by Egelstaff et al., disagree with the analysis of other workers.[43,47,55]

The neutron data of Egelstaff et al. have also been analyzed by Gubbins, Gray, and Egelstaff[61] and by Lowden and Chandler.[55] Calculations by Gubbins et al. are based on a perturbation treatment, using as a reference a system of molecules with isotropic forces. The authors conclude that for CCl_4 attractive forces from octopolar interactions must be considered in a description of the structure function in the region $2 \text{ Å}^{-1} \gtrsim Q \gtrsim 3.8 \text{ Å}^{-1}$. Their conclusions agree with an analysis by Narten,[47] to be discussed below, but are disputed by Lowden and Chandler.[55] The latter authors[55] compare structure functions computed for a fused sphere model of CCl_4 and multicentered, nonspherical repulsive forces with the neutron results of Egelstaff et al. The agreement is of similar quality to that shown in Fig. 6 for the case of CS_2 and CSe_2, and Lowden and Chandler argue that the quality of the Egelstaff data does not justify a more elaborate model.

The analysis and interpretation of diffraction data for linear and tetrahedral molecules described so far has been based on the total structure function. We have argued that the molecular part $H_m(Q)$ should be subtracted from the total structure function $H(Q)$ to yield the distinct part, $H_d(Q)$, and we have shown that this can be done in practice. The functions $H_d(Q)$ for CCl_4, shown in Fig. 4, deviate from zero value only in the region $Q \gtrsim 6 \text{ Å}^{-1}$. Narten[47] has shown that the functions $H_d(Q)$ below $Q \gtrsim 2 \text{ Å}^{-1}$

are almost completely determined by the spherical part, $|a_0{}^0|^2 h_c(Q)M(Q)$, of the expansion (4.8), and that the very crude approximation for $h_c(Q)$ obtained from a one-center hard-sphere model is adequate to describe this region of the functions $H_d(Q)$ for CCl_4. He has further shown that the coefficients $a_0{}^0[a_0{}^3]^*$ and $|a_0{}^3|^2$, the latter descriptive of octopolar interactions, are most pronounced in the region $2 \text{ Å}^{-1} \gtrsim Q \gtrsim 3.8 \text{ Å}^{-1}$. Hence we speculate that the distinct structure function $H_d(Q)$ for liquid CCl_4 can be approximated by the first three nonzero terms in the expansion (4.8).

We conclude our discussion of CCl_4 with the observation that the function $H_d(Q)$ below $Q \sim 2 \text{ Å}^{-1}$ is described equally well by the one-center hard-sphere model[47] and by the multicenter fused-sphere model of Lowden and Chandler[55] Molecular center correlation functions, $g_c(R)$, for the two models are compared in Fig. 7. Also shown are results obtained for a more realistic model of CCl_4 tested by Fowler[62] in a computer experiment using the Monte Carlo method. Both models agree poorly with the Monte Carlo results in the region of the first peak in $g_c(R)$. However, the near-neighbor distance region in $g_c(R)$ affects the distinct structure function $H_d(Q)$ predominantly at values $Q > 2 \text{ Å}^{-1}$ where, for CCl_4, the scattering coefficients $|a_0{}^0|^2$ in (4.8) decay rapidly to rather small values (see Fig. 1 of Ref. 47). Hence, the diffraction

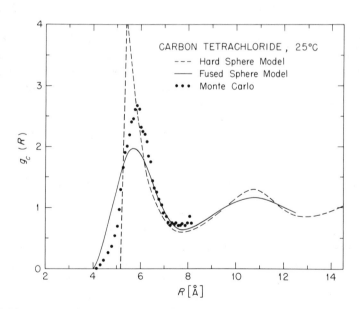

Fig. 7. Correlation functions for molecular centers, $g_c(R)$, computed for a one-center hard sphere model (Ref. 47), a multi-center fused sphere model (Ref. 55), and a computer simulation model (Ref. 61).

data predominantly sample the longer range tail of the molecular center correlation (and potential) function for liquid CCl_4.

A very interesting and special class consists of liquids with molecules such as H_2O, NH_3, HF, which have a very nearly spherical distribution of electron density.[22] For these molecules the nonspherical terms in the expansion (3.42) for the X-ray and electron (but not neutron) scattering factors contribute only about 1 % to the total sum. Hence, only the spherical term needs to be considered and structure functions for molecular centers can be obtained from X-ray and electron diffraction data.

Liquid water (H_2O) has been studied with X-rays by Narten and Levy,[62] and with high-energy electrons by Kálmán, Lengyel, Pálinkás, and Haklik.[46] Structure and correlation functions for H_2O at 4°C obtained by the two methods agree with each other within the experimental errors, which are as yet somewhat uncertain for the very difficult electron diffraction experiment.[46] Molecular center structure functions, $h_c(Q)$, for H_2O at 25°C, taken from Table II of Ref. 63, are shown in Fig. 8. Liquid water (D_2O) has been studies with neutrons by Page and Powles[64] and by Narten.[65] The two data sets are in fair agreement (see Fig. 1 of Ref. 19), and we choose to discuss only the data of Narten because of their consistency with the X-ray data and the wider range of momentum transfer covered in the experiment. The distinct

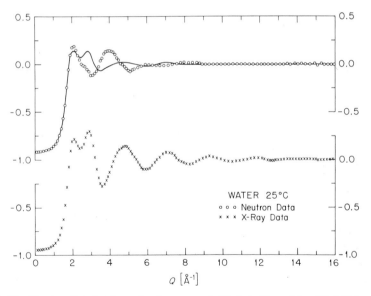

Fig. 8. Structure function for molecular centers, $h_c(Q)$, from X-ray data of Ref. 62 (× × ×). Distinct structure function, $S_d(Q)$, from neutron data of Ref. 64 (circles). Solid line represents the spherical part, $|a_0^0|^2 h_c(Q)M(Q)$, with $M(Q)$ from (4.10). (Reproduced from Ref. 47).

structure function, $H_d(Q)$, for D_2O is shown in Fig. 8 (circles). Also shown (solid line) is the computed spherical part, $|a_0{}^0|^2 h_c(Q)M(Q)$, using the molecular center structure function (xxx) derived from the X-ray data, and the factor $M(Q)$ defined in (4.10). As was found for CCl_4 (Fig. 4), the distinct neutron structure function for D_2O is completely determined by its spherical part below $Q \sim 2 \text{ Å}^{-1}$. However, in the case of water the expansion for the neutron scattering coefficients, and hence for $S_d(Q)$, converges slowly and extraction of the nonspherical coefficients $\mathcal{H}_{\mu\nu,0}^{mn}(Q)$ from the available diffraction data alone appears hopeless. The diffraction data for water have not yet been successfully analyzed in terms of realistic theoretical models. They have, however, been used to test intermolecular potential functions for H_2O molecules in computer experiments by Rahman and Stillinger[66] and by Popkie, Kistenmacher, and Clementi.[67]

VI. SUMMARY AND CONCLUSIONS

The determination of all structure function coefficients $h_{\mu\nu}^{mnl}(Q)$, and hence all correlation function coefficients $h_{\mu\nu}^{mnl}(R)$ necessary to construct the molecular pair correlation function $g(\mathbf{X}_1, \mathbf{X}_2)$, is not possible from diffraction data alone. Only in special cases, for example CCl_4, is there hope to obtain a small number of the coefficients $\mathcal{H}_{\mu\nu,0}^{mn}(Q)$ from scattering experiments. In other special cases (H_2O, NH_3, HF, CH_4, H_2S) the extraction of structure and correlation functions for molecular centers from a single X-ray or electron diffraction experiment is feasible. In general we have to make use of theoretical considerations to interpret the data.

It has now become clear that diffraction data of high accuracy are necessary to obtain information about correlations in molecular liquids or to test theoretical models. An overall precision of $\sim 1\%$ can be achieved in present-day experiments, and this is barely adequate for the purpose. We believe that, at this level of accuracy, departures from the SFA-approximation, convenient for the computation of molecular X-ray scattering factors from atomic form factors, need only be considered for a few molecules with light nuclides. For a better understanding of the structure of liquids, it would also be useful to perform diffraction experiments at high pressures in order to disentangle the different effects of temperature and density.

On the theoretical side, realistic and computationally convenient models for the prediction of structure and correlation functions from intermolecular potentials have been developed. The fused sphere or, more precisely, reduced interaction site model (RISM) of Chandler has been shown to predict[55] the structure of a variety of liquids in satisfactory agreement with diffraction data. Another model, the mean spherical approximation (MSA), has been successful in the prediction[68] of the structure of atomic liquids, and the expansions for the properties of molecular liquids in their irreducible representations appear tractable in the MSA.[69] The outlook for these procedures as a viable theory for liquids such as water has brightened considerably with the results of Verlet and Weiss,[70] who showed that rather simple corrections to the MSA give good agreement with computer experiments in the case of dipolar liquids.

Another powerful, if expensive, method of testing in intermolecular potential functions against diffraction data is computer simulation by the Monte Carlo or molecular dynamics method. In practice, theoretical models will most likely be used to establish the force law. The computer experiments can then be used to predict not only the structure but many other thermodynamic and transport properties of molecular fluids.

APPENDIX. EVALUATION OF ENERGY MOMENTS FOR MOLECULAR SYSTEMS

We will follow Placzek's original derivation in our evaluation of the first and second moments, $\langle \omega \rangle$ and $\langle \omega^2 \rangle$. Inverting the Fourier transform (3.6), we get

$$\Pi(\mathbf{Q}, t) = \int_{-\infty}^{\infty} dt \, e^{i\omega t} \mathscr{S}(\mathbf{Q}, \omega)$$

$$= \sum_{\alpha, \beta} \langle f_\alpha^* f_\beta e^{-i\mathbf{Q}\cdot\mathbf{R}_\alpha(0)} e^{i\mathbf{Q}\cdot\mathbf{R}_\beta(t)} \rangle \qquad (A.1)$$

Then, using a known property of the Fourier transforms,[25]

$$\langle \omega^n \rangle = (-i)^n \frac{d^n}{dt^n} \Pi(\mathbf{Q}, t) \qquad (A.2)$$

Since thermal neutrons are quantum particles, we need to calculate the quantum mechanical ensemble average of (3.55), rather than the classical average.

The first moment is

$$\langle \omega \rangle = -i \frac{\partial}{\partial t} \sum_{\alpha, \beta} \langle f_\alpha^* f_\beta \exp[-i\mathbf{Q} \cdot \mathbf{R}_\alpha(0)] \exp[i\mathbf{Q} \cdot \mathbf{R}_\beta(t)] \rangle_Q \Big|_{t=0}$$

$$= -i \frac{\partial}{\partial t} \sum_{\alpha, \beta} f_\alpha^* f_\beta \, \mathrm{Tr} \exp -\left[\frac{\hat{H}}{kT}\right] \exp[-i\mathbf{Q} \cdot \mathbf{R}_\alpha(0)] \exp[i\mathbf{Q} \cdot \mathbf{R}_\beta(t)] \Big|_{t=0}$$

$$(A.3)$$

where $\hat{H}$ is the quantum mechanical Hamiltonian operator and the trace is taken in an appropriate, time-invariant basis.[13] Then, we make use of the property of time-dependent operators

$$\frac{\partial \hat{O}}{\partial t} = \frac{1}{i\hbar} [\hat{O}, \hat{H}] \qquad (A.4)$$

where the symbol $[A, B]$ is the commutator of the operators A and B.

Since the Hamiltonian is

$$\hat{H} = -\frac{\hbar^2}{2} \sum \frac{1}{M_\alpha} \nabla_\alpha{}^2 + U(\mathbf{R}_1 \ldots \mathbf{R}_N) \tag{A.5}$$

a short calculation yields

$$\langle \omega \rangle = \sum_{\alpha, \beta} \left\langle f_\alpha^* f_\beta e^{-\mathbf{Q} \cdot (\mathbf{R}_\alpha - \mathbf{R}_\beta)} \left[\frac{\hbar Q^2}{2M_\alpha} - i\frac{\hbar}{M_\alpha} \mathbf{Q} \cdot \mathbf{V}_\alpha \right] \right\rangle \tag{A.6}$$

But, from the hermiticity of $\Pi(\mathbf{Q}, t)$, we find

$$\langle e^{i\mathbf{Q} \cdot (\mathbf{R}_\alpha - \mathbf{R}_\beta)} (\mathbf{Q} \cdot \mathbf{V}_\alpha) \rangle = -\frac{\hbar Q}{M_\alpha} \langle e^{-i\mathbf{Q} \cdot (\mathbf{R}_\alpha - \mathbf{R}_\beta)} \rangle \quad \text{For } \alpha \neq \beta \tag{A.7}$$

$$= \langle \mathbf{Q} \cdot \mathbf{v}_\alpha \rangle = 0 \quad \text{For } \alpha = \beta$$

Here we have made use of the fact that in the Schrödinger representation,

$$-\frac{i\hbar}{M_\alpha} \nabla_\alpha$$

is the operator for $\mathbf{v}_\alpha$, and also that the average speed in any direction $(\mathbf{Q})$ must be zero, for any atom, no matter what the interactions are. This is a consequence of equilibrium: in an equilibrium system the atoms do not drift in any direction whatsoever, and therefore the average velocity must be zero. Substituting (A.7) into (A.6) we get at once

$$\langle \omega \rangle = \frac{N\hbar Q^2}{2} \left[\sum_{\alpha=1}^{\nu} \frac{\overline{f_\alpha^2}}{M_\alpha} \right] \tag{A.8}$$

where ν is the number of atoms in the molecule. Notice that the average of the square scattering amplitude is taken, rather than the square of the average. This is an important distinction for neutrons. Also, $\langle \omega \rangle$ coherent is equal to $\langle \omega \rangle$ incoherent, as required. We remark also that the first moment is independent of the interactions.

The second moment can be obtained by a similar method. The algebra is considerably more complex. Starting from

$$\langle \omega^2 \rangle = \left\langle \left| \frac{d}{dt} \sum_\alpha f_\alpha e^{i\mathbf{Q} \cdot \mathbf{R}_\alpha} \right|^2 \right\rangle$$

$$= \sum_{\alpha, \beta} \left\langle f_\alpha^* f_\beta \left\{ -\frac{i}{\hbar} [\hat{H}, e^{-i\mathbf{Q} \cdot \mathbf{R}_\alpha}] \right\} \left\{ \frac{i}{\hbar} [\hat{H}, e^{i\mathbf{Q} \cdot \mathbf{R}_\beta}] \right\} \right\rangle \tag{A.9}$$

We get

$$\langle \omega^2 \rangle = \frac{\hbar^2 Q^4}{4} \sum_{\alpha,\beta} \left\langle \frac{f_\alpha^* f_\beta}{M_\alpha M_\beta} (2\delta_{\alpha\beta} - e^{-i\mathbf{Q}\cdot\mathbf{R}_{\alpha,\beta}}) \right\rangle$$

$$+ \sum_{\alpha,\beta} \langle f_\alpha^* f_\beta e^{-i\mathbf{Q}\cdot\mathbf{R}_{\alpha,\beta}} (\mathbf{Q}\cdot\mathbf{v}_\alpha)(\mathbf{Q}\cdot\mathbf{v}_\beta) \rangle \qquad (A.10)$$

with $\mathbf{R}_{\alpha,\beta} = \mathbf{R}_\alpha - \mathbf{R}_\beta$

We will neglect the terms of $O(\hbar^2)$ or higher, and therefore, we keep only the second term of (A.10). Furthermore, we separate the contributions to this term into two categories [see (3.14)]:

$$\langle \omega^2 \rangle = \langle \omega^2 \rangle_m + \langle \omega^2 \rangle_d \qquad (A.11)$$

In the first, the atoms α, β belong to the same molecule, whereas in the second $\langle \omega \rangle_d$, they belong to different molecules.

The calculation of both these terms is, in principle, quantum mechanical. The first term is due to the interactions of the neutron with isolated translating, rotating, and vibrating molecules. If the kinetic energy of the neutrons E_0 is larger than the level spacing of the levels of the molecular energy modes, then we can perform the averages classically without serious error. For the translational modes, this condition is always satisfied. For the rotational levels of molecules containing hydrogen this condition is not satisfied and the classical calculation should, perhaps, be corrected.

The condition is never true for the vibrational modes, but in this case since the harmonic oscillator has no dispersion, the contribution to the second moment is zero. Coriolis forces and anharmonicity effects are negligible. Let us also mention that in the incoherent neutron scattering experiments the semiclassical theory,[25] which uses classical averages, is moderately successful for molecules such as methane and ammonia at high values of Q. The complete quantum mechanical calculation[29] is quite complex, and we do not believe that it is necessary for the corrections of the scattering of most molecular liquids.

Using Rayleigh expansion (3.21), and tensor algebra notation,[30] we get

$$\langle \omega^2 \rangle_m = \sum_{\alpha,\beta} \langle f_\alpha^* f_\beta e^{-i\mathbf{Q}\cdot\mathbf{r}_{\alpha,\beta}} (\mathbf{Q}\cdot\mathbf{v}_\alpha)(\mathbf{Q}\cdot\mathbf{v}_\beta) \rangle$$

$$= \frac{Q^2}{3} \sum \overline{f_\alpha^2} \langle v_\alpha^2 \rangle + \frac{Q^2}{3} \sum_{\alpha \neq \beta} \bar{f}_\alpha \bar{f}_\beta \langle j_0(Q r_{\alpha,\beta})(\mathbf{v}_\alpha \cdot \mathbf{v}_\beta) \rangle \qquad (A.12)$$

$$- Q^2 \sqrt{\tfrac{2}{3}} \sum_{\alpha \neq \beta} \bar{f}_\alpha \bar{f}_\beta \langle j_2(Q r_{\alpha\beta})\{[\mathbf{v}_\alpha \otimes \mathbf{v}_\beta]^2 \odot \mathbf{C}^2(\hat{\mathbf{r}}_{\alpha\beta})\} \rangle$$

where we have used spherical tensor notation.[30] The scalar product of two tensors of rank m is

$$\mathbf{a}^m \odot \mathbf{b}^m = \sum a_\mu{}^m b_{-\mu}{}^m (-)^\mu \qquad (A.13)$$

and the direct or Kronecker product of two tensors of rank m, n is

$$[\mathbf{a}^m \otimes \mathbf{b}^n]_\lambda^l = \sum_{\mu,\,\nu} \langle m\mu n\nu | l\lambda \rangle a_\mu{}^m b_\nu{}^n \qquad (A.14)$$

where $\langle m\mu n\nu | l\lambda \rangle$ is the vector coupling coefficient.[5] The quantities involved in these operations are expressed in polar tensor form. So, the velocity $\mathbf{v}_\alpha$ in polar form is

$$v_{\pm 1}^1 = \mp \frac{1}{\sqrt{2}} (v_x \pm i v_y), \qquad v_0{}^1 = v_z \qquad (A.15)$$

The Racah spherical harmonic $C_\mu{}^m(\mathbf{R})$ is defined in Ref. 30, where also the other definitions af angular coupling coefficients can be found.

Since we ignore vibrations, the velocity of atom α in molecule i is

$$\mathbf{v}_\alpha = \mathbf{r}_i + \boldsymbol{\omega}_i \times \mathbf{r}_{\alpha i} \qquad (A.16)$$

where $\mathbf{r}_i$ is the velocity of the center of mass of the molecule, and $\boldsymbol{\omega}_i$ is the angular velocity, in a frame that diagonalizes the moment of inertia tensor (usually, the axes of this frame coincide with the symmetry axes of the molecule).

Substituting (A.16) (in polar form) into (A.12), we get after much algebra

$$\frac{1}{N} \langle \omega \rangle = \frac{Q^2 \frac{\hbar}{2} \sum_\alpha \overline{f_\alpha^2}}{M} \qquad (A.17)$$

$$\frac{1}{N} \langle \omega^2 \rangle_m = kTQ^2 \left\{ C + \sum_{\alpha \neq \beta} \overline{f_\alpha} \overline{f_\beta} [j_0(QR_{\alpha\beta}) A_{\alpha\beta} + j_2(QR_{\alpha\beta}) B_{\alpha\beta}] \right\} \qquad (A.18)$$

where we have used the following abbreviations:

$$A_{\alpha\beta} = \frac{1}{M_{\text{mol}}} + \frac{1}{3} [\text{Tr}\,(\mathbf{r}^\alpha \mathbf{r}^\beta)\,\text{Tr}\,(\mathbf{I}^{-1}) - (\mathbf{r}^\alpha \mathbf{r}^\beta : \mathbf{I}^{-1})] \qquad (A.19)$$

$$B_{\alpha\beta} = A_{\alpha\beta} - \frac{1}{M_{\text{mol}}} - (\mathbf{r}^\alpha \times \hat{\mathbf{r}}^{\alpha\beta}) \cdot \mathbf{I}^{-1} \cdot (\mathbf{r}^\beta \times \hat{\mathbf{r}}^{\alpha\beta}) \qquad (A.20)$$

$$C = \sum_\alpha \overline{f_\alpha^2} A_{\alpha\alpha} \qquad (A.21)$$

where we have used Cartesian coordinates and standard vector–tensor notation. M_α is the atomic mass, M_{mol} the molecular mass, and $\mathbf{I}$ the momentum of inertia tensor.

For (3.25) we need the derivatives of (A.18), which can be computed and substituted into (3.52). The algebraic expressions are long, and are easily programmable in a digital computer. Therefore, they will be omitted here.

We still have to evaluate $\langle\omega\rangle_d$ in (A.11). This can be shown[28] to give a quantum correction term of order $O[\hbar(\partial^2 U/\partial r^2)]$. Presently we would have no way of calculating this term, but for most liquids it should give only a small contribution. For a liquid like water, where the interactions are strong, we suspect that this term is of importance, but we lack ways of computing it.

Acknowledgments

We are grateful to S. I. Sandler, D. Chandler, R. H. Fowler, R. W. Hendricks, H. A. Levy, and Inez Hodge for their cooperation. Dr. Sandler and Dr. Chandler kindly sent us, prior to publication, preprints of Refs. 51 and 55 as well as tables of their calculated scattering functions from which Figs. 5 and 6 were prepared. Dr. Fowler made available the unpublished data of Ref. 62, used in Fig. 7. Dr. Hendricks and Dr. Levy made many helpful suggestions during the preparation of this manuscript which was typed with great patience by Ms. Hodge.

References

1. D. W. Jepsen and H. L. Friedman, *J. Chem. Phys.*, **38**, 846 (1963); W. A. Steele, *J. Chem. Phys.*, **39**, 3197 (1963).
2. L. Blum and A. J. Torruella, *J. Chem. Phys.*, **56**, 303 (1972).
3. L. Blum, *J. Chem. Phys.*, **57**, 1862 (1972); **58**, 3295 (1973).
4. A more comprehensive historical account can be found in L. Blum, A. H. Narten, and R. H. Fowler, *J. Chem. Phys.*, to be submitted.
5. A. R. Edmonds, *Angular Momenta in Quantum Mechanics*, Princeton, N.J., 1960.
6. M. Hamermesh, *Group Theory*, Addison-Wesley, Reading, Mass., 1962.
7. S. L. Altmann and A. P. Cracknell, *Rev. Mod. Phys.*, **37**, 19 (1965).
8. M. S. Wertheim, *J. Chem. Phys.*, **55**, 4291 (1971).
9. R. M. Hochstrasser, *Molecular Aspects of Symmetry*, W. A. Benjamin, New York, 1966.
10. T. L. Hill, *Statistical Mechanics*, McGraw-Hill, New York, 1956.
11. See, for example, the excellent review by J. M. Deutsch, *Ann. Rev. Phys. Chem.*, **24**, 301 (1973).
12. J. D. Ramshaw, *J. Chem. Phys.*, **57**, 2684 (1972).
13. L. J. Schiff, *Quantum Mechanics*, 3rd ed., McGraw-Hill, New York, N.Y., 1968, p. 324.
14. See for example, V. F. Turchin, *Slow Neutrons*, Israel Program for Scientific Translations, Jerusalem, 1965, and references cited therein.
15. M. I. Davis, *Electron Diffraction in Gases*, Dekker, New York, N.Y., 1971.
16. L. Van Hove, *Phys. Rev.*, **95**, 249 (1954).
17. J. G. Powles, *Adv. Physics*, **22**, 1 (1973).
18. G. Placzek, *Phys. Revs.*, **86**, 377 (1952).
19. A. H. Narten, *J. Chem. Phys.*, **56**, 5681 (1972).
20. W. A. Steele and R. Pecora, *J. Chem. Phys.*, **42**, 1863 (1965). See also R. McWeeney, *Acta Cryst.*, **7**, 180 (1954).

21. R. McWeeney, *Acta Cryst.*, **6**, 631 (1953); C. Tavard, M. Rouault, M. Roux, and M. Cornille, *J. Chim. Physique*, **61**, 1324, 1330 (1964); R. Stewart, *J. Chem. Phys.*, **42**, 3175 (1965); **51**, 4569 (1969).

22. L. Blum, *J. Comput. Phys.*, **7**, 592 (1971); P. F. Morrison and C. J. Pings, *J. Chem. Phys.*, **56**, 280 (1972).

23. *International Tables for X-Ray Crystallography*, The Kynoch Press, Birmingham, England, 1974, Vol. IV.

24. S. K. Sinha, and G. Venkataraman, *Phys. Rev.*, **149**, 1 (1966).

25. W. Marshall and S. W. Lovesey, *Theory of Thermal Neutron Scattering*, Oxford University Press (1971).

26. S. J. Cyvin, *Molecular Vibrations and Mean Square Amplitudes*, Elsevier, Amsterdam, 1968.

27. G. Placzek, *Phys. Rev.*, **86**, 377 (1952); A clear treatment of this correction can be found in the article by J. L. Yarnell, M. J. Katz, R. G. Wenzel and S. H. Koenig, *Phys. Rev.*, **A7**, 2130 (1973). Our account follows this reference closely.

28. P. G. De Gennes, *Physica*, **25**, 825 (1959); A. Rahman, K. S. Singwi and A. Sjölander, *Phys. Rev.*, **126**, 986 (1962).

29. G. W. Griffing, *Inelastic Scattering of Neutrons in Solids and Liquids*, Vol. I, International Atomic Energy Commission, Vienna, 1963.

30. A clear exposition of these techniques can be found in B. W. Shore and D. H. Menzel, *Principles of Atomic Spectra*, Wiley, New York, N.Y., 1968.

31. See for example, F. J. Webb, *Proc. Phys. Soc.*, **92**, 912 (1967).

32. R. W. Hendricks, P. G. Mardon, and L. B. Shaffer, *J. Chem. Phys.*, **61**, 319 (1974).

33. J. H. Clarke, J. C. Dore, and R. N. Sinclair, *Mol. Phys.*, **29**, 581 (1975).

34. The description and analysis of a time-of-flight diffractometer for use with a pulsed neutron source has been given by J. G. Powles, *Mol. Phys.*, **26**, 1325 (1973).

35. J. E. Enderby, in *Physics of Simple Liquids*, H. N. V. Temperley, J. S. Rowlinson, G. S. Rushbrooke, Eds., North-Holland Publishing Co., Amsterdam, 1968, p. 611.

36. F. Hajdu and G. Pálinkás, *J. Appl. Cryst.*, **5**, 395 (1972).

37. A. H. Narten and H. A. Levy, in *Water: A Comprehensive Treatise*, Vol. I, F. Franks, Ed., Plenum Press, New York, 1972, p. 311.

38. C. J. Pings, in Ref. 35, p. 422.

39. W. A. Schlup, *Phys. Chem. Liquids*, **1**, 73 (1968).

40. C. Lanczos, *Applied Analysis*, Prentice-Hall, Englewood Cliffs, N.J., 1956, p. 336.

41. A. Rahman, *J. Chem. Phys.*, **42**, 3540 (1965).

42. R. D. Mountain, *J. Chem. Phys.*, **57**, 4346 (1972).

43. A. H. Narten, unpublished data; to be submitted to *J. Chem. Phys.*

44. For a review of experimental work up to 1960, see R. F. Kruh, *Chem. Rev.*, **62**, 319 (1962); earlier attempts of interpreting diffraction data on molecular liquids are summarized by J. A. Prins and W. Prins, *Physica*, **23**, 253 (1957).

45. For a review of experimental work up to 1960, see G. E. Bacon, *Neutron Diffraction*, Clarendon Press, Oxford, 1962, p. 380.

46. E. Kálmán, E. Lengyel, G. Pálinkás, L. Haklik, and E. Eke, in *Structure of Water and Aqueous Solutions*, W. A. P. Luck, Ed., Verlag Chemie/Physik, Weinheim, 1974, p. 366.

47. A. H. Narten, *Trans. Am. Cryst. Assoc.*, **10**, 19 (1974).

48. S. I. Sandler, A. Das Gupta, and W. A. Steele, *J. Chem. Phys.*, **61**, 1326 (1974).

49. H. C. Andersen, J. D. Weeks, and D. Chandler, *Phys. Rev.*, **A4**, 1597 (1971).

50. W. A. Steele and S. I. Sandler, *J. Chem. Phys.*, **61**, 1315 (1974).

51. A. Das Gupta, S. I. Sandler, and W. A. Steele, *J. Chem. Phys.*, **62**, 1769 (1975).

52. D. C. Henshaw, D. C. Hurst, and N. K. Pope, *Phys. Rev.*, **92**, 1229 (1953).

53. For a convenient tabulation of diffraction data on homonuclear diatomics, see P. W. Schmidt and C. W. Tompson, in *Simple Dense Fluids*, H. L. Frisch and Z. W. Salsburg, Eds., Academic Press, New York, 1968, p. 78.
54. K. Suzuki and P. A. Egelstaff, *Can. J. Phys.*, **52**, 241 (1974).
55. L. J. Lowden and D. Chandler, *J. Chem. Phys.*, **61**, 5228 (1974).
56. A. H. Narten, M. D. Danford, and H. A. Levy, *J. Chem. Phys.*, **46**, 4875 (1967).
57. G. Reichelt, Dissertation, Universität Freiburg, Germany, 1973.
58. P. R. Ireland, R. Mason, and A. I. M. Rae, *Mol. Phys.*, **24**, 17 (1972).
59. K. R. Rao, *J. Chem. Phys.*, **48**, 2395 (1968).
60. P. A. Egelstaff, D. I. Page, and J. G. Powles, *Mol. Phys.*, **20**, 881 (1971).
61. K. E. Gubbins, C. G. Gray, P. A. Egelstaff, and M. S. Ananth, *Mol. Phys.*, **25**, 1353 (1973).
62. R. H. Fowler, private communication. The calculations were done with a relatively small system of particles interacting with atom-atom pair potentials derived from an analysis of crystalline CCl_4 by W. R. Busing. The form of the potentials as well as the parameters are the same as those derived for hexachlorobenzene: J. B. Bates and W. R. Busing, *J. Chem. Phys.*, **60**, 2414 (1974).
63. A. H. Narten and H. A. Levy, *J. Chem. Phys.*, **55**, 2263 (1971).
64. D. I. Page and J. G. Powles, *Mol. Phys.*, **21**, 901 (1971).
65. The neutron data presented in Ref. 19 have since been extended to values of $Q \sim 16$ Å^{-1}, using 0.75 Å neutrons. The extended data, shown in Fig. 4 of Ref. 47, will be discussed in more detail in a future report.
66. F. H. Stillinger and A. Rahman, *J. Chem. Phys.*, **60**, 1545 (1974).
67. H. Popkie, H. Kistenmacher, and E. Clementi, *J. Chem. Phys.*, **59**, 1325 (1973).
68. L. Blum and A. H. Narten, *J. Chem. Phys.*, **56**, 5197 (1972); A. H. Narten, L. Blum, and R. H. Fowler, *J. Chem. Phys.*, **60**, 3378 (1974).
69. L. Blum, *J. Chem. Phys.*, **61**, 2129 (1974).
70. L. Verlet and J. J. Weiss, *Mol. Phys.*, **28**, 665 (1974).

THE EXPANSION OF
THE MASTER EQUATION

N. G. VAN KAMPEN

Institute of Theoretical Physics of the University, Utrecht, Netherlands

CONTENTS

I. INTRODUCTION

Macroscopic physics deals with macroscopic quantities q_v, such as the positions and velocities of bodies, electrical charges and currents, amounts or concentrations of chemical compounds, temperatures at various points in a material, or local density and velocity of a fluid. They obey macroscopic laws

$$\dot{q}_v = F_v(q_1, q_2, \ldots) \qquad (1.1)$$

which on a macroscopic level can be derived from general principles such as conservation laws, together with some specific assumptions of phenomenological nature, for example, those of Fourier and Fick. On the one hand this macroscopic picture is incomplete, because it has to introduce phenomenological coefficients, as many as there are phenomena, but this is of no concern to us here. On the other hand the macroscopic laws (1.1) are merely an approximation, valid when so many particles are involved that fluctuations are negligible. The present work is concerned with improving on this macroscopic approximation by taking into account the fluctuations. Of course the quantities that characterize the discreteness are essential: Boltzmann's constant, the elementary charge, and the masses of individual particles.

In order to study the corrections to (1.1) caused by the discrete nature of matter one must view the macroscopic phenomena as the outcome of the collective behavior of many particles. This does not merely require an investigation of the phenomenological assumptions mentioned above, but a reappraisal of the very definitions of the q_v is needed.

In principle all information is contained in the microscopic equations of motion of all particles, but it hardly needs saying that an exact solution of these equations is beyond human means, excepting a small number of simple models.[1] Even the macroscopic laws (1.1) can only be derived from them with the help of simplifications and assumptions, which are no more reliable than the phenomenological assumptions used in the purely macroscopic approach. It is therefore sensible to embark upon a less ambitious program and to develop a theory which goes beyond the macroscopic description in that it includes fluctuations, but short-cuts the connection with the microscopic equations by an appeal to some suitably chosen semiphenomenological assumptions. This is the customary approach in noise theory; we propose to call it the *mesoscopic* level of description.

One popular mesoscopic approach consists in adding to (1.1) a fluctuating term

$$\dot{q}_v = F_v(q_1, q_2, \ldots) + l_v(t) \qquad (1.2)$$

and making suitable assumptions concerning the statistical properties of the random functions $l_v(t)$ (see Section XI). It should be clear that this device changes the nature of the q_v; they are now also stochastic quantities. The macroscopic values that enter into (1.1) are identified with the averages of the q_v in (1.2). This approach was first used by Langevin in his treatment of the Brownian movement, and his success has led many authors to apply the same device to other systems.[2] However, we shall show in Section XI that in many cases it leads to wrong results.

A second approach starts out by introducing the probability distribution $P(q_1, q_2, \ldots; t)$ defined as follows: $P(q_1, q_2, \ldots; t)\, dq_1\, dq_2 \ldots$ = the joint probability that at time t the first quantity has a value between q_1 and $q_1 + dq_1$, and the second one between q_2 and $q_2 + dq_2$, etc. Note that the q's have changed their nature again and are merely coordinates in some q-space on which the probability density P is defined. One then assumes that P obeys the Fokker-Planck equation

$$\frac{\partial P}{\partial t} = -\sum_v \frac{\partial}{\partial q_v} F_v(q)P + \sum_{v\mu} \frac{\partial^2}{\partial q_v \partial q_\mu} D_{v\mu}P \qquad (1.3)$$

The F_v are the same as in (1.1), and the new coefficients $D_{v\mu}$ are found from the fluctuation-dissipation theorem. Although this looks quite different from the Langevin approach, it is actually equivalent to it, and therefore subject to the same criticism (Section XXII).

A third mesoscopic approach is the basis of this article. It also starts out from the probability density P, but merely assumes that it obeys an equation of type

$$\dot{P} = \mathbf{W}P \qquad (1.4)$$

where $\mathbf{W}$ is a linear operator acting on the q-dependence. Let $W(q|q')$ be the integral kernel of $\mathbf{W}$; then the requirement that the total probability must remain equal to unity tells

$$\int W(q|q')\,dq = 0$$

(Each q stands for the whole set of q_v, and dq is a volume element in q-space.) Hence one may write (1.4) in the physically more transparent form

$$\dot{P}(q, t) = \int \{W(q|q')P(q', t) - W(q'|q)P(q, t)\}\,dq' \qquad (1.5)$$

The kernel $W(q|q')$ for $q \neq q'$ represents the transition probability per unit time from q' to q and must be nonnegative. The second term represents the decrease of $P(q, t)$ due to transitions to other values q'. Equation (1.4) or (1.5) is called the "master equation".*

The assumption (1.4) implies that the stochastic process described by q is a Markov process. This is a strong assumption, which in most applications is only approximately true and with the conditions that a suitable coarse-grained time scale is used, and that the correct set of variables $q = \{q_1, q_2, \ldots\}$ is chosen. On the other hand, it is weaker than the assumptions needed in the two previously mentioned mesoscopic approaches. Moreover it is easier to assess on physical grounds. The transition probabilities $W(q|q')$ usually have a direct physical interpretation in terms of the microscopic quantities

---

* Throughout this article the term is used in its original sense:[3] an equation of the type (1.4) for a probability distribution.

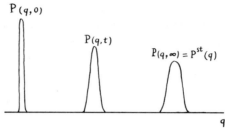

Fig. 1. The evolution of the probability density towards equilibrium.

describing the system, for instance, collision cross-sections or quantum mechanical matrix elements. We shall meet cases where the master equation (1.5) obviously holds, but neither the Langevin approach, nor the Fokker-Planck equation (1.3) leads to correct results.

The master equation purports to describe the entire behavior of the q and hence also the macroscopic equation (1.1) should follow from it. How is it possible for an equation governing the probability density in q-space to give rise to a deterministic set of equations for the q_v? The idea is that $P(q, t)$ is a sharp peak located at a rather well-defined point in q-space (Fig. 1). If the width of the peak may be neglected it is possible to consider its position in q-space as the macroscopic value of q_v. While P varies in time according to (1.5) the peak moves through q-space according to (1.1). Note that there is no contradiction between the fact that (1.5) is linear whereas (1.1) may well be nonlinear. The situation is analogous to the way in which a linear Schroedinger equation gives rise to a nonlinear classical equation of motion in the approximation in which the particle is heavy enough to neglect the spreading of the wave function.

The mathematical scheme describing this state of affairs was developed some time ago.[4-6] The present review is confined to the special but frequently occurring class of master equations in which the variable q takes only integral values. We therefore write n rather than q and the master equation is

$$\dot{P}(n, t) = \sum_{n'} \{W(n|n')P(n', t) - W(n'|n)P(n, t) \qquad (1.6)$$

It may happen that n runs from $-\infty$ to $+\infty$, or from 0 to ∞, or only takes values in some finite range. The transition probabilities $W(n|n')$ need only be defined for $n \neq n'$ and are nonnegative. They are properties of the system and, of course, independent of the $P(n, t)$, which describe the special mesoscopic state considered.[7] It is possible to include the case that W depends on time (nonautonomous systems), but we shall not do so. Equation (1.6) may also be written by means of a matrix $\mathbf{W}$

$$\dot{P}(n, t) = \sum_{n'} \mathbf{W}_{nn'} P(n', t)$$

The master equation is "solved" if one can find the $P(n, t)$ that obey (1.6) and take arbitrarily prescribed initials values at $t = 0$. Obviously it suffices to consider the initial condition

$$P(n, 0) = \delta_{n, m} \tag{1.7}$$

for each m. The corresponding solution is

$$P(n, t \,|\, m, 0) = (e^{\mathbf{W}t})_{nm} \tag{1.8}$$

In order to evaluate this formal expression one has to diagonalize $\mathbf{W}$, but only in rare cases can that be done exactly. Hence it is necessary to have a systematic approximation scheme in the form of a power series expansion in some physical parameter. It appeared that the appropriate quantity is $\Omega^{-1/2}$, where Ω is a measure for the size of the system or the total number of particles involved. This scheme is demonstrated on a simple example in Section III, formulated in general in Sections IV and V, and subsequently applied to various problems.

It will appear that most of the problems treated in the literature can be readily handled with the Ω-expansion method. Many of the existing controversies and paradoxes[5] are caused by unsystematic approximations, in which terms are neglected according to the taste of the author. In addition it will be shown that the popular Langevin approach may lead to wrong results even in simple cases (Section XI), and the limitations of the Fokker–Planck equation are discussed in Section XXII. On the other hand, it must be stressed that the expansion is essentially based on the smallness of fluctuations and has only limited validity in unstable situations (Section XVIII) or phase transitions (Section XX).

II. PRELIMINARIES

The jump moments or derivate moments[8] are defined by

$$a_p(n) = \sum_{n'} (n' - n)^p W(n' \,|\, n) \qquad (p = 1, 2, \ldots) \tag{2.1}$$

Multiply (1.6) with n and sum

$$\frac{d}{dt} \langle n \rangle = \sum_{nn'} \{ nW(n \,|\, n')P(n') - nW(n' \,|\, n)P(n) \}$$

$$= \sum_{nn'} (n' - n)W(n' \,|\, n)P(n)$$

$$= \langle a_1(n) \rangle \tag{2.2}$$

If $a_1(n)$ is a linear function this is identical with

$$\frac{d}{dt} \langle n \rangle = a_1(\langle n \rangle) \tag{2.3}$$

which permits us to determine $\langle n \rangle$ as a function of t. If, however, $a_1(n)$ is not linear, (2.3) is at best an approximation, which amounts to neglecting all fluctuations. We shall see that (2.3) is, indeed, the zeroth approximation in the Ω-expansion scheme, and is therefore to be identified with the macroscopic equation (1.1).* The exact identity (2.2) is not a closed equation for $\langle n \rangle$ but involves higher moments of n as well. To improve on the approximation (2.3) we expand in (2.2) the function $a_1(n)$ in $n - \langle n \rangle$ and break off after the second derivative:

$$\frac{d}{dt} \langle n \rangle = a_1(\langle n \rangle) + \tfrac{1}{2} a_1''(\langle n \rangle) \sigma_n^2 \tag{2.4}$$

where $\sigma_n^2 = \langle (n - \langle n \rangle)^2 \rangle$. As this equation involves $\langle n^2 \rangle$ we also multiply (1.6) with n^2 and sum

$$\frac{d}{dt} \langle n^2 \rangle = \sum_{nn'} (n'^2 - n^2) W(n'|n) P(n)$$

$$= \langle a_2(n) \rangle + 2 \langle n a_1(n) \rangle \tag{2.5}$$

Combination with (2.2) yields the exact identity

$$\frac{d}{dt} \sigma_n^2 = \langle a_2(n) \rangle + 2\{ \langle n a_1(n) \rangle - \langle n \rangle \langle a_1(n) \rangle \} \tag{2.6}$$

Making somewhat loose approximations similar to (2.3) we write for this

$$\frac{d}{dt} \sigma_n^2 = a_2(\langle n \rangle) + 2 a_1'(\langle n \rangle) \sigma_n^2 \tag{2.7}$$

In Section V it will be shown that the pair of equations (2.4) and (2.7) together actually constitute a consistent approximation.[9,10]

The upshot is that in order to improve on (2.3) two coupled equations [(2.4) and (2.7)] are needed (unless $a_1(n)$ happens to be linear). That means that it is no longer possible to determine $\langle n \rangle$ from its initial value; one also needs to know the initial value of σ_n^2. All this is subject to the condition that σ_n^2 remains finite (of order n), otherwise there is no justification for omitting higher moments. This condition amounts to $a_1'(\langle n \rangle) < 0$, that is, the system must be stable (compare Section V).

---

* This statement requires a minor modification, see Section IV.

A special but important class of discrete Markov processes are the one-step or birth-and-death processes. They are defined by $W(n|n') = 0$ unless $n = n' \pm 1$, that is,

$$W(n|n') = r(n')\delta_{n, n' - 1} + g(n')\delta_{n, n' + 1}$$

$r(n)$ and $g(n)$ may be any two nonnegative functions, usually analytic; their names stem from recombination and generation of charge carriers in semi-conductors.[11]

The master equation of a one-step process has the following form:

$$\dot{P}(n, t) = r(n + 1)P(n + 1, t) + g(n - 1)P(n - 1, t) - \{r(n) + g(n)\}P(n, t) \tag{2.8}$$

It is convenient to define the difference operator $\mathbf{E}$ by[12]

$$\mathbf{E}f(n) = f(n + 1), \qquad \mathbf{E}^{-1}f(n) = f(n - 1) \tag{2.9}$$

With its aid the master equation (2.8) may be written

$$\dot{P} = (\mathbf{E} - 1)r(n)P + (\mathbf{E}^{-1} - 1)g(n)P \tag{2.10}$$

The jump moments are

$$a_p(n) = (-1)^p r(n) + g(n) \tag{2.11}$$

The macroscopic rate equation (2.3) takes the form

$$\frac{d}{dt}\langle n \rangle = -r(\langle n \rangle) + g(\langle n \rangle) \tag{2.12}$$

and the coupled equations [(2.4) and (2.7)] are

$$\frac{d}{dt}\langle n \rangle = g(\langle n \rangle) - r(\langle n \rangle) + \tfrac{1}{2}\{g''(\langle n \rangle) - r''(\langle n \rangle)\}\sigma_n^2$$

$$\frac{d}{dt}\sigma_n^2 = g(\langle n \rangle) + r(\langle n \rangle) + 2\{g'(\langle n \rangle) - r'(\langle n \rangle)\}\sigma_n^2$$

Some general properties of one-step processes are listed in Section VI.

III. FIRST EXAMPLE: SPREADING OF AN EPIDEMIC

As a first example for demonstrating the expansion we choose a simple nonlinear one-step process, which describes the spreading of an epidemic in a population of Ω individuals.[6,13] If n is the number of infected individuals, the probability per unit time for a new infection to occur is proportional to n, and to the number $\Omega - n$ of uninfected. Thus $g(n) = \beta n(\Omega - n)$ with constant

β. Furthermore we take $r(n) = 0$, that is, no cure is possible. Hence

$$W(n|n') = \beta\delta_{n,n'+1}n'(\Omega - n') \tag{3.1}$$

The master equation is

$$\begin{aligned}\dot{P}(n, t) &= \beta(n - 1)(\Omega - n + 1)P(n + 1, t) - \beta n(\Omega - n)P(n, t)\\ &= \beta(\mathbf{E}^{-1} - 1)n(\Omega - n)P\end{aligned} \tag{3.2}$$

The more general problem with arbitrary $g(n)$ and $r(n) = 0$ has been treated by Weiss as a model for superradiance.[14] Actually such problems can be solved without approximations in a more or less closed form, but the result is too involved to be of much use, unless $g(n)$ is sufficiently simple.

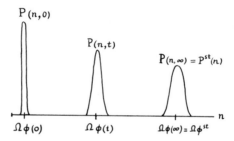

Fig. 2.   The definition of the macroscopic part of a variable n.

One expects n to consist of a macroscopic part $\Omega\varphi(t)$ plus fluctuations of order $\Omega^{1/2}$. That is, $P(n, t)$ will be a sharp peak located roughly at $\Omega\varphi(t)$ with a width of order $\Omega^{1/2}$ (see Fig. 2). Hence we set

$$n = \Omega\varphi(t) + \Omega^{1/2}x \tag{3.3}$$

where x is the new variable and $\varphi(t)$ will be chosen presently. We shall call $\Omega\varphi(t)$ the "macroscopic part" and $\Omega^{1/2}x$ the "fluctuating part" of n, and refer to the new variables as the "Ω language." Accordingly the probability distribution of n now becomes a probability distribution Π of x,

$$P(n, t)\Delta n = \Pi(x, t)\Delta x$$

$$\Pi(x, t) = \Omega^{1/2}P(\Omega\varphi(t) + \Omega^{1/2}x, t) \tag{3.4}$$

The following transformation formulas apply

$$\frac{\partial\Pi}{\partial x} = \Omega^{1/2}\frac{\partial P}{\partial n}, \qquad \frac{\partial\Pi}{\partial t} = \Omega^{1/2}\left\{\Omega\frac{d\varphi}{dt}\frac{\partial P}{\partial n} + \frac{\partial P}{\partial t}\right\}$$

Hence*

$$\Omega^{1/2} \frac{\partial P}{\partial t} = \frac{\partial \Pi}{\partial t} - \Omega^{1/2} \frac{d\varphi}{dt} \frac{\partial \Pi}{\partial x} \tag{3.5}$$

Moreover one has

$$\mathbf{E} = 1 + \Omega^{-1/2} \frac{\partial}{\partial x} + \frac{1}{2} \Omega^{-1} \frac{\partial^2}{\partial x^2} + \cdots \tag{3.6}$$

Substitute the new variables in the master equation (3.2) and cancel an overall factor $\Omega^{-1/2}$,

$$\frac{\partial \Pi}{\partial t} - \Omega^{1/2} \frac{d\varphi}{dt} \frac{\partial \Pi}{\partial x} = \beta\Omega^2 \left\{ -\Omega^{-1/2} \frac{\partial}{\partial x} + \frac{1}{2} \Omega^{-1} \frac{\partial^2}{\partial x^2} \right\}$$
$$\times (\varphi + \Omega^{-1/2}x)(1 - \varphi - \Omega^{-1/2}x)\Pi \tag{3.7}$$

We absorb one factor Ω into the time variable (and for convenience also the β) by setting

$$\beta\Omega t = \tau \tag{3.8}$$

Then the largest terms are

$$-\Omega^{1/2} \frac{d\varphi}{d\tau} \frac{\partial \Pi}{\partial x} = -\varphi(1 - \varphi) \frac{\partial \Pi}{\partial x}$$

They can be made to cancel by subjecting φ to the equation

$$\frac{d\varphi}{d\tau} = \varphi(1 - \varphi) \tag{3.9}$$

This equation determines how the macroscopic part of n varies with time. Translating back to the original variables it takes the form

$$\frac{dn}{dt} = \beta n(\Omega - n) \tag{3.10}$$

which appears to be identical with the macroscopic rate equation (2.12).

The terms of order Ω^0 in (3.7) yield an equation for Π,

$$\frac{\partial \Pi}{d\tau} = -(1 - 2\varphi) \frac{\partial}{\partial x} x\Pi + \frac{1}{2} \varphi(1 - \varphi) \frac{\partial^2 \Pi}{\partial x^2} \tag{3.11}$$

This is a Fokker–Planck equation whose coefficients involve φ and therefore depend on time. Observe, however, that the coefficient of the first term is

---

* It is possible to arrive at (3.5) without the intervention of the dubious symbol $\partial P/\partial n$. Let t in (3.4) vary by δt and simultaneously x by $-\Omega^{1/2}\dot\varphi(t)\delta t$; this leads immediately to (3.5).

linear in x, and that the second term does not depend on x; we shall indicate these features of a Fokker–Planck equation by calling it *linear*. (Of course, all Fokker–Planck equations are linear in the unknown function—in this case Π.) Equation (3.11) governs the fluctuations in n of order $\Omega^{1/2}$ about the macroscopic part.

The strategy for solving the master equation (3.2) with initial condition (1.7) now emerges. First solve (3.9) with initial value $\varphi(0) = m/\Omega$. Then solve (3.11) with initial $\Pi(x, 0) = \delta(x)$. Then

$$P(n, t \,|\, m, 0) = \Omega^{-1/2}\Pi\!\left(\frac{n - \Omega\varphi(\tau)}{\Omega^{1/2}}, \tau\right)$$

In this solution terms of relative order $\Omega^{-1/2}$ have been neglected.

IV. THE GENERAL EXPANSION METHOD

The basic idea is that there is a parameter Ω measuring the size of the system, such that for large Ω the fluctuations are relatively small. It is then possible to expand in descending powers of Ω, as will be outlined in five steps.*

First step: specifying the dependence of the transition probabilities on Ω. It is assumed that the way in which $W(n|n')$ depends on Ω has the following form:

$$W(n|n') = f(\Omega)\left[\Phi_0\!\left(\frac{n'}{\Omega}; n - n'\right) + \Omega^{-1}\Phi_1\!\left(\frac{n'}{\Omega}; n - n'\right) + \cdots\right] \quad (4.1)$$

Each function Φ_j has a Taylor expansion with respect to its first argument, but is of course a discrete function of its second argument, which is the jump size. The factor $f(\Omega)$, usually some power of Ω, is innocuous because it can be absorbed in the time variable. The jump moments (2.1) are transformed accordingly,

$$a_p(n) = f(\Omega)\alpha_p\!\left(\frac{n}{\Omega}\right) \quad (4.2)$$

In the following we suppose for simplicity that $\Phi_1, \Phi_2, \ldots$ vanish. They are not hard to include when they occur, as in Section IX, but cumbersome in the general treatment. When they do not vanish it is not strictly true that (2.3) is identical with the macroscopic law, inasmuch as a_1 involves higher orders in $1/\Omega$, which do not belong to a macroscopic description. The macroscopic law is determined by the first jump moment of Φ_0 alone,[16] but in the next approximation (2.4) both Φ_0 and Φ_1 have to be used for a_1.

---

*Previously we have used the Kramers–Moyal expansion as a convenient intermediate step,[4,5] but we shall avoid it here, since its role has been misconstrued.[15]

Substituting (4.1) in (1.6) and changing the summation variable from n' to $v = n - n'$ one obtains for the master equation

$$\frac{\partial P(n, t)}{\partial t} = f(\Omega) \sum_v \left\{ \Phi_0\left(\frac{n - v}{\Omega}; v\right) P(n - v, t) - \Phi_0\left(\frac{n}{\Omega}; -v\right) P(n, t) \right\} \qquad (4.3)$$

Second step: postulating the way in which P depends on Ω. One expects $P(n, t)$ to be a sharp peak located at some point $\Omega\varphi(t)$ with a width of order $\Omega^{1/2}$. Hence one transforms the variable n to a new variable x as in (3.3). This transforms $P(n, t)$ into $\Pi(x, t)$ according to (3.4). Substitute this in the master equation:

$$\frac{\partial \Pi}{\partial t} - \Omega^{1/2} \frac{d\varphi}{dt} \frac{\partial \Pi}{\partial x} = f(\Omega) \left[\sum_v \Phi_0(\varphi(t) + \Omega^{-1/2}x - \Omega^{-1/2}v; v) \right.$$

$$\left. \times \Pi(x - \Omega^{-1/2}v, t) - \sum_v \Phi_0(\varphi(t) + \Omega^{-1/2}x; -v)\Pi(x, t) \right]$$

$$(4.4)$$

The factor [ ] vanishes to lowest order in $\Omega^{-1/2}$. To obtain the next order it is convenient to write it in the form

$$\left[\; \right] = \sum_v \left\{ -\Omega^{-1/2}v \frac{\partial}{\partial x} + \frac{1}{2}\Omega^{-1}v^2 \frac{\partial^2}{\partial x^2} - \cdots \right\} \Phi_0(\varphi + \Omega^{-1/2}x; v)\Pi(x, t)$$

$$(4.5)$$

Third step: extracting the largest terms to obtain the macroscopic equation. The lowest order in [ ] is $\Omega^{-1/2}$; it can be combined with the term of order $\Omega^{1/2}$ on the left if we define a scaled time τ by*

$$f(\Omega)t = \Omega\tau \qquad (4.6)$$

Then the largest terms are of order $\Omega^{1/2}$ on both sides,

$$-\Omega^{1/2} \frac{d\varphi}{d\tau} \frac{\partial \Pi}{\partial x} = \Omega\left(-\Omega^{-1/2} \frac{\partial \Pi}{\partial x} \right) \sum_v v\Phi_0(\varphi; v)$$

Since both terms involve Π only through the factor $\partial\Pi/\partial x$, it is possible to satisfy this equation by choosing for φ a solution of

$$\frac{d\varphi}{d\tau} = \sum_v v\Phi_0(\varphi, v) = \alpha_1(\varphi) \qquad (4.7)$$

* In many of our examples we shall find $f(\Omega) = \Omega$ so that $t = \tau$.

This is the equation for the macroscopic part of n, that is, the macroscopic rate equation.

Fourth step: the next order determines the fluctuations. The terms of order Ω^0 in (4.4) are

$$
\frac{\partial \Pi}{\partial \tau} = - \left\{ \sum_\nu \nu \Phi_0'(\varphi; \nu) \right\} \frac{\partial}{\partial x} x \Pi + \frac{1}{2} \left\{ \sum_\nu \nu^2 \Phi_0(\varphi; \nu) \right\} \frac{\partial^2 \Pi}{\partial x^2}
$$

$$
= -\alpha_1'(\varphi) \frac{\partial}{\partial x} x \Pi + \frac{1}{2} \alpha_2(\varphi) \frac{\partial^2 \Pi}{\partial x^2} \tag{4.8}
$$

The prime indicates differentiation with respect to φ. This is again a linear Fokker–Planck equation with time-dependent coefficients, which governs the fluctuations in n of order $\Omega^{1/2}$ about the macroscopic part $\Omega\varphi$.

Final step: collecting the results in order to solve (1.6) with initial condition (1.7). First solve (4.7) with initial condition

$$
\Omega\varphi(0) = m \tag{4.9}
$$

and call the solution $\varphi(\tau \,|\, m/\Omega)$. Next solve (4.8) with initial condition

$$
\Pi(x, 0) = \delta(x) \tag{4.10}
$$

and call the solution $\Pi(x, \tau \,|\, 0, 0)$. Then

$$
P(n, t \,|\, m, 0) = \Omega^{-1/2} \Pi\left(\frac{n - \Omega\varphi(\tau \,|\, m/\Omega)}{\Omega^{1/2}}, \tau \,\middle|\, 0, 0 \right) \tag{4.11}
$$

Note that one has to the same order

$$
P(n, t \,|\, m, 0) = \Omega^{-1/2} \Pi\left(\frac{n - \Omega\varphi(\tau \,|\, m/\Omega - c\Omega^{1/2})}{\Omega^{1/2}}, \tau \,\middle|\, c, 0 \right) \tag{4.12}
$$

where c is an arbitrary number of order 1.

This program can be carried out by a number of integrations (see Appendix). It is simpler, however, and in many cases sufficient to determine only $\langle n \rangle$ and $\sigma_n^2 = \langle n^2 \rangle - \langle n \rangle^2$ as functions of t. The relevant formulas are derived in the next section.

Higher orders can be added and have the effect of modifying the equation for $\Pi(x, t)$ (see Section VIII). However, we shall be mainly concerned with the approximation to order Ω^0 as given here. This will be called the *linear noise approximation* since to this order the fluctuations are governed by the linear Fokker–Planck equation (4.8). It is the approximation on which the familiar theory of noise in electrical networks[11] is based.

V. THE EQUATIONS FOR THE MOMENTS

Without actually solving (4.8) one may deduce directly from it (by multiplying with x and integrating)

$$\frac{d}{d\tau}\langle x\rangle = \alpha'_1(\varphi)\langle x\rangle \tag{5.1}$$

Observe that this is identical with the "variational equation" belonging to (4.7), that is, the equation for the difference between two neighboring solutions of (4.7). The fact that this must be so can be gleaned from (4.12); a slight variation (of order $\Omega^{-1/2}$) in the initial value of φ can be compensated by the initial value of $\langle x\rangle$. An important consequence is the following: Since the variational equation of (4.7) determines the stability of the macroscopic solution $\varphi(t)$, it follows that the macroscopic stability also determines whether or not the average $\langle x\rangle$ of the fluctuations grows with time.

One also deduces directly from (4.8)

$$\frac{d}{d\tau}\langle x^2\rangle = 2\alpha'_1(\varphi)\langle x^2\rangle + \alpha_2(\varphi) \tag{5.2}$$

In both (5.1) and (5.2) terms of order $\Omega^{-1/2}$ have been neglected. With the choice of (4.9) or (4.10) for the initial values one has at $t = 0$

$$\langle x\rangle_0 = 0, \qquad \langle x^2\rangle_0 = 0 \tag{5.3}$$

Hence $\langle x\rangle$ remains zero at all $t > 0$, so that

$$\langle n\rangle_t = \Omega\varphi(t) + \mathcal{O}(1) \tag{5.4}$$

To the present order, therefore, the macroscopic part of n is also its average.

Furthermore, if the macroscopic solution $\varphi(\tau)$ is stable, and therefore also (5.1), it follows from (5.2) that $\langle x^2\rangle$ remains finite as well. Consequently x remains of order unity at all times, which constitutes the *a posteriori* justification of the Ansatz (3.3). Note that the stability is crucial for our approximation scheme: if $\langle x^2\rangle$ grows exponentially in time, the separation of powers of Ω becomes invalid after a time of order long Ω. In Section XVIII we shall meet an example where $\langle x^2\rangle$ grows linearly with time.

It is possible to improve the equation for $\langle n\rangle$ by one order without going beyond the linear noise approximation. To this end we rewrite the exact equation (2.2) in the Ω language in order to display the powers of Ω,

$$\frac{d}{d\tau}(\varphi + \Omega^{-1/2}\langle x\rangle) = \langle \alpha_1(\varphi + \Omega^{-1/2}x)\rangle$$

$$= \alpha_1(\varphi) + \Omega^{-1/2}\alpha'_1(\varphi)\langle x\rangle + \tfrac{1}{2}\Omega^{-1}\alpha''_1(\varphi)\langle x^2\rangle + \mathcal{O}(\Omega^{-3/2})$$

Since φ obeys (4.7) by definition,

$$\frac{d}{d\tau}\langle x\rangle = \alpha_1'(\varphi)\langle x\rangle + \tfrac{1}{2}\Omega^{-1/2}\alpha_1''(\varphi)\langle x^2\rangle + \mathcal{O}(\Omega^{-1}) \tag{5.5}$$

Although this equation for $\langle x\rangle$ involves $\langle x^2\rangle$ there is a factor $\Omega^{-1/2}$; hence the approximation (5.2) for $\langle x^2\rangle$ suffices. Rewriting the result in the original variables we see that the two coupled equations (2.4) and (2.7) determine $\langle n\rangle$ to order Ω^0 and σ_n^2 to order Ω^1.

The conclusion reached here in a slightly devious manner can also be obtained by simply adding the next order correction to (4.8) and then computing the first and second moments to the desired order (compare Section VIII).

VI. ONE-STEP PROCESSES

One-step processes have been defined by processes that obey the master equation (2.8) or (2.10). However, it is necessary to specify in addition the range of n. There are three possibilities: (a) all integers, $-\infty < n < \infty$; (b) half-infinite range, $n = 0, 1, 2, \ldots$; (c) finite range, $n = 0, 1, 2, \ldots, N$. If the range consists of several intervals with gaps between them, a one-step process cannot have transitions between them, so that the process decomposes into several independent processes.

If $r(n)$ and $g(n)$ are constants and n ranges from $-\infty$ to $+\infty$, the one-step process is identical with the (unsymmetric) random walk. The master equation can then easily be solved explicitly and no Ω expansion is needed. If r and g are constant and n has a limited range, for example, $n = 0, 1, 2, \ldots, \infty$, then (2.8) cannot be valid for all n. It can at best hold for $n = 1, 2, \ldots$, whereas for $n = 0$ it must have a slightly different form. We shall then call the boundary at $n = 0$ *artificial*. The random walk with one or two artificial boundaries can still be solved explicitly and will therefore not be considered.

If $r(n)$ and $g(n)$ are linear functions of n, there must be at least one boundary to prevent them from becoming negative. Again this makes a modification of (2.8) necessary. The following particular case is of special interest and will be called a *natural boundary*. Suppose again $n = 0, 1, 2, \ldots$. Then $n = 0$ is a natural boundary if (Fig. 3)

(a)
$$r(0) = 0 \tag{6.1a}$$

(b) the modified equation at $n = 0$ is

$$\dot{P}(0, t) = r(1)P(1, t) - g(0)P(0, t) \tag{6.1b}$$

Note that this is identical with (2.8) for $n = 0$ if one knows that $P(-1, t) = 0$.

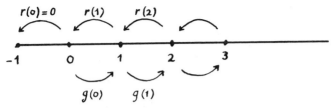

Fig. 3. The one-step process with an absorbing boundary. The boundary is "natural" when
$r(0) = 0$.

Although artificial boundaries often create considerable difficulties for solving the master equation,[17] natural boundaries do not—for the following reason. Consider the master equation (2.8) with natural boundary (6.1b). Now solve (2.8) for $-\infty < n < \infty$, paying no attention to the boundary. Then take an initial state in which $P(n, 0) = 0$ for $n < 0$, for instance (1.7) with some nonnegative m. It will now be true that $P(n, t)$ for $n < 0$ remains zero at all $t > 0$; owing to (6.1a) no probability spills over to negative n. Hence (6.1b) is automatically satisfied, since it is implied in (2.8) if one has $P(-1, t) = 0$.

An upper boundary $n = N$ is called natural if $g(N) = 0$ and the modified equation for $P(N, t)$ is obtained from (2.8) by setting $P(N + 1, t) = 0$. (In a way infinity might also be considered as a natural boundary.) The one-step process with linear or constant $r(n)$ and $g(n)$ and no other than natural boundaries can be solved explicitly, for example with the aid of generating functions.

If $r(n)$ or $g(n)$ or both are nonlinear functions, for instance polynomials, the definition (6.1) of a natural boundary remains valid. Explicit solutions of such master equations are rare, but it is always possible to find the stationary, that is, time-independent solution. For this purpose write (2.10) in the form

$$0 = (\mathbf{E} - 1)\{r(n)P^{\mathrm{st}}(n) - \mathbf{E}^{-1}g(n)P^{\mathrm{st}}(n)\} \tag{6.2}$$

It follows that $\{\ \}$ must be constant

$$r(n)P^{\mathrm{st}}(n) - g(n - 1)P^{\mathrm{st}}(n - 1) = J \tag{6.3}$$

J is the net probability flow from n to $n - 1$. Using (6.3) one can construct the successive $P^{\mathrm{st}}(n)$, starting from a single one, for instance $P^{\mathrm{st}}(0)$, which then serves as a normalizing factor.

If there is a natural boundary, for instance at $n = 0$, one finds on substituting $n = 0$ in (6.3) that J must vanish:

$$r(n)P^{\mathrm{st}}(n) = g(n - 1)P^{\mathrm{st}}(n - 1) \tag{6.4}$$

It then follows directly that

$$P^{st}(n) = \frac{g(n-1)g(n-2)\cdots g(0)}{r(n)r(n-1)\cdots r(1)} P^{st}(0) \tag{6.5}$$

The normalizing factor $P^{st}(0)$ is subsequently found from

$$[P^{st}(0)]^{-1} = \sum_{n=0}^{N} \frac{g(n-1)g(n-2)\cdots g(0)}{r(n)r(n-1)\cdots r(1)} \tag{6.6}$$

When the upper bound N is infinite it may happen that the sum does not converge. In that case every solution $P(n, t)$ continues to spread out indefinitely, in the same way as in the familiar random walk.

It should be emphasized that (6.4) is simply a mathematical identity for one-step processes. It has to be distinguished from detailed balance, which for one-step processes reads

$$r(n)P^{eq}(n) = g(n-1)P^{eq}(n-1) \tag{6.7}$$

Here P^{eq} is the thermal equilibrium distribution and is known *a priori* from the familiar phase space argument of equilibrium statistical mechanics. On the one hand, detailed balance is not restricted to one-step processes; on the other hand it only applies to closed physical systems, without magnetic field or overall rotation.[18] The identity (6.4) also holds for open systems, for example, the photoconductor mentioned in the next section, and for population problems.

VII. SEMICONDUCTOR

As a second example of a nonlinear one-step process consider the following model of an intrinsic semiconductor. A crystal has a nearly empty conduction band and a nearly full valence band. Let n denote the number of electrons that by thermal fluctuations have been excited into the conduction band. The probability per unit time for an excitation to occur is $g(n) = \beta\Omega$, where Ω is the volume of the crystal and β a constant (see Fig. 4). The probability for a recombination is proportional to the number of excited electrons and to the density n/Ω of the available holes: $r(n) = \gamma n^2/\Omega$. Thus the macroscopic rate equation is

$$\frac{dn}{dt} = \beta\Omega - \frac{\gamma}{\Omega} n^2 \tag{7.1}$$

On the mesoscopic level the process is specified by the transition probabilities:

$$W(n|n') = \beta\Omega\delta_{n, n'+1} + \frac{\gamma}{\Omega} n'^2\delta_{n, n'-1} \tag{7.2}$$

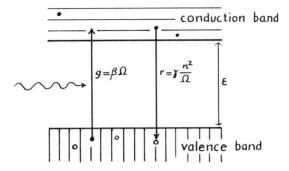

Fig. 4. Model for a semiconductor or a photoconductor.

This is actually of the form stipulated in (4.1)—with $f(\Omega) = \Omega$ and vanishing $\Phi_1, \Phi_2, \ldots$. The master equation for the probability $P(n, t)$ of having n excited electrons is

$$\dot{P}(n, t) = \beta\Omega(\mathbf{E}^{-1} - 1)P + \frac{\gamma}{\Omega}(\mathbf{E} - 1)n^2 P \qquad (7.3)$$

The range of n is $\{0, 1, 2, \ldots\}$ and the equation has a natural boundary at $n = 0$. The stationary solution is found from (6.5) to be

$$P^{\text{st}}(n) = \frac{\Omega^{2n}}{n!^2}\left(\frac{\beta}{\gamma}\right)^n P^{\text{st}}(0) \qquad (7.4)$$

For large Ω this represents a sharp peak whose position and width are determined by

$$\langle (n - \langle n \rangle^{\text{st}})^2 \rangle^{\text{st}} = \tfrac{1}{2}\langle n \rangle^{\text{st}} = \tfrac{1}{2}\Omega\sqrt{\beta/\gamma} \qquad (7.5)$$

The distribution (7.4) must be the same as the Boltzmann distribution given by equilibrium statistical mechanics. Indeed, Ω^{2n} is the phase volume of the n electrons and n holes, and each has a Gibbs factorial. Thus the remaining factor in (7.4) must be the energy exponential:

$$\frac{\beta}{\gamma} = e^{-\epsilon/kT} \qquad (7.6)$$

where ϵ is the energy gap between both bands.

The same master equation (7.3), however, also describes a photoconductor. In that case β has an additional term due to excitation by incident photons, so that (7.6) is no longer true. We shall therefore not make us of it. Note, however, that it is necessary that the arrival times of the photons are un-correlated (shot noise); otherwise their effect cannot be taken into account by a simple addition to the excitation probability but a much more elaborate treatment is necessary.[19]

The jump moments are in the Ω language:

$$\alpha_p(\varphi) = \beta + (-1)^p \gamma \varphi^2 \tag{7.7}$$

Hence the macroscopic equation is, according to (4.7),

$$\dot{\varphi} = \beta - \gamma \varphi^2 \tag{7.8}$$

This equation can readily be solved, but all we shall need to know is that every solution tends to

$$\varphi^{st} = \sqrt{\beta/\gamma} \tag{7.9}$$

which is immediately seen from (7.8) [compare (7.5)]. The Fokker-Planck equation for the fluctuations is, according to (4.8),

$$\frac{\partial \Pi}{\partial t} = 2\gamma \varphi \frac{\partial}{\partial x} x\Pi + \frac{1}{2}(\beta + \gamma \varphi^2) \frac{\partial^2 \Pi}{\partial x^2} \tag{7.10}$$

One finds from this [compare (5.1) and (5.2)]

$$\frac{d}{dt}\langle x \rangle = -2\gamma \varphi \langle x \rangle \tag{7.11}$$

$$\frac{d}{dt}\langle x^2 \rangle = -4\gamma \varphi \langle x^2 \rangle + \beta + \gamma \varphi^2 \tag{7.12}$$

The extended equation (5.5) for $\langle x \rangle$ is

$$\frac{d}{dt}\langle x \rangle = -2\gamma \varphi \langle x \rangle - \gamma \Omega^{-1/2} \langle x^2 \rangle \tag{7.13}$$

which together with (7.12) determines $\langle x \rangle$ to order $\Omega^{-1/2}$ and therefore $\langle n \rangle$ to order Ω^0.

As the solution of these equations is rather laborious we shall merely study the fluctuations in the stationary state. On substituting (7.9) in (7.11) one finds

$$\langle x(t) \rangle = \langle x(0) \rangle e^{-2t\sqrt{\beta \gamma}} \tag{7.14}$$

From (7.12) one obtains, in agreement with (7.5),

$$\langle x^2 \rangle^{st} = \frac{\beta + \gamma \varphi^2}{4\gamma \varphi} = \frac{1}{2}\sqrt{\frac{\beta}{\gamma}} \tag{7.15}$$

These data suffice to find the autocorrelation function of x,

$$\langle x(0)x(t)\rangle^{\text{st}} = \frac{1}{2}\sqrt{\frac{\beta}{\gamma}}\, e^{-2t\sqrt{\beta\gamma}} \tag{7.16}$$

The autocorrelation function of n is therefore

$$\langle\{n(0) - \langle n\rangle^{\text{st}}\}\{n(t) - \langle n\rangle^{\text{st}}\}\rangle^{\text{st}} = \frac{1}{2}\Omega\sqrt{\frac{\beta}{\gamma}}\, e^{-2t\sqrt{\beta\gamma}} \tag{7.17}$$

where terms of order $\Omega^{1/2}$ have been neglected.

This autocorrelation function consists of a single exponential, so that the spectral density of the fluctuations consists of a single Debye term with relaxation time $[2\sqrt{\beta\gamma}]^{-1}$. That result is not surprising since to this order the noise is treated in linear approximation. In the next section it will be seen that the higher orders give rise to additional exponentials involving other relaxation times.

VIII. HIGHER-ORDER CORRECTIONS

It is easy to work out the higher orders of (4.4) beyond the power Ω^0 displayed in (4.8). The result is

$$
\begin{aligned}
\frac{\partial\Pi}{\partial t} &= -\frac{\partial}{\partial x}\left\{\alpha_1' x + \frac{1}{2}\Omega^{-1/2}\alpha_1'' x^2 + \frac{1}{3!}\Omega^{-1}\alpha_1''' x^3 + \cdots\right\}\Pi \\
&\quad + \frac{1}{2}\frac{\partial^2}{\partial x^2}\left\{\alpha_2 + \Omega^{-1/2}\alpha_2' x + \frac{1}{2}\Omega^{-1}\alpha_2'' x^2 + \cdots\right\}\Pi \\
&\quad - \frac{1}{3!}\Omega^{-1/2}\frac{\partial^3}{\partial x^3}\{\alpha_3 + \Omega^{-1/2}\alpha_3' x + \cdots\}\Pi \\
&\quad + \frac{1}{4!}\Omega^{-1}\frac{\partial^4}{\partial x^4}\{\alpha_4 + \cdots\}\Pi + \cdots .
\end{aligned}
\tag{8.1}
$$

All functions α have the argument φ, the prime indicates differentiation with respect to φ. All terms of order Ω^{-1} have been written down, and it is clear how the expansion continues: each order of $\Omega^{-1/2}$ adds one power of x to the coefficients and at the same time an additional derivative appears. The nonlinear Fokker–Planck equation, which is so often used for describing nonlinear random processes in physics, contains all terms on the first two lines and ignores all other lines; clearly that is an inconsistent approximation, unfit to describe anything beyond the linear noise approximation (see Section XXII).

One cannot expect to solve the full equation (8.1), but it is remarkable that one can find the moments to any given order explicitly. For example, to find $\langle x \rangle$ to order $\Omega^{-1/2}$ we write

$$\frac{d}{dt}\langle x \rangle = \alpha_1'\langle x \rangle + \tfrac{1}{2}\Omega^{-1/2}\alpha_1''\langle x^2 \rangle + \mathcal{O}(\Omega^{-1}) \tag{8.2a}$$

$$\frac{d}{dt}\langle x^2 \rangle = 2\alpha_1'\langle x^2 \rangle + \alpha_2 + \mathcal{O}(\Omega^{-1/2}) \tag{8.2b}$$

These are the two equations already discussed in Sections II and V.

We shall now calculate the next correction to the autocorrelation function (7.16) or (7.17) of the model in the previous section. It is readily seen that the approximation (8.2) does not contribute owing to symmetry between positive and negative values of x. Hence we have to go one step further to find how the nonlinearity affects the autocorrelation function and thereby the fluctuation spectrum. To save writing we rescale the variables,

$$2t\sqrt{\beta\gamma} = \tau, \qquad (\gamma/\beta)^{1/4}x = \xi, \qquad (\gamma/\beta)^{1/4}\Omega^{-1/2} = \epsilon \tag{8.3}$$

Then (8.1) to order ϵ^2 takes the form

$$\frac{\partial\Pi}{\partial\tau} = \frac{\partial}{\partial\xi}\left\{\xi + \frac{1}{2}\epsilon\xi^2\right\}\Pi + \frac{1}{2}\frac{\partial^2}{\partial\xi^2}\left\{1 + \epsilon\xi + \frac{1}{2}\epsilon^2\xi^2\right\}\Pi$$
$$+ \frac{1}{3}\epsilon^2\frac{\partial^3}{\partial\xi^3}\xi\Pi + \frac{1}{24}\epsilon^2\frac{\partial^4\Pi}{\partial\xi^4} \tag{8.4}$$

The equations for the first four moments are to the required order

$$\frac{d}{d\tau}\langle\xi\rangle = -\langle\xi\rangle - \frac{1}{2}\epsilon\langle\xi^2\rangle + \mathcal{O}(\epsilon^3) \tag{8.5a}$$

$$\frac{d}{d\tau}\langle\xi^2\rangle = -2\langle\xi^2\rangle - \epsilon\langle\xi^3\rangle + 1 + \epsilon\langle\xi\rangle + \frac{1}{2}\epsilon^2\langle\xi\rangle + \mathcal{O}(\epsilon^3) \tag{8.5b}$$

$$\frac{d}{d\tau}\langle\xi^3\rangle = -3\langle\xi^3\rangle - \frac{3}{2}\epsilon\langle\xi^4\rangle + 3\langle\xi\rangle + 3\epsilon\langle\xi^2\rangle + \mathcal{O}(\epsilon^2) \tag{8.5c}$$

$$\frac{d}{d\tau}\langle\xi^4\rangle = -4\langle\xi^4\rangle + 6\langle\xi^2\rangle + \mathcal{O}(\epsilon) \tag{8.5d}$$

From this we first conclude

$$\langle\xi^4\rangle^{st} = \tfrac{3}{2}\langle\xi^2\rangle^{st} = \tfrac{3}{4} + \mathcal{O}(\epsilon) \tag{8.6a}$$

$$\langle\xi^3\rangle^{st} = -\tfrac{3}{8}\epsilon - \tfrac{1}{4}\epsilon + \tfrac{1}{2}\epsilon = -\tfrac{1}{8}\epsilon + \mathcal{O}(\epsilon^2) \tag{8.6b}$$

$$\langle\xi^2\rangle^{st} = \tfrac{1}{2} - \tfrac{1}{16}\epsilon^2 + \mathcal{O}(\epsilon^3) \tag{8.6c}$$

$$\langle\xi\rangle^{st} = -\tfrac{1}{4}\epsilon + \mathcal{O}(\epsilon^3) \tag{8.6d}$$

Notice that to order Ω^{-1} it is no longer true that $\langle\xi\rangle = 0$, so that the macroscopic part $\Omega\varphi$ of n is no longer identical with the average $\langle n\rangle$.

Subsequently solve the equations for the first three moments as functions of time to the desired order in ϵ. The initial conditions are $\langle\xi\rangle_0 = \xi_0$, $\langle\xi^2\rangle_0 = \xi_0{}^2$, $\langle\xi^3\rangle_0 = \xi_0{}^3$. Finally multiply $\langle\xi(t)\rangle_{\xi_0}$ with ξ_0 and average ξ_0 over the stationary distribution. The result is the autocorrelation function:

$$\langle\xi_0\langle\xi(t)\rangle_\xi\rangle^{st} - \{\langle\xi\rangle^{st}\}^2 = \tfrac{1}{2}(1 - \tfrac{1}{2}\epsilon^2)e^{-(1-(1/4)\epsilon^2)\tau} + \tfrac{1}{8}\epsilon^2 e^{-(2+\epsilon^2)\tau}$$

The coefficients and the exponents are correct to order ϵ^2.

From this result one sees that the order Ω^{-1} lowers the rate of decay of the leading term, and modifies its coefficient. More strikingly, however, another exponential term appears which decays roughly twice as fast. Thus the nonlinearity gives rise to an additional Debye term in the fluctuation spectrum, which in principle could be observed. Higher orders give rise to a sequence of such terms.

The microscopic treatment of Bernard and Callen[20] led them to conclude that the nonlinearity does not affect the fluctuation spectrum at all. Admittedly, the present mesoscopic treatment is founded on the assumption that the master equation applies, but there are doubts about the applicability of the microscopic results as well.[5,21] It would be of great interest to perform an *experimentum crucis*.

IX. THE MALTHUS–VERHULST PROBLEM

In a population of n individuals each individual has a probability $\alpha\,dt$ to die in the next dt, and a probability $\beta\,dt$ to give birth to an additional individual. Moreover the struggle for life gives rise to an additional death rate, which is proportional to the density $(n-1)/\Omega$ of other individuals, where Ω is the amount of space or food available. The growth of the population is described by the Malthus–Verhulst equation,[22]

$$\frac{dn}{dt} = (\beta - \alpha)n - \frac{\gamma}{\Omega}n(n-1) \tag{9.1}$$

On the mesoscopic level one has a transition probability

$$W(n|n') = r(n')\delta_{n,\,n'-1} + g(n')\delta_{n,\,n'+1}$$

$$= \left\{\alpha n' + \frac{\gamma}{\Omega}n'(n'-1)\right\}\delta_{n,\,n'-1} + \beta n'\delta_{n,\,n'+1} \tag{9.2}$$

It has the form (4.1) with $f(\Omega) = \Omega$ including a term

$$\Phi_1(n'/\Omega;\,n-n') = \gamma(n'/\Omega)\delta_{n,\,n'-1} \tag{9.3}$$

which has been underlined in (9.2). For this reason we shall carry out the expansion once more explicitly. The result will be that this term neither contributes to the macroscopic equation, nor to the linear noise approximation, but only to higher orders.

The master equation is

$$\dot{P}(n, t) = \alpha(\mathbf{E} - 1)nP + \beta(\mathbf{E}^{-1} - 1)nP + \frac{\gamma}{\Omega}(\mathbf{E} - 1)n(n - 1)P \qquad (9.4)$$

Substitute the transformation (3.3), (3.4) using (3.6)

$$\frac{\partial \Pi}{\partial t} - \Omega^{1/2}\frac{d\varphi}{dt}\frac{\partial \Pi}{\partial x} = \alpha\Omega\left\{\Omega^{-1/2}\frac{\partial}{\partial x} + \frac{1}{2}\Omega^{-1}\frac{\partial^2}{\partial x^2}\right\}(\varphi + \Omega^{-1/2}x)\Pi$$

$$+ \beta\Omega\left\{-\Omega^{-1/2}\frac{\partial}{\partial x} + \frac{1}{2}\Omega^{-1}\frac{\partial^2}{\partial x^2}\right\}(\varphi + \Omega^{-1/2}x)\Pi$$

$$+ \gamma\Omega\left\{\Omega^{-1/2}\frac{\partial}{\partial x} + \frac{1}{2}\Omega^{-1}\frac{\partial^2}{\partial x^2}\right\}$$

$$\times (\varphi + \Omega^{-1/2}x)(\varphi + \Omega^{-1/2}x - \underline{\Omega^{-1}})\Pi \qquad (9.5)$$

Collecting the terms of order $\Omega^{1/2}$ one obtains the macroscopic equation

$$-\dot{\varphi} = (\alpha - \beta)\varphi + \gamma\varphi^2 \qquad (9.6)$$

It is the same equation as (9.1) with $n(n - 1)$ replaced by n^2. This difference is meaningless on the macroscopic level, because on this level one also ignores the difference between $\langle n^2 \rangle$ and $\langle n \rangle^2$, which is of the same order. It is therefore permissible, and in fact more consistent, to write n^2 in (9.1), as is usually done.

The solution of (9.6) is

$$\varphi(t) = \frac{\varphi(0)e^{(\beta - \alpha)t}}{1 + \varphi(0)\dfrac{\gamma}{\beta - \alpha}\{e^{(\beta - \alpha)t} - 1\}} \qquad (9.7)$$

For $\beta < \alpha$ it tends to zero: the population dies out because the death rate exceeds the birth rate. For $\beta > \alpha$ it tends to

$$\varphi^{st} = \frac{\beta - \alpha}{\gamma} \qquad (9.8)$$

which is a balance between the natural growth and the mutual interference.

The terms of order Ω^0 in (9.5) yield the Fokker-Planck equation

$$\frac{\partial \Pi}{\partial t} = (\alpha - \beta + 2\gamma\varphi)\frac{\partial}{\partial x}x\Pi + \frac{1}{2}(\alpha\varphi + \beta\varphi + \gamma\varphi^2)\frac{\partial^2 \Pi}{\partial x^2} \qquad (9.9)$$

Note that the underlined term in (9.2) and (9.5) still does not contribute. From (9.9) one obtains for the first two moments

$$\frac{d}{dt}\langle x \rangle = (\beta - \alpha - 2\gamma\varphi)\langle x \rangle \tag{9.10}$$

$$\frac{d}{dt}\langle x^2 \rangle = 2(\beta - \alpha - 2\gamma\varphi)\langle x^2 \rangle + (\alpha\varphi + \beta\varphi + \gamma\varphi^2) \tag{9.11}$$

In particular, after φ has reached its stationary value (9.8) the fluctuations are determined by

$$\frac{d}{dt}\langle x \rangle = -(\beta - \alpha)\langle x \rangle \tag{9.12}$$

$$\frac{d}{dt}\langle x^2 \rangle = -2(\beta - \alpha)\langle x^2 \rangle + 2\beta(\beta - \alpha)/\gamma \tag{9.13}$$

Notice that whereas in the macroscopic equation (9.6) the quantities α and β only occur in the combination $\alpha - \beta$, (9.13) contains β separately. This demonstrates how the observation of fluctuations permits one to find more information about a system than mere macroscopic measurements. Of course this is well known; observation of Brownian motion or critical opalescence permits one to determine Avogadro's number.

In order to check the validity of the expansion in Ω we have to investigate whether the fluctuations remain small of the order anticipated in (3.3). First, for $\beta < \alpha$ there is one stationary value $\varphi^{\text{st}} = 0$, and it is asymptotically stable since all other solutions tend to zero [see (9.7)]. It is evident from Fig. 5a that the fluctuations tend to decrease rather than to grow, and this is confirmed by the explicit calculation in the Appendix.

When $\beta > \alpha$ the stationary solution $\varphi = 0$ is unstable; all other solutions tend to (9.8). It is clear from Fig. 5b that the fluctuations may be magnified for some time, but will ultimately decrease again. Hence they cannot grow such that they violate the assumption they are of order $\Omega^{1/2}$.

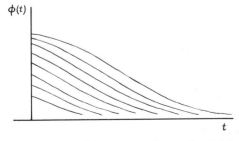

Fig. 5a.   The solutions of the macroscopic equation (9.6) for $\beta < \alpha$.

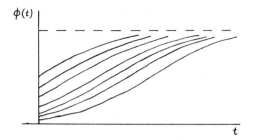

Fig. 5b.   The solutions of the same equation for $\beta > \alpha$.

X. THE TRANSITION REGION

The fact that for different values of the parameters different stationary states prevail may be considered as a phase transition. This point of view will be amplified in Section XX, but it is useful to work out some details in the context of the present example.

In the previous section it was tacitly understood that $|\alpha - \beta|$ was of order unity. There is a transition region, however, where it is small of order $\Omega^{-1/2}$. In that case the stable and unstable stationary states are not clearly separated, since their distance is of the same order as the fluctuations. In order to treat this we set $\beta - \alpha = \Omega^{-1/2}\Delta$, with constant Δ, either positive or negative. Without loss of generality we also take for convenience $\beta + \alpha = 1$, $\gamma = 1$.

The transformed master equation (9.5) may now be written

$$\frac{\partial \Pi}{\partial t} - \Omega^{1/2} \frac{d\varphi}{dt} \frac{\partial \Pi}{\partial x} = -\Delta\varphi \frac{\partial \Pi}{\partial x} + \frac{1}{2} \varphi \frac{\partial^2 \Pi}{\partial x^2}$$
$$+ \Omega\left(\Omega^{-1/2} \frac{\partial}{\partial x} + \frac{1}{2} \Omega^{-1} \frac{\partial^2}{\partial x^2}\right)(\varphi + \Omega^{-1/2}x)^2\Pi \quad (10.1)$$

The terms of order $\Omega^{1/2}$ yield

$$\frac{d\varphi}{dt} = -\varphi^2 \tag{10.2}$$

This agrees with (9.6) since α and β are now equal from a macroscopic point of view. The terms of order Ω^0 yield

$$\frac{\partial \Pi}{\partial t} = \varphi \frac{\partial}{\partial x} (-\Delta + 2x)\Pi + \frac{1}{2} (\varphi + \varphi^2) \frac{\partial^2 \Pi}{\partial x^2} \tag{10.3}$$

Note that this equation is essentially different from (9.7) and cannot be obtained from it as a limiting case.

The solution of (10.2) with initial value φ_0 is

$$\varphi(t) = \frac{\varphi_0}{1 + t\varphi_0}$$

Thus the macroscopic part of n tends to zero, but merely as t^{-1} rather than exponentially. On the other hand, the average of the fluctuations obeys according to (10.3)

$$\frac{d}{dt}\langle x \rangle = \varphi(t)(\Delta - 2\langle x \rangle)$$

This is no longer identical with the variational equation associated with (10.2). The solution with initial value $\langle x \rangle_0 = 0$ is

$$\langle x \rangle_t = \frac{\Delta}{2}\left\{ 1 - \frac{1}{(1 + t\varphi_0)^2} \right\}$$

Hence it slowly tends to $\frac{1}{2}\Delta$ rather than to zero. In this case therefore, already in order Ω^0 the average $\langle n \rangle$ does not coincide with the macroscopic part $\Omega\varphi$ of n.

One also finds after some algebra

$$\langle x^2 \rangle_t - \langle x \rangle_t^2 = \frac{1}{4}\left\{ 1 - \frac{1}{(1 + t\varphi_0)^4} \right\} + \frac{\varphi_0}{3}\frac{1}{1 + t\varphi_0}\left\{ 1 - \frac{1}{(1 + t\varphi_0)^3} \right\}$$

which tends to $\frac{1}{4}$. Thus when $t \gg \varphi_0^{-1}$ the distribution $P(n, t)$ consists of a peak with center $\frac{1}{2}\Omega^{1/2}\Delta$ and width $\frac{1}{2}\Omega^{1/2}$.

This cannot be the whole story, however, because according to this picture n would also take negative values. The present model has the additional complication that one of the two stationary states also happens to be a boundary. The above solution can therefore only be trusted as long as the fluctuations do not reach that boundary:

$$\Omega\varphi(t) \gg \Omega^{1/2}\sqrt{\langle x^2 \rangle_t}$$

This condition amounts to $t \ll \Omega^{1/2}$. Thus in this case the expansion in powers of $\Omega^{-1/2}$ is no longer uniformly valid for all t, and cannot provide information about the behavior for $t \to \infty$.

It is obvious what that behavior is. Whenever by a fluctuation n reaches the value zero the population has died out, so that n remains zero ever after. No spontaneous generation occurs in the Malthus–Verhulst equation, in contrast with the semiconductor in Section VII. Thus $n = 0$ is an absorbing state and the Fokker–Plank equation (10.1) for Π should be solved with an absorbing boundary condition at $n = 0$. In terms of x the boundary moves as determined by

$$x = -\Omega^{1/2}\varphi(t)$$

Absorption at the boundary is represented by the condition that Π must vanish, hence one has to solve (10.1) with the boundary condition $\Pi(-\Omega^{1/2}\varphi(t), t) = 0$. Of course in principle this is also true for (9.9) in the previous section, but in that case $P(0, t)$ is small of order $\exp(-\Omega)$, so that error is negligible (compare Section XX).

XI. THE LANGEVIN APPROACH

The success of Langevin's treatment of Brownian motion has led to the erroneous idea that all fluctuations can be treated in that way. The method consists of three steps.

(*i*) One supplements the macroscopic equation with a stochastic driving force. In the notation of Section II

$$\frac{dn}{dt} = a_1(n) + l(t) \tag{11.1}$$

This equation determines a stochastic process $n(t)$, whose mean and variance are supposed to provide the macroscopic value and the fluctuations of n, respectively. Of course (11.1) is moot unless one says something about $l(t)$. First the usually tacit assumption is made that the statistical properties of l are independent of $n(t)$.

(*ii*) Subsequently two more explicit assumptions are made:

$$\langle l(t) \rangle = 0, \qquad \langle l(t_1)l(t_2) \rangle = \Gamma\delta(t_1 - t_2) \tag{11.2}$$

with constant Γ. When $a_1(n)$ is linear in n these properties are sufficient for computing the first two moments of $n(t)$.[23] If one is interested in higher moments, or when $a_1(n)$ is nonlinear, one needs the higher moments of $l(t)$ as well. They are ordinarily chosen by assuming $l(t)$ to be Gaussian, that is, all higher cumulants of $l(t)$ vanish, or all higher moments of $l(t)$ factorize, for example,

$$\langle l(t_1)l(t_2)l(t_3)l(t_4) \rangle = \langle l(t_1)l(t_2) \rangle \langle l(t_3)l(t_4) \rangle$$
$$+ \langle l(t_1)l(t_3) \rangle \langle l(t_2)l(t_4) \rangle + \langle l(t_1)l(t_4) \rangle \langle l(t_2)l(t_3) \rangle \tag{11.3}$$

Whereas (11.2) makes physical sense for the force exerted by gas molecules on a Brownian particle, in other cases it is often doubtful, and of course (11.3) goes far beyond all physical intuition.

(iii) The macroscopic stationary value of n is found from $a_1(n^{st}) = 0$. It is stable if $a'_1(n^{st}) < 0$. One then obtains for the fluctuations in linear approximation:

$$\frac{d\Delta n}{dt} = a'_1(n^{st})\Delta n + l(t) \tag{11.4}$$

On solving this equation[23] one finds in the limit $t \to \infty$:

$$\langle \Delta n^2 \rangle_\infty = \frac{\Gamma}{2} |\alpha'_1(n^{st})| \tag{11.5}$$

If one is dealing with a system in thermal equilibrium, (11.5) must be identical with the equilibrium fluctuation $\langle \Delta n^2 \rangle^{eq}$ as known from equilibrium statistical mechanics. This identification determines Γ, supposing that the macroscopic $a_1(n)$ is known—which is the fluctuation-dissipation theorem. If one is dealing with a stationary state other than thermal equilibrium, no *a priori* knowledge of the mean square fluctuations is available and the value of Γ must be found in another way.

We shall now test the various assumptions on which this approach is based by comparing its results with those of the systematic expansion. Throughout we are only concerned with the weaker version (11.2) rather than (11.3). Moreover by not using (11.5) the discussion is not confined to thermal equilibrium.

On averaging (11.4) one obtains for the average of the fluctuations near the stationary state

$$\frac{d\langle \Delta n \rangle}{dt} = a'_1(n^{st})\langle \Delta n \rangle \tag{11.6}$$

This is identical with (5.1) when in the latter the stationary value of φ is substituted. Also one finds in the usual way

$$\frac{\langle (\Delta n)^2 \rangle}{\Delta t} = 2a'_1(n^{st})\langle (\Delta n)^2 \rangle + \Gamma \tag{11.7}$$

This is identical with (5.2) if one takes $\Gamma = a_2(n^{st})$. Thus the Langevin approach yields equations for $\langle \Delta n \rangle$ and $\langle (\Delta n)^2 \rangle$ in the stationary state that are correct to the same approximation as (2.3), (2.7), that is, with neglect of $\mathcal{O}(\Omega^0)$ and $\mathcal{O}(\Omega^{1/2})$, respectively.

However, let us consider the fluctuations around a nonstationary solution. On averaging (11.1) one obtains (2.2), which is exact, but not identical with the macroscopic equation (2.3) unless $a_1(n)$ happens to be a linear function.

An equation for $\langle(\Delta n)^2\rangle$ cannot even be deduced from (11.1) with (11.2). Actually the difficulty can be demonstrated on a simple linear example.

A radioactive sample consists of identical nuclei, each having a probability λ per unit time to decay. The transition probability per unit time for the number n of active nuclei is

$$W(n|n') = \lambda n' \delta_{n, n'-1} \tag{11.8}$$

Hence

$$a_p(n) = (-1)^p \lambda n \tag{11.9}$$

so that the exact equations (2.2) and (2.5) take the form

$$\frac{d}{dt} \langle n \rangle = -\lambda \langle n \rangle$$

$$\frac{d}{dt} \sigma_n^2 = \lambda \langle n \rangle - 2\lambda \sigma_n^2 \tag{11.10}$$

Their solutions with the initial condition (1.7) are

$$\langle n \rangle_t = me^{-\lambda t}$$

$$\sigma_n^2(t) = me^{-\lambda t}(1 - e^{-\lambda t}) = \langle n \rangle_t (1 - e^{-\lambda t}) \tag{11.11}$$

The Langevin approach, on the other hand, starts from the equation

$$\frac{dn}{dt} = -\lambda n + l(t) \tag{11.12}$$

with the solution

$$n(t) = me^{-\lambda t} + e^{-\lambda t} \int_0^t e^{\lambda t'} l(t')\, dt \tag{11.13}$$

On averaging one duly finds (11.10). However, on squaring (11.13)

$$\langle n(t)^2 \rangle = m^2 e^{-2\lambda t} + \frac{\Gamma}{2\lambda} (1 - e^{-2\lambda t})$$

so that

$$\sigma_n(t)^2 = \frac{\Gamma}{2\lambda} (1 - e^{-2\lambda t}) \tag{11.14}$$

Clearly it is not possible to choose the constant Γ in order to find agreement with (11.11). A fortiori the fluctuation–dissipation value for Γ cannot be correct; in fact for the present case it would give $\Gamma = 0$. It is true that (11.14) reduces to (11.11) if one puts

$$\Gamma = 2\lambda m(e^{\lambda t} + 1)^{-1}$$

but that would make Γ depend not only on time, but also on the special solution considered. The conclusion is therefore that fluctuations around a time-dependent state cannot be treated by the Langevin method.

XII. THE GENERATION AND RECOMBINATION CURRENTS

Consider the one-step master equation [(2.8) or (2.10)]. The solution $P(n, t)$ is the probability distribution of n at time t, resulting from the random up and down jumps of n. We now ask how these jumps are distributed in time. Rather than treating the general case we shall concentrate on the recombination events in the stationary state of the semiconductor model of Section VII. In a photoconductor these events may emit photons, which can be observed. In population problems the result might be of interest to morticians.

The recombination events constitute a random sequence of dots on the time axis. An appropriate tool for describing the statistical properties of such sequences is furnished by the hierarchy of distribution functions,[24]

$$f_n(t_1, t_2, \ldots, t_n) \qquad (n = 1, 2, 3, \ldots)$$

They are defined as follows. First $f_1(t_1) \, dt_1$ is the probability for having an event in the time interval $(t_1, t_1 + dt_1)$, regardless of what happens outside that interval. The probability for having two or more events in $(t_1, t_1 + dt_1)$ is of higher order in dt_1 and therefore negligible. Next $f_2(t_1, t_2) \, dt_1 \, dt_2$ is the joint probability for having one event in $(t_1, t_1 + dt_1)$ and another in $(t_2, t_2 + dt_2)$, regardless of all other events. Since this definition only holds for $t_1 \neq t_2$ we have to add that on integrating f_2 no extra contribution arises from $t_1 = t_2$; that is, f_2 contains no terms with $\delta(t_1 - t_2)$.

The higher f_n are defined in the same way, but we shall only compute f_1 and f_2. It is often convenient to express the result in terms of the correlation functions g_n, which are related to the f_n by the familiar cluster expansion,

$$g_1(t_1) = f_1(t_1), \qquad g_2(t_1, t_2) = f_2(t_1, t_2) - f_1(t_1)f_1(t_2)$$

For independent events g_2 and the higher g_n vanish. By definition f_2 and g_2 are symmetric functions of t_1, t_2. When a sequence of events is stationary its f_1 is independent of time, and f_2 only depends on $t_1 - t_2$.

Now let $P(n, t)$ be a solution of (2.10) and therefore describe a mesoscopic state of the system. The first term describes the recombination events. The probability for the occurrence of a jump from n to $n - 1$ in $(t_1, t_1 + dt_1)$ is

$$r(n_1) \, dt_1 \, P(n_1, t_1)$$

The probability per unit time for a recombination to occur is therefore

$$f_1(t_1) = \sum_{n_1=1}^{\infty} r(n_1)P(n_1, t_1) = \langle r(n_1) \rangle_{t_1} \tag{12.1}$$

This is the average recombination current at time t_1.

In order to compute $f_2(t_1, t_2)$ we have to find the joint probability for having a jump from n_1 to $n_1 - 1$ in $(t_1, t_1 + dt_1)$ and subsequently a jump from n_2 to $n_2 - 1$ in $(t_2, t_2 + dt_2)$. This is given by

$$r(n_1) \, dt_1 P(n_1, t) \cdot r(n_2) \, dt_2 \, P(n_2, t_2 | n_1 - 1, t_1)$$

The last factor is the conditional probability (1.8), which describes the evolution from t_1 to t_2. It starts from the initial value $n_1 - 1$, because a recombination took place at t_1. This detail cannot be ignored as was first remarked by Ubbink.[18] Thus we find

$$f_2(t_1, t_2) = \sum_{n_1, n_2} r(n_1)r(n_2)P(n_1, t_1)P(n_2, t_2 | n_1 - 1, t_1) \tag{12.2}$$

Of course similar expressions obtain for the f_1 and f_2 of the generation events and it is easy to see how they generalize to higher f_n. They are exact, but in order to evaluate them one needs to know the solutions of the master equation.

For more explicit results take the stationary state (7.4) of the semiconductor. According to (12.1) and (3.3)

$$f_1 = \frac{\gamma}{\Omega} \sum_{n=1}^{\infty} n^2 P^{\text{st}}(n) = \beta\Omega + \mathcal{O}(\Omega^{-1}) \tag{12.3}$$

To show that the term of order Ω^0 vanishes use has been made of

$$\langle x \rangle^{\text{st}} = -\tfrac{1}{4}\Omega^{-1/2} \tag{12.4}$$

which follows from (8.6d) with the aid of (8.3).

The evaluation of (12.2) takes some algebra. We go to order Ω and are careful to include all the necessary powers of Ω. The symbol φ throughout denotes $\varphi^{\text{st}} = \sqrt{\beta/\gamma}$, and $\Pi(x)$ is the stationary distribution.

$$f_2(t_1, t_2) = \gamma^2\Omega^2 \iint \{\varphi + \Omega^{-1/2}x_2\}^2 \Pi(x_2, t_2 | x_1 - \Omega^{-1/2}, t_1)$$

$$\times \{\varphi + \Omega^{-1/2}x_1\}^2 \Pi(x_1) \, dx_1 \, dx_2 \tag{12.5}$$

Begin by working out the first quadratic factor.

$$f_2(t_1, t_2) = \gamma^2 \Omega^2 \varphi^2 \int \{\varphi + \Omega^{-1/2} x_1\}^2 \Pi(x_1)\, dx_1$$

$$+ 2\gamma^2 \Omega^{3/2} \varphi \iint x_2 \Pi(x_2, t_2 | x_1 - \Omega^{-1/2}, t_1)$$

$$\times \{\varphi + \Omega^{-1/2} x_1\}^2 \Pi(x_1)\, dx_1\, dx_2$$

$$+ \gamma^2 \Omega \iint x_2{}^2 \Pi(x_2, t_2 | x_1, t_1) \varphi^2 \Pi(x_1)\, dx_1\, dx_2$$

$$= \gamma^2 \Omega^2 \varphi^4 + 2\gamma^2 \Omega^{3/2} \varphi^3 \langle x \rangle^{\mathrm{st}} + \gamma^2 \Omega \varphi^2 \langle x^2 \rangle^{\mathrm{st}}$$

$$+ 2\gamma^2 \Omega^{3/2} \varphi^3 \iint x_2 \Pi(x_2, t_2 | x_1 - \Omega^{-1/2}, t_1) \Pi(x_1)\, dx_1\, dx_2$$

$$+ 4\gamma^2 \Omega \varphi^2 \iint x_2 \Pi(x_2, t_2 | x_1, t_1) x_1 \Pi(x_1)\, dx_1$$

$$+ \gamma^2 \Omega \varphi^2 \langle x^2 \rangle^{\mathrm{st}} \tag{12.6}$$

The integral on the third line is given to sufficient approximation by (7.16); we set $t_2 - t_1 = 2\tau \sqrt{\beta \gamma}$. The integral on the second line can be split into two parts. The first is with the aid of (12.4):

$$\iint x_2 \Pi(x_2, t_2 | x_1, t_1) \Pi(x_1)\, dx_1\, dx_2 = \langle x \rangle^{\mathrm{st}} = -\tfrac{1}{4}\Omega^{-1/2}$$

The second part is to sufficient approximation:

$$\int x_2 \Pi(x_2, t_2 | -\Omega^{-1/2}, t_1) \int \Pi(x_1)\, dx_1\, dx_2 = -\Omega^{-1/2} \exp\left[-2\sqrt{\beta \gamma}(t_2 - t_1)\right]$$

Collecting these results one obtains surprisingly

$$f_2(t_1, t_2) = \gamma^2 \Omega^2 \varphi^4 = \beta^2 \Omega^2 \tag{12.7}$$

Hence $g_2 = 0$; the recombination events are uncorrelated! It is not clear (to me) whether this could have been predicted on *a priori* grounds.

XIII. EXAMPLE OF A MULTIVARIATE MASTER EQUATION

As a simple example in which two variables n, m occur, consider the following modification of the Malthus–Verhulst equation. A population consists of m males and n females. They have the same probability per unit time to die, both due to natural death and to the competition with the surrounding population density $(n + m)/\Omega$. Their birth probability is for both sexes proportional with the number n of females.

$$\dot{n} = \beta n - \alpha n - \gamma n(n + m)/\Omega \tag{13.1a}$$

$$\dot{m} = \beta n - \alpha m - \gamma m(n + m)/\Omega \tag{13.1b}$$

It is possible, but unnecessarily cumbersome, to use different values for α, β, γ in both equations.

The master equation for the joint distribution $P(n, m, t)$ is

$$\dot{P} = \beta(\mathbf{E}_n^{-1} - 1)nP + \alpha(\mathbf{E}_n - 1)nP + \frac{\gamma}{\Omega}(\mathbf{E}_n - 1)n(n + m)P$$

$$+ \beta(\mathbf{E}_m^{-1} - 1)nP + \alpha(\mathbf{E}_m - 1)mP + \frac{\gamma}{\Omega}(\mathbf{E}_m - 1)m(n + m)P \tag{13.2}$$

$\mathbf{E}_n$ and $\mathbf{E}_m$ are difference operators (2.9) acting on n and m, respectively. The ready-made equations derived in Sections IV and V cannot be used, but it is easy to adapt the Ω expansion to this multivariate case.

Apply again the information (3.3) combined with

$$m = \Omega\psi(t) + \Omega^{1/2}y$$

The new distribution $\Pi(x, y, t)$ then obeys to order Ω^0

$$\frac{\partial \Pi}{\partial t} - \Omega^{1/2}\frac{d\varphi}{dt}\frac{\partial \Pi}{\partial x} - \Omega^{1/2}\frac{d\psi}{dt}\frac{\partial \Pi}{\partial y}$$

$$= \beta\Omega\left(-\Omega^{-1/2}\frac{\partial}{\partial x} + \frac{1}{2}\Omega^{-1}\frac{\partial^2}{\partial x^2}\right)(\varphi + \Omega^{-1/2}x)\Pi$$

$$+ \alpha\Omega\left(\Omega^{-1/2}\frac{\partial}{\partial x} + \frac{1}{2}\Omega^{-1}\frac{\partial^2}{\partial x^2}\right)(\varphi + \Omega^{-1/2}x)\Pi$$

$$+ \gamma\Omega\left(\Omega^{-1/2}\frac{\partial}{\partial x} + \frac{1}{2}\Omega^{-1}\frac{\partial^2}{\partial x^2}\right)$$

$$\times (\varphi + \Omega^{-1/2}x)(\varphi + \psi + \Omega^{-1/2}x + \Omega^{-1/2}y)\Pi$$

$$+ \beta\Omega\left(-\Omega^{-1/2}\frac{\partial}{\partial y} + \frac{1}{2}\Omega^{-1}\frac{\partial^2}{\partial y^2}\right)(\varphi + \Omega^{-1/2}x)\Pi$$

$$+ \alpha\Omega\left(\Omega^{-1/2}\frac{\partial}{\partial y} + \frac{1}{2}\Omega^{-1}\frac{\partial^2}{\partial y^2}\right)(\psi + \Omega^{-1/2}y)\Pi$$

$$+ \gamma\Omega\left(\Omega^{-1/2}\frac{\partial}{\partial y} + \frac{1}{2}\Omega^{-1}\frac{\partial^2}{\partial y^2}\right)$$

$$\times (\psi + \Omega^{-1/2}y)(\varphi + \psi + \Omega^{-1/2}x + \Omega^{-1/2}y)\Pi \tag{13.3}$$

All terms of order $\Omega^{1/2}$ are either proportional to $\partial\Pi/\partial x$ or to $\partial\Pi/\partial y$. The coefficient of each of these derivatives vanishes if

$$-\frac{d\varphi}{dt} = -\beta\varphi + \alpha\varphi + \gamma\varphi(\varphi + \psi) \tag{13.4a}$$

$$-\frac{d\psi}{dt} = -\beta\varphi + \alpha\psi + \gamma\psi(\varphi + \psi) \tag{13.4b}$$

These are again the macroscopic rate equations (13.1). They can be solved explicitly; the result is sketched in Fig. 6. For our purpose it is sufficient to notice that, supposing $\beta > \alpha$, there is one stable stationary point

$$\varphi^{\text{st}} = \psi^{\text{st}} = \frac{\beta - \alpha}{2\gamma} \tag{13.5}$$

The terms of order Ω^0 yield a bivariate linear Fokker–Planck equation with time-dependent coefficients for Π

$$\frac{\partial\Pi}{\partial t} = \frac{\partial}{\partial x}\{-\beta x + \alpha x + \gamma(\varphi + \psi)x + \gamma\varphi(x + y)\}\Pi$$

$$+ \frac{\partial}{\partial y}\{-\beta x + \alpha x + \gamma(\varphi + \psi)y + \gamma\psi(x + y)\}\Pi$$

$$+ \frac{1}{2}\{\beta\varphi + \alpha\varphi + \gamma\varphi(\varphi + \psi)\}\frac{\partial^2\Pi}{\partial x^2}$$

$$+ \frac{1}{2}\{\beta\varphi + \alpha\psi + \gamma\psi(\varphi + \psi)\}\frac{\partial^2\Pi}{\partial y^2} \tag{13.6}$$

The general solution of such equations will be given in the next section.

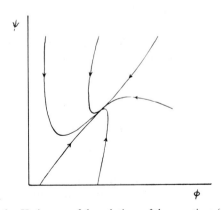

Fig. 6. Hodogram of the solutions of the equations (13.4).

Here we shall only study the fluctuations in the stationary state. Substitution of (13.5) reduces (13.6) to

$$\frac{\partial \Pi}{\partial t} = \frac{1}{2}(\beta - \alpha)\frac{\partial}{\partial x}(x + y)\Pi + \frac{\partial}{\partial y}\left\{-\frac{1}{2}(\beta + \alpha)x + \left(\frac{3}{2}\beta - \frac{1}{2}\alpha\right)y\right\}\Pi$$
$$+ \frac{\beta}{2\gamma}(\beta - \alpha)\left(\frac{\partial^2 \Pi}{\partial x^2} + \frac{\partial^2 \Pi}{\partial y^2}\right) \tag{13.7}$$

The resulting equations for the first moments are

$$\frac{d}{dt}\langle x \rangle = -\frac{1}{2}(\beta - \alpha)(\langle x \rangle + \langle y \rangle) \tag{13.8a}$$

$$\frac{d}{dt}\langle y \rangle = \frac{1}{2}(\beta + \alpha)\langle x \rangle - \left(\frac{3}{2}\beta - \frac{1}{2}\alpha\right)\langle y \rangle \tag{13.8b}$$

The eigenvalues of the coefficient matrix are $-\beta$ and $-\beta + \alpha$, which confirms that the stationary solution (13.5) is asymptotically stable. The equations for the second moments are

$$\frac{d}{dt}\langle x^2 \rangle = -(\beta - \alpha)(\langle x^2 \rangle + \langle xy \rangle) + \frac{\beta}{\gamma}(\beta - \alpha) \tag{13.9a}$$

$$\frac{d}{dt}\langle xy \rangle = \frac{1}{2}(\beta + \alpha)\langle x^2 \rangle + (2\beta - \alpha)\langle xy \rangle - \frac{1}{2}(\beta - \alpha)\langle y^2 \rangle \tag{13.9b}$$

$$\frac{d}{dt}\langle y^2 \rangle = (\beta + \alpha)\langle xy \rangle - (3\beta - \alpha)\langle y^2 \rangle + \frac{\beta}{\gamma}(\beta - \alpha) \tag{13.9c}$$

They can be used for instance to find the equilibrium fluctuations

$$\langle x^2 \rangle = \frac{7\beta^2 - 6\beta\alpha + \alpha^2}{4\gamma(2\beta - \alpha)} \tag{13.10a}$$

$$\langle xy \rangle = \frac{\beta^2 + 2\beta\alpha - \alpha^2}{4\gamma(2\beta - \alpha)} \tag{13.10b}$$

$$\langle y^2 \rangle = \frac{3\beta^2 - 2\beta\alpha + \alpha^2}{4\gamma(2\beta - \alpha)} \tag{13.10c}$$

This result, together with the solutions of (13.8) permits us to determine the two-time correlations; they will involve both exponential factors $e^{-\alpha t}$ and $e^{-(\beta - \alpha)t}$.

XIV. SOLUTION OF THE MULTIVARIATE FOKKER–PLANCK EQUATION

The general linear Fokker–Planck equation with K variables and time-dependent coefficients has the form

$$\frac{\partial \Pi}{\partial t} = -\sum_{i,j} \frac{\partial}{\partial x_i} A_{ij}(t) x_j \Pi + \frac{1}{2} \sum_{i,j} B_{ij}(t) \frac{\partial^2 \Pi}{\partial x_i \partial x_j} \tag{14.1}$$

The matrix B is symmetric but in general A is not. The first moments obey the equations

$$\frac{d}{dt} \langle x_i \rangle = \sum_j A_{ij}(t) \langle x_j \rangle \tag{14.2}$$

It is convenient to define an evolution matrix $V(t|t')$ by

$$\frac{dV(t|t')}{dt} = A(t) V(t|t') \qquad V(t'|t') = 1 \tag{14.3}$$

Then the solution of (14.2) with given $\langle x_i \rangle$ at $t = 0$ is

$$\langle x_i \rangle_t = \sum_j V_{ij}(t|0) \langle x_j \rangle_0 \tag{14.4}$$

The second moments obey the set of equations

$$\frac{d}{dt} \langle x_i x_j \rangle = \sum_k A_{ik} \langle x_k x_j \rangle + \sum_k A_{jk} \langle x_i x_k \rangle + B_{ij} \tag{14.5}$$

Introducing the matrix $X_{ij}(t) = \langle x_i x_j \rangle$ one may write this equation in matrix notation:

$$\dot{X} = AX + X\tilde{A} + B \tag{14.6}$$

where $\tilde{A}$ is the transpose of A. Note that by construction X must be symmetric and positive definite (or at least semidefinite). The problem is to find such an X obeying (14.6) with arbitrary (symmetric, positive definite) initial value.

The resemblance of (14.6) with the Neumann equation for a quantum-mechanical density matrix suggests that one transforms X by putting

$$X(t) = V(t|0) Y(t) \tilde{V}(t|0) \tag{14.7}$$

One then finds after some algebra

$$\begin{aligned} \dot{Y}(t) &= V(t|0)^{-1} (\dot{X} - AX - X\tilde{A}) \tilde{V}(t|0)^{-1} \\ &= V(t|0)^{-1} B(t) \tilde{V}(t|0)^{-1} \end{aligned} \tag{14.8}$$

Hence

$$Y(t) = X(0) + \int_0^t V(t'|0)^{-1}B(t')\tilde{V}(t'|0)^{-1}\,dt'$$

Thus we have found the solution of (14.6)

$$X(t) = V(t|0)X(0)\tilde{V}(t|0) + \int_0^t V(t|t')B(t')\tilde{V}(t|t')\,dt' \qquad (14.9)$$

Clearly it is symmetric and positive definite when $X(0)$ has these properties and when $B(t)$ is positive definite. Thus a condition for our solution of (14.1) is that B is positive definite or at least semidefinite at all t. What goes wrong when that condition is violated is exemplified by

$$\frac{\partial \Pi(x, t)}{\partial t} = - \frac{\partial^2 \Pi(x, t)}{\partial x^2}$$

This amounts to solving the heat conduction equation towards negative times, which is not a correctly set problem in the sense of Hadamard.[25]

Finally we have to find the solution of (14.1). The transformation (14.7) amounts to introducing new variables,

$$x_i = \sum_j V_{ij}(t|0)y_j \qquad (14.10)$$

We transform the whole equation (14.1) to these new variables. The probability distribution of y is

$$\Xi(y, t) = \Pi(x, t)\,\mathrm{Det}\,V \qquad (14.11)$$

It is not necessary to carry out the transformation in detail, because we know beforehand that Ξ obeys an equation of the same type (14.1) and that the first-order derivatives vanish since the $\langle y_i \rangle$ are constant by construction. Hence the equation for Ξ must have the form

$$\frac{\partial \Xi}{\partial t} = \frac{1}{2}\sum_{i,j} C_{ij}(t)\frac{\partial^2 \Xi}{\partial y_i\,\partial y_j} \qquad (14.12)$$

The second-order derivatives can only stem from those in (14.1) and one must therefore have

$$\sum_{ij} C_{ij} V_{ki} V_{lj} = B_{kl} \quad \text{or} \quad B = VC\tilde{V} \qquad (14.13)$$

Note that C is identical with the matrix in (14.8) so that

$$\dot{Y} = C \qquad (14.14)$$

The form of (14.12) suggests that Ξ is a Gaussian

$$\Xi(y, t) = (2\pi)^{-1/2K}(\text{Det } R)^{-1/2} \exp\left[-\frac{1}{2}\sum_{k,l} R_{kl}(t)(y_k - c_k)(y_l - c_l)\right] \quad (14.15)$$

If that is so one knows that R^{-1} is the correlation matrix of the y's and hence identical with Y. To verify that (14.15) is a solution we substitute in (14.12)

$$-\frac{1}{2}\frac{d}{dt}\log \text{Det } R - \frac{1}{2}\sum_{k,l} \dot{R}_{kl}(t)(y_k - c_k)(y_l - c_l)$$

$$= \frac{1}{2}\sum_{i,j} C_{ij}\left\{\sum_{k,l} R_{ik}(y_k - c_k)R_{jl}(y_l - c_l) - R_{ij}\right\}$$

For this to be an identity it is necessary that

$$-\dot{R} = RCR$$

$$\frac{d}{dt}\log \text{Det } R = \text{Tr } CR$$

The first one easily follows from (14.14). The second one is obtained by means of the Wronski identity:

$$\text{Det } (Y + \dot{Y} dt) - \text{Det } Y = \text{Det } Y\{\text{Det}(1 + Y^{-1}\dot{Y}dt) - 1\}$$
$$= \text{Det } Y(\text{Tr } Y^{-1}\dot{Y}) dt$$

This completes the solution of (14.1) or at least reduces it to solving (14.3). The final formula is obtained by restoring the variables x_i,

$$\Pi(x, t) = (2\pi)^{-1/2K} (\text{Det } X)^{1/2} \exp\left[-\frac{1}{2}\sum_{ij} (X^{-1})_{ij}(x_i - \langle x_i\rangle)(x_j - \langle x_j\rangle)\right]$$
$$(14.16)$$

where $\langle x_i\rangle$ is obtained from (14.2) and $X(t)$ from (14.9). Of course this result can be checked by direct substitution in (14.1).* In the case of a single variable $(K = 1)$ the solution is the same as in the Appendix and in Ref. 4. In the case that A and B are independent of time the solution can be made more explicit as is shown in the next section. We have to add a remark here on stability.

Suppose $A(t)$ is such that all solutions of (14.2) tend to zero, strongly enough for the time integral of $\langle x_i\rangle_t$ to converge.† According to (14.3) that implies that the operator $V(t|t')$ vanishes for $t \to \infty$ and fixed t'. Hence the

---

* An alternative way of solving (14.1) is by Fourier transformation, i.e., with the use of the characteristic function of Π.

† The precise condition is presumably that $\sum \langle x_i\rangle_t^2$ can be integrated from $t = 0$ to $t = \infty$.

first term on the right of (14.9) tends to zero, which means that the correlations become independent of their initial values. Moreover, if $B(t)$ is bounded for $t \rightarrow \infty$, one may write for large t

$$X(t) = \int_0^\infty V(t|t - \tau)B(t - \tau)\tilde{V}(t|t - \tau)\,d\tau \qquad (14.17)$$

Not only the initial value but also the initial time has disappeared from the formula; the fluctuations depend exclusively on the coefficient matrices of (14.1). This fact is of course taken for granted in many physical applications, but it should be borne in mind that it requires that all solutions of (14.2) tend to zero, which is tantamount to saying that the macroscopic solution is asymptotically stable. What happens when this is not so will be investigated in Section XVIII.

XV. FOKKER–PLANCK EQUATION WITH CONSTANT COEFFICIENTS

Take the same equation (14.1) but with A and B independent of t. The solution of (14.3) may then be written

$$V(t|t') = V(t - t') = e^{A(t - t')} \qquad (15.1)$$

Hence the expression (14.9) for X becomes

$$X(t) = e^{At}X(0)e^{At} + \int_0^t e^{A(t - t')}Be^{A(t - t')}\,dt' \qquad (15.2)$$

It is clear that the stability conditions mentioned in the previous paragraph are obeyed if, and only if, each eigenvalue of A has a negative real part. In that case $X(t)$ tends to the constant matrix

$$X(\infty) = \int_0^\infty e^{A\tau}Be^{A\tau}\,d\tau \qquad (15.3)$$

This is a general formula for the stationary state fluctuations in the linear noise approximation, applicable to all systems that have a stable stationary state.

A method for actually evaluating (15.2), and hence also (15.3), is suggested by the following observation. If A were symmetric one would perform an orthogonal transformation which diagonalizes A; as a result the matrix equation (15.2) would decompose in a set of uncoupled equations for the separate matrix elements of X in this new representation. A similar approach is possible now, using a nonorthogonal transformation, owing to the following fact. When Q is a nonsingular matrix, (15.2) is invariant for the transformation

$$X' = QX\tilde{Q}, \qquad B' = QB\tilde{Q}, \qquad A' = QAQ^{-1} \qquad (15.4)$$

If a Q can be found such that A' is diagonal the problem is again solved.

Denote the eigenvalues of A by $-\lambda_k$ and the corresponding *left* eigenvectors by $q_i^{(k)}$. There is no guarantee that the $q^{(k)}$ form a complete set of K linearly independent vectors, but we assume that they do, because that suffices for our applications.* Then the diagonalizing matrix is $Q_{ki} = q_i^{(k)}$ whereas Q^{-1} consists of the suitably normalized right eigenvectors $p_i^{(k)}$

$$(Q^{-1})_{ik} = p_i^{(k)}, \qquad \sum_i q_i^{(k)} p_i^{(l)} = \delta_{kl} \qquad (15.4)$$

Thus we have the transformation formulas

$$A'_{kl} = \sum_{ij} q_i^{(k)} A_{ij} p_j^{(l)} = -\lambda_k \delta_{kl} \qquad (15.5)$$

$$X'_{kl} = \sum_{ij} q_i^{(k)} X_{ij} q_j^{(l)} = \tilde{q}^{(k)} \cdot X \cdot q^{(l)} \qquad (15.6)$$

In the new representation (15.2) reads

$$X'_{kl}(t) = X'_{kl}(0)e^{-(\lambda_k + \lambda_l)t} + \int_0^t e^{-\lambda_k(t-t')} B'_{kl} e^{-\lambda_l(t-t')} \, dt' \qquad (15.7)$$

The solution is simple:

$$X'_{kl}(t) = X'_{kl}(0)e^{-(\lambda_k + \lambda_l)t} + B'_{kl} \frac{1 - e^{-(\lambda_k + \lambda_l)t}}{\lambda_k + \lambda_l} \qquad (15.8)$$

When the denominator vanishes, the fraction is replaced by its limiting value, t. This cannot occur, however, when (15.1) is stable, since in that case all λ_k have a positive real part and (15.8) tends to the limit

$$X'_{kl}(\infty) = \frac{B'_{kl}}{\lambda_k + \lambda_l} \qquad (15.9)$$

Finally one has to transform back to $X_{ij} \equiv \langle x_i x_j \rangle$. The result is an algebraic expression for the integral (15.3), or also a solution of (14.6) with the left-hand side zero.[27] A special example of this result was obtained in (13.10).

---

* If they do not, A can be transformed to the Jordan form, which makes the solution still possible but quite a bit more cumbersome.[26]

XVI. CHEMICAL REACTIONS

Chemical reactions provide a wide variety of applications.[28,29] Consider a volume Ω containing a mixture of different species of molecules, and suppose (*i*) that the mixture is sufficiently dilute to constitute an ideal gas, (*ii*) that it is homogeneously distributed* in Ω, and (*iii*) that the momenta are distributed according to Boltzmann with temperature T. The concentrations of the various compounds, however, are not in thermal equilibrium and vary as a result of reactive collisions.

The simplest possible reaction is the transition between two isomers

$$A + X \underset{\beta}{\overset{\alpha}{\rightleftharpoons}} A + Y \tag{16.1}$$

where A is a fixed catalyzer needed for the collisions. The total number of molecules X and Y is a constant N. If n is the number of X, the number of reactive collisions from left to right is proportional with n and the fixed concentration of A. Hence the macroscopic rate equation has the form

$$\frac{dn}{dt} = -\alpha n + \beta(N - n) \tag{16.2}$$

The transition probability per unit time is

$$W(n|n') = \alpha n' \delta_{n, n'-1} + \beta(N - n')\delta_{n, n'+1} \tag{16.3}$$

The master equation is therefore

$$\dot{P}(n, t) = \alpha(\mathbf{E} - 1)nP + \beta(\mathbf{E}^{-1} - 1)(N - n)P \tag{16.4}$$

Note that there is a natural boundary at $n = 0$ (depletion of X) and another one at $n = N$ (depletion of Y). This master equation is linear and can readily be solved by means of a generating function. Rather than writing the full solution we only note that according to (6.4) and (6.5)

$$P^{\text{st}}(n) = (\alpha + \beta)^{-N}\binom{N}{n}\alpha^{N-n}\beta^n \tag{16.5}$$

This is identical with the thermal equilibrium provided that

$$\beta/\alpha = \exp(\sigma_x - \sigma_y) \tag{16.6}$$

where σ_x, σ_y are the internal entropies of X and Y.

The mixture of X and Y in the volume Ω constituted a closed system since no molecules could enter or escape. In order to extend the theory to open systems one supposes that it is possible to inject molecules at a constant rate,

---

* Reactions that are not homogeneous in space are of considerable interest[28,30–33] but will not be studied in this work.

or to extract them at a rate proportional to their number. This adds two new terms to the rate equation for n,

$$\frac{dn}{dt} = b - an + \cdots \tag{16.7}$$

They correspond to two additional terms in the transition probability

$$W(n|n') = b\delta_{n,n'+1} + an'\delta_{n,n'-1} + \cdots \tag{16.8}$$

In the master equation they give rise to the terms

$$\dot{P}(n,t) = b(\mathbf{E}^{-1} - 1)P + a(\mathbf{E} - 1)nP + \cdots \tag{16.9}$$

Of course it is understood that the injection and extraction do not violate assumptions (i), (ii), and (iii).

Another way of arriving at these equations is by taking a closed system which includes a reservoir of molecules that produce or absorb X. To show this consider the reaction (16.1) in the limit $N \to \infty$, $\beta \to 0$ with constant $\beta N = b$. Then (16.2) tends to (16.7), (16.3) to (16.8) and (16.4) to (16.9). Thus the molecules Y serve as a reservoir which creates, and annihilates molecules X according to the same equations as used for describing injection and extraction. The Y do not otherwise enter the picture and may be ignored. The remaining species X alone constitutes an open system, although X and Y together were a closed system.

This way of creating and annihilating X is not uniquely determined by the macroscopic equation. For instance the reaction

$$Y \xrightarrow{\frac{1}{2}\beta} X + X$$

would produce the same term b in (16.7) but not the same terms in (16.8) and (16.9). For this reason one usually includes in the reaction scheme the reactions with the reservoir molecules; they are indicated by the earlier letters of the alphabet A, B, ... and the numbers n_A, n_B, ... are then taken constant.

The Malthus–Verhulst problem can be interpreted as the following chemical reaction in an open system:

$$X \xrightarrow{k_1} A$$
$$B + X \underset{k_2'}{\overset{k_2}{\rightleftharpoons}} X + X \tag{16.10}$$

The amounts of A and B are supposed fixed, so that the only variable is the number n of molecules X. The first line simulates the natural death: $k_1 = \alpha$. The reverse process A $\to$ X is supposed to be negligible, for instance because A is drained or undergoes further decays. The second line is an autocatalytic process. The arrow to the right simulates natural births proportional to the existing population, $k_2 n_B/\Omega = \beta$. The reverse process corresponds to the Verhulst term: $k_2' = \gamma$. This process was studied by Schlögl.[30]

A reaction in a closed system having the same rate equation but a different master equation is[34]

$$X + Y \mathrel{\mathop{\rightleftharpoons}^{k}_{k'}} 2X \qquad (16.11)$$

The concentrations of both compounds vary in time, but there is a conserved quantity $n_X + n_Y = C$. Hence there is a single rate equation for $n_X = n$,

$$\frac{dn}{dt} = \frac{k}{\Omega} n(C - n) - \frac{k'}{\Omega} n^2 = \frac{kC}{\Omega} n - \frac{k + k'}{\Omega} n^2 \qquad (16.12)$$

Ω is the volume in which the reacting mixture is enclosed. We have written n^2 rather than $n(n - 1)$ in agreement with the remark in Section IX that the difference is negligible on the macroscopic scale. The equation is identical with the Malthus–Verhulst equation (9.1) with the identification

$$kC/\Omega = \beta - \alpha, \qquad k + k' = \gamma$$

However, the master equation for (16.11) is

$$\dot{P}(n, t) = \frac{k}{\Omega} (\mathbf{E}^{-1} - 1)n(C - n)P + \frac{k'}{\Omega} (\mathbf{E} - 1)n(n - 1)P \qquad (16.13)$$

Obviously this differs from (9.4) and we may expect that the fluctuations are different. In fact one finds

$$\frac{\partial \Pi}{\partial t} = \{-kc + 2(k - k')\varphi\} \frac{\partial}{\partial x} x\Pi + \frac{1}{2} \{kc\varphi + (k' - k)\varphi^2\} \frac{\partial^2 \Pi}{\partial x^2} \qquad (16.14)$$

with $c = C/\Omega$. This is manifestly different from (9.5). The variance of the fluctuations in the stationary state is

$$\sigma_n^2 = \Omega \langle x^2 \rangle^{\mathrm{st}} = \frac{k'kc}{(k' + k)(3k' - k)} \qquad (16.15)$$

which cannot even be expressed in α, β, γ.

XVII. TWO-STEP CHEMICAL REACTIONS

Consider the reaction scheme[31,35]

$$\begin{aligned} A &\xrightarrow{k^{(1)}} X \\ 2X &\xrightarrow{k^{(2)}} B \end{aligned} \qquad (17.1)$$

The amounts of A and B are fixed and the reverse reactions are supposed to be negligible. The rate equation for the number of molecules X is

$$\frac{dn}{dt} = k^{(1)}n_A - 2k^{(2)}\frac{n^2}{\Omega} \tag{17.2}$$

This is the same as for the semiconductor, see (7.1) if

$$k^{(1)}n_A/\Omega = \beta, \qquad 2k^{(2)} = \gamma \tag{17.3}$$

The master equation, however, has to take into account that the X can only disappear in pairs,

$$\dot{P}(n, t) = k^{(1)}n_A(\mathbf{E}^{-1} - 1)P + \frac{k^{(2)}}{\Omega}(\mathbf{E}^2 - 1)n(n - 1)P \tag{17.4}$$

(Of course the number of pairs is $\frac{1}{2}n(n - 1)$, but this $\frac{1}{2}$ has been incorporated in the definition of $k^{(2)}$.)

Expanding as usual and utilizing (17.3) as convenient abbreviations we get

$$\frac{\partial \Pi}{\partial t} - \Omega^{1/2}\frac{d\varphi}{dt}\frac{\partial \Pi}{\partial x} = \Omega\beta\left(-\Omega^{-1/2}\frac{\partial}{\partial x} + \frac{1}{2}\Omega^{-1}\frac{\partial^2}{\partial x^2}\right)\Pi$$

$$+ \frac{1}{2}\gamma\Omega\left(2\Omega^{-1/2}\frac{\partial}{\partial x} + 2\Omega^{-1}\frac{\partial^2}{\partial x^2}\right)$$

$$\times (\varphi + \Omega^{-1/2}x)(\varphi + \Omega^{-1/2}x - \Omega^{-1})\Pi \tag{17.5}$$

The terms of order $\Omega^{1/2}$ reproduce the macroscopic equation and those of order Ω^0 yield

$$\frac{\partial \Pi}{\partial t} = 2\gamma\varphi\frac{\partial}{\partial x}x\Pi + \left(\frac{1}{2}\beta + \gamma\varphi^2\right)\frac{\partial^2 \Pi}{\partial x^2} \tag{17.6}$$

Compare this width (7.10). In particular, to find the variance in the stationary state we substitute $\varphi^{st} = \sqrt{\beta/\gamma}$ and obtain

$$\sigma_n^2 = \Omega\langle x^2\rangle^{st} = \frac{3}{4}\Omega\sqrt{\frac{\beta}{\gamma}} = \frac{3}{4}\langle n\rangle^{st} \tag{17.7}$$

The factor $3/4$ shows that the stationary distribution is narrower than a Poisson distribution. This fact has caused some misgivings about the correctness, but is actually easy to understand. The Poisson distribution is expected for independent particles, but the X molecules are not independent as they disappear in pairs. If at any time their number is slightly higher than the average they disappear more rapidly not only because there are more, but also because the probability for each molecule to disappear is enhanced by the higher number of others available to form a pair. A similar argument applies when their number happens to be lower than average. Hence it was to be expected that the distribution should be narrower than for molecules that appear and disappear independently from each other.

On the other hand, when the molecules are created in pairs one expects that the stationary distribution will be broader than Poisson. As an example consider the reaction scheme

$$
\begin{array}{ccc}
A & \xrightarrow{k^{(1)}} & 2X \\
X & \xrightarrow{k^{(2)}} & B
\end{array}
\tag{17.8}
$$

n_A and n_B are again fixed, and the master equation is with the abbreviations $k^{(1)}n_A/\Omega = \alpha$ and $k^{(2)} = \beta$

$$
\dot{P}(n, t) = \alpha\Omega(\mathbf{E}^{-2} - 1)P + \beta(\mathbf{E} - 1)nP
\tag{17.9}
$$

The expansion in Ω yields

$$
\frac{\partial\Pi}{\partial t} - \Omega^{1/2}\frac{d\varphi}{dt}\frac{\partial\Pi}{\partial x} = \alpha\Omega\left(-2\Omega^{-1/2}\frac{\partial}{\partial x} + 2\Omega^{-1}\frac{\partial^2}{\partial x^2}\right)\Pi
$$

$$
+ \beta\Omega\left(\Omega^{-1/2}\frac{\partial}{\partial x} + \frac{1}{2}\Omega^{-1}\frac{\partial^2}{\partial x^2}\right)(\varphi + \Omega^{-1/2}x)\Pi \tag{17.10}
$$

The terms of order $\Omega^{1/2}$ give the macroscopic equation

$$
\frac{d\varphi}{dt} = 2\alpha - \beta\varphi
\tag{17.11}
$$

The terms of order Ω^0 yield

$$
\frac{\partial\Pi}{\partial t} = \beta\frac{\partial}{\partial x}x\Pi + \left(2\alpha + \frac{1}{2}\beta\varphi\right)\frac{\partial^2\Pi}{\partial x^2}
\tag{17.12}
$$

It is again easy to find $\varphi^{st} = 2\alpha/\beta$ and subsequently

$$
\sigma_n^2 = \Omega\langle x^2\rangle^{st} = \frac{3\Omega\alpha}{\beta} = \frac{3}{2}\langle n\rangle^{st}
\tag{17.13}
$$

This confirms the expectation.*

The reader will have no difficulty in treating the case that p molecules are created simultaneously, while q are annihilated simultaneously. The master equation is

$$
\dot{P} = \alpha\Omega(\mathbf{E}^{-p} - 1)P + \beta\Omega^{-q+1}(\mathbf{E}^q - 1)(n)_q P
\tag{17.14}
$$

---

* Actually the master equation (17.9) happens to be linear in n and can therefore be solved exactly.

Here $(n)_q$ is an abbreviation (Pochhammer's symbol) for

$$(n)_q = n(n - 1)(n - 2) \cdots (n - q + 1) \tag{17.15}$$

but in our approximation no error is committed on replacing it with n^q. The result for the stationary distribution is

$$\sigma_n{}^2 = \frac{p + q}{2q} \langle n \rangle^{st}$$

In particular for $p = q$ one recovers the Poisson-like result. In fact one can verify by direct substitution in (17.14) that the Poisson distribution with $\langle n \rangle = \Omega(\alpha/\beta)^p$ is an exact solution.

XVIII. FLUCTUATIONS ABOUT A LIMIT CYCLE*

As an example we take a reaction scheme whose macroscopic behavior has been analyzed by Glansdorff and Prigogine,[28] Nicolis,[36] and others[37,38]

$$
\begin{array}{rcl}
A & \longrightarrow & X \\
2X + Y & \longrightarrow & 3X \\
B + X & \longrightarrow & Y + D \\
X & \longrightarrow & E
\end{array}
\tag{18.1}
$$

There are two variables n_X and n_Y; they transport matter from the reservoir A into E and from B into D. These flows are supposed to be sufficiently strong for the reverse reactions to be negligible. With a suitable choice of units of t and Ω the master equation takes the form

$$\dot{P}(n_X, n_Y, t) = \Omega\alpha(\mathbf{E}_X{}^{-1} - 1)P + \Omega^{-2}(\mathbf{E}_X{}^{-1}\mathbf{E}_Y - 1)n_X{}^2 n_Y P$$
$$+ (\mathbf{E}_X - 1)n_X P + \beta(\mathbf{E}_X \mathbf{E}_Y{}^{-1} - 1)P \tag{18.2}$$

In order to expand in Ω we set

$$n_X = \Omega\varphi(t) + \Omega^{1/2}x, \qquad n_Y = \Omega\psi(t) + \Omega^{1/2}y \tag{18.3}$$

The macroscopic equations so obtained are

$$\dot{\varphi} = \alpha + \varphi^2\psi - \beta\varphi - \varphi \tag{18.4a}$$

$$\dot{\psi} = \beta\varphi - \varphi^2\psi \tag{18.4b}$$

A careful study of these equations[36,37] shows that the solutions do not tend to a stationary point but to a limit cycle (see Fig. 7). Thus there is a periodic

---

* Note added in proof. After this article was submitted, a paper by K. Tomita, T. Ohta, and H. Tomita appeared [Progr. Theor. Phys., **52**, 1744, (1974)], which covers the same ground as our Section XVIII.

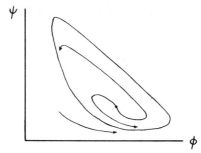

Fig. 7. Hodogram of the solutions of the equations (18.4) according to Glansdorff and Prigogine.[28]

solution with period T, or rather a family of such solutions because the phase is arbitrary. All other solutions tend to one of these as $t \to \infty$. We propose to investigate the fluctuations about this periodic solution of the macroscopic equations.

The terms of Ω^0 in the expansion of (18.2) yield a Fokker–Planck equation for Π, which is of the form (14.1) with

$$A(t) = \begin{pmatrix} 2\varphi\psi - 1 - \beta & \varphi^2 \\ -2\varphi\psi + \beta & -\varphi^2 \end{pmatrix} \tag{18.5}$$

$$B(t) = \begin{pmatrix} \varphi^2\psi + \beta\varphi + \varphi + \alpha & -\varphi^2\psi - \beta\varphi \\ -\varphi^2\psi - \beta\varphi & \varphi^2\psi + \beta\varphi \end{pmatrix} \tag{18.6}$$

According to the general scheme of Section XIV one obtains the correlation matrix $X(t)$ from (14.9) after solving the linear equation (14.3) for $V(t|t')$. The condition that B should be positive definite is duly satisfied. The behavior for $t \to \infty$, however, will be different from the behavior in the case of a limit point.

Suppose that the system has evolved long enough, so that (φ, ψ) has approached the limit cycle within a distance of order $\Omega^{-1/2}$. Thus we have to substitute in $A(t)$ and $B(t)$ the periodic solution with some value of the phase. It follows that $A(t)$ is periodic with period T, rather than constant. This fact by itself does not exclude the possibility that the solution $V(t|t')$ of (14.3) tends to zero but we shall show that in fact it does not.

It is convenient to write φ_1, φ_2 instead of φ, ψ and to employ the symbol φ for this two-component vector. The macroscopic equations (18.4) may then be written

$$\dot{\varphi}_i(t) = f_i(\varphi(t)) \tag{18.7}$$

Of course it is also true that

$$\dot{\varphi}_i(t + \delta t) = f_i(\varphi(t + \delta t)) \tag{18.8}$$

and therefore

$$\ddot{\varphi}_i(t) = \sum_j \frac{\partial f_i(\varphi)}{\partial \varphi_j} \dot{\varphi}_j(t) \tag{18.9}$$

As mentioned in Section V the matrix of the derivatives is just A_{ij}. Hence we have found one solution of (14.2), namely, $\dot{\varphi}_j(t)$. Now substitute for φ a periodic solution φ^{lc} with some prescribed phase. Then (18.9) shows that there is one solution of (14.2) that does not go to zero since it is periodic in t. Hence the equation (14.17) for large t cannot be correct for fluctuations about φ^{lc}.

In order to investigate the modifications needed to deal with this lack of stability we construct a new coordinate system for the neighborhood of the limit cycle. In each point P of the limit cycle a natural base vector $\vartheta^{(1)}$ is defined by the tangent (Fig. 8):

$$\vartheta_j^{(1)} = \dot{\varphi}_j \tag{18.10}$$

This equation may also serve to define the length of $\vartheta^{(1)}$, since $\|\dot{\varphi}_j\|$ has a positive lower bound along the cycle. By construction we have according to (18.9)

$$\dot{\vartheta}_i^{(1)} = \sum_j A_{ij} \vartheta_j^{(1)} \tag{18.11}$$

Next a suitable direction for the other base vector $\vartheta^{(2)}$ has to be found. From each point Q in the vicinity of P, but not on the cycle itself, a solution curve starts out that approaches the cycle. Compare this solution with the one that starts out from P at the same time. If the direction PQ is almost

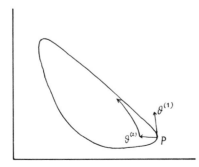

Fig. 8. The new base vectors along the limit cycle.

parallel to $\vartheta^{(1)}$ the former will end up by being slightly ahead in time with respect to the latter. If PQ is almost antiparallel to $\vartheta^{(1)}$ it will end up by being behind. In between there is a direction where both solutions will ultimately coincide, not merely their orbits. We take $\vartheta^{(2)}$ in this direction and choose some arbitrary length for it. Then we know

$$\varphi(t\,|\,P) - \varphi(t\,|\,P + \epsilon\vartheta^{(2)}) \to 0 \tag{18.12}$$

where the notation introduced below (4.9) is employed.

The variables x_j, which describe the fluctuations about the macroscopic value $\varphi_j^{lc}(t)$, are now transformed by expressing them as linear combinations of the base vectors $\vartheta^{(\alpha)}$ attached to the point $P = \varphi_j^{lc}(t)$,

$$x_k = \sum_{\alpha=1}^{2} x'_\alpha \vartheta_j^{(\alpha)} = \sum_\alpha \Theta_{j\alpha} x'_\alpha \tag{18.13}$$

The moments transform accordingly,

$$\langle x_j \rangle = \sum_\alpha \Theta_{j\alpha} \langle x'_\alpha \rangle \qquad X = \Theta X' \tilde{\Theta} \tag{18.14}$$

In the primed variables (14.2) becomes

$$\frac{d}{dt} \langle x'_\alpha \rangle = \sum_\beta A'_{\alpha\beta}(t) \langle x'_\beta \rangle \tag{18.15}$$

where the matrix A' is

$$A' = \Theta^{-1} A \Theta - \Theta^{-1} \dot{\Theta} \tag{18.16}$$

The solution of (18.15) may again be expressed in an evolution matrix $V'(t\,|\,t')$, which is easily seen to be

$$V'(t\,|\,t') = \Theta(t)^{-1} V(t\,|\,t') \Theta(t') \tag{18.17}$$

Owing to the choice of the new base, A' and V' have special properties. The primed components of $\vartheta^{(1)}$ are simply $(1, 0)$; hence (18.10) translates into $0 = A'(t)(1, 0)$. This says that $(1, 0)$ is invariant in time and therefore $V'(t\,|\,t')(1, 0) = (1, 0)$, so that V' must have the form

$$V'(t\,|\,t') = \begin{pmatrix} 1 & V'_{12} \\ 0 & V'_{22} \end{pmatrix} \tag{18.18}$$

Moreover, according to (18.12) one has $V'(t\,|\,t')(0, 1) \to 0$ so that

$$\lim_{t\to\infty} V'_{12}(t\,|\,t') = \lim_{t\to\infty} V'_{22}(t\,|\,t') = 0 \tag{18.19}$$

How fast these elements go to zero depends on how fast the solutions of the macroscopic equations tend to the limit cycle; we shall assume here that they are integrable.

In the new variables equation (14.9) for the second moments takes the form

$$X'(t) = V'(t|0)X'(0)\tilde{V}'(t|0) + \int_0^t V'(t|t')B'(t')\tilde{V}'(t|t')\, dt \qquad (18.20)$$

where we have defined $B'(t)$ by

$$B(t) = \Theta(t)B'(t)\tilde{\Theta}(t) \qquad (18.21)$$

As $t \to \infty$ one obtains for the (1, 1)-element

$$X'_{11}(t) = X'_{11}(0) + \int_0^t B'_{11}(t')\, dt$$

$$+ 2\int_0^\infty V'_{12}(t|t-\tau)B'_{12}(t-\tau)\, d\tau + \int_0^\infty V'_{12}(t|t-\tau)^2 B'_{12}(t-\tau)\, d\tau$$

$$(18.22)$$

Thus the initial value does not disappear, but even a term that grows roughly linearly with t appears, and some finite additional terms. That means that fluctuations directed along the limit cycle are not damped but grow roughly like the displacement of a Brownian particle. When X'_{11} has become of order Ω the whole limit cycle is covered; the time needed for that is also of order Ω.

The other matrix elements of $X(t)$ obey equations of the same type as (14.17) and therefore tend to finite values. This means that fluctuations away from the limit cycle do not grow.

Thus the probability distribution $P(n, t)$ for large t spreads out over a narrow strip covering the limit cycle. As it is no longer a sharp peak, no position can be ascribed to it, and no macroscopic part $\Omega\varphi$ of n exists. Hence no macroscopic equation of motion can be formulated that describes the behavior during times of order Ω, because after such times the position on the limit cycle is no longer determined. The same effect is known in electrical engineering as "phase slip."

XIX. THE RANDOM WALK PICTURE

The following way of visualizing a given one-step process is suggestive but, as will be shown presently, somewhat misleading. Take a line with a coordinate s on it and mark the points $s = n\epsilon$. Define a "potential energy" $U(s)$ by putting

$$P^{st}(n) = \text{const } e^{-U(n\epsilon)} \qquad (19.1)$$

The given one-step process may then be visualized as a random walk on the point set $\{n\epsilon\}$ under influence of a force with potential $U(s)$; the factor kT has been absorbed in U. Of course this model does not uniquely specify the

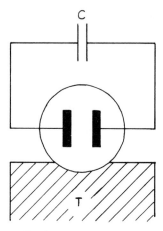

Fig. 9. The diode circuit.

jump probabilities $r(n)$ and $g(n)$, but expresses them in terms of a single function $k(n)$,

$$r(n) = k(n)e^{U(n\epsilon)}, \qquad g(n) = k(n + 1)e^{U(n\epsilon)} \tag{19.2}$$

For example, in the much studied diode circuit[40,41] of Fig. 9, where n is the number of excess electrons on the left condenser plate, one has $U = e^2n^2/2kTC$. When $n > 0$, the electrons that jump from left to right only face the potential threshold formed by the work function of the diode plate so that $r(n)$ is a constant given by Richardson's formula. The electrons jumping from right to left, however, face both the work function and the potential difference; one then has

$$r(n) = A, \qquad g(n) = A \exp\left[-\frac{e^2}{kTC}\left(n + \frac{1}{2}\right) \right] \tag{19.3}$$

For $n < 0$ the roles of r and g are interchanged.

The function U does not always have such a concrete meaning, but this analogy with a potential energy makes the picture suggestive. If it has a single minimum, as in the diode example, one sees how the particles jump back and forth with a preference for lower energies, and in so doing establish a one-humped P^{st} (see Fig. 10a). Now suppose that U has two minima (Fig. 10b). Then P^{st} has one hump in the lower minimum, and an additional one in the other minimum only if it is as low or almost as low. If it is possible to alter the relative depths of the two minima by varying an external parameter, one may make the hump of P^{st} jump from one minimum to the other, and so create a phase transition. If the two minima have equal depth, and one varies a second parameter in order to reduce the maximum between them,

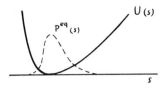

Fig. 10a. The stationary distribution with one maximum.

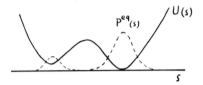

Fig. 10b. A stationary distribution with two maxima.

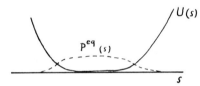

Fig. 10c. The maxima coalesce into a critical point.

eventually a critical point will be reached where both minima merge into a single flat minimum (Fig. 10c). The fluctuations in the flat minimum are exceptionally large, as computed in Section XXI. In cases with more than one variable a great variety of phenomena may be visualized in this way.[10, 41]

Of course all this is nothing but interpreting (19.1). In addition, however, the picture also demonstrates some well-known features that are not implied in the stationary distribution. A metastable state occurs when the lower minimum is gradually raised above the other one, because the probability has to be transferred across the potential maximum. The rate at which this happens is mainly determined by the density of the initial distribution at the maximum, which is roughly

$$\exp\left[-(U_{\max} - U_{\min})\right] \qquad (19.4)$$

This may easily be less than the rate at which the external parameter varies. Landauer[43] has pointed out that flip-flop circuits, tunnel diodes, and in fact all memory-storing elements are examples of this situation and only retain their memory thanks to the smallness of (19.4).

Another time-dependent phenomenon that can be visualized by the random walk picture is the critical slowing down. When the minimum of U

is flat the difference between the jump probabilities towards and away from the minimum is small, which makes the distribution slow in moving to its equilibrium.

It has already been tacitly assumed that ϵ is small and that U can be expressed as a smooth function of a continuous variable s. We now also set

$$r(n) = \rho(s), \qquad g(n) = \gamma(s), \qquad k(n) = \kappa(s), \qquad \epsilon P(n, t) = p(s, t) \quad (19.5)$$

Then the master equation (2.8) may be expanded in ϵ:

$$\frac{\partial p(s)}{\partial t} = \epsilon \frac{\partial}{\partial s} \{\rho(s) - \gamma(s)\}p + \frac{1}{2} \epsilon^2 \frac{\partial^2}{\partial s^2} \{\rho(s) + \gamma(s)\}p + \mathcal{O}(\epsilon^3) \quad (19.6)$$

Using (19.2) and setting $\epsilon^2 t = \tau$ one obtains the nonlinear Fokker–Planck equation:

$$\frac{\partial p(s)}{\partial \tau} = -\frac{\partial}{\partial s} \kappa'(s)e^{U(s)}p + \frac{\partial^2}{\partial s^2} \kappa(s)e^{U(s)}p \quad (19.7a)$$

Instead of κ one may also use the function $\lambda(s) = \kappa e^U$ to specify the jump probabilities,

$$\frac{\partial p(s)}{\partial \tau} = \frac{\partial}{\partial s} (\lambda U' - \lambda')p + \frac{\partial^2 p}{\partial s^2} \quad (19.7b)$$

Notice that the average of s obeys

$$\frac{d}{d\tau} \langle s \rangle = \langle \kappa'(s)e^{U(s)} \rangle = -\langle \lambda(s)U'(s) \rangle + \langle \lambda'(s) \rangle \quad (19.8)$$

The fact that this is not equal to $-\langle U'(s) \rangle$, let alone $-U'(\langle s \rangle)$, as one might expect from a macroscopic point of view, has given rise to much discussion.[40,49]

This confusion is caused by the inherent weakness of the random walk model. Although the model is appropriate for describing an actual random walk in an external potential field,[45] in general it does *not* apply to the fluctuations of a macroscopic quantity in a many-body system. The formal expansion in ϵ is not an expansion in an actual physical quantity like $1/\Omega$. This is already apparent because for small ϵ the width of the stationary distribution does not become small, since it is solely determined by $U(s)$ according to (19.1). The correct expansion is obtained by first identifying ϵ with $1/\Omega$ (so that s becomes the intensive variable corresponding to the extensive variable n), and then defining the function $U(s)$ by

$$P^{st}(n) = \exp\left[-\frac{1}{\epsilon} U(s)\right] = \exp\left[-\Omega U(s)\right] \quad (19.9)$$

rather than by (19.1). For instance in the diode circuit[40] the capacity C has the role of $\Omega = 1/\epsilon$, $s = n/C = V/e$ represents the voltage, and

$$P^{st}(n) = \text{const} \exp \left[-\frac{1}{\epsilon} \frac{e^2 s^2}{2kT} \right]$$

The reason why (19.9) is correct rather than (19.1) can be understood as follows. The mesoscopic description of a many-body system in terms of a master equation is based on the assumption that on a suitable coarse-grained level the state of the system can be described by a single variable n or s (or by a number of such variables, but much less than the number of microscopic variables). The other degrees of freedom merely cause the random jumps of this single variable, called the "order parameter." The probability distribution (19.1) is the partition function for fixed s. Hence $U(s)$ is the free energy, which is of course proportional to Ω. To obtain a correct expansion in $1/\Omega$ one must display this factor explicitly as in (19.9).*

The modification (19.9) changes the expansion radically, undermines the validity of (19.6) and (19.7), and largely spoils the suggestive random walk picture, but it gives the results correctly. As an additional benefit we found that the Ω expansion leads to a Fokker–Planck equation that can be solved generally, whereas (19.7) can be solved only in very few cases.

Kubo has applied the same idea to the time-dependent solutions of the master equation.[10] He substituted

$$P(n, t) = \exp \left[-\Omega U(s, t) \right]$$

and solved the resulting equation for U in successive powers of $1/\Omega$. The results were, of course, the same as those obtained by the Ω expansion of P itself, as described in the present article.

XX. PHASE TRANSITIONS

The random walk picture is particularly attractive for illustrating phase transitions, as in Fig. 10b. In this section we describe a few examples. They have been amply discussed in the literature; our purpose is merely to emphasize the role of the parameter Ω.

The Weiss theory of ferromagnetism† provides a simple model for a phase transition. The free energy per particle, at a fixed value s of the magnetization, is, in proper units,

$$N\{(\tfrac{1}{2} + s) \log (\tfrac{1}{2} + s) + (\tfrac{1}{2} - s) \log (\tfrac{1}{2} - s) - 2Ks^2 - 2Hs\} \quad (20.1)$$

---

* The new function U as it appears in (19.9) may involve higher orders in $1/\Omega$, but we shall ignore this complication here.

† Also called Bragg–Williams approximation (see Ref. 46).

N is the number of dipoles and serves as size parameter Ω. H is the external field and the interaction energy among the particles is represented by the Weiss field Ks. Thus (19.9) holds, where $U(s)$ is the expression $\{\quad\}$ in (20.1). If for simplicity one takes $H = 0$ it is easily seen that $U(s)$ looks like Fig. 10a for $|K| < 1$ and like Fig. 10b for $|K| > 1$.

In order to endow this model with a temporal behavior, Ruijgrok and Tjon[47] considered a collection of N spins, each having an up and a down state. The number of up spins minus the number of down spins is $2n = 2Ns$. The canonical distribution is, supposing N even,

$$P^{\mathrm{eq}}(n) = \mathrm{const}\left(\frac{N}{\frac{1}{2}N + n}\right) \exp\left[\frac{2K}{N} n^2 + 2Hn\right] \tag{20.2}$$

which for large N is the same as (20.1). The variable n performs a random walk between $\frac{1}{2}N$ and $-\frac{1}{2}N$. The jump probabilities are taken in agreement with (20.2) and (6.4) to be

$$r(n) = (\tfrac{1}{2}N + n) \exp\left[-\frac{2K}{N} n - H\right]$$

$$g(n) = (\tfrac{1}{2}N - n) \exp\left[\frac{2K}{N} n + H\right]$$

It should be clear that they are not derived from any microscopic Hamiltonian, but are postulated as part and parcel of the model, in the spirit of Glauber's model.[48] Otherwise one would rather expect to find transition probabilities similar to (19.3).

If one now lets N grow, the function $U(s)$ remains the same, while the steps in the s-scale tend to zero. However, the equilibrium distribution (19.9) also changes and becomes very sharp. As a consequence the fluctuations of s become small and therefore $U(s)$ may well be approximated near its minima by parabolas. This explains why the linear noise approximation holds and why the low-lying eigenvalues of the master equation are equidistant.[47] Thus in the systematic expansion the nonlinear Fokker–Planck equation (19.7) never comes in, and efforts to solve it appear to be irrelevant.

As a second example we take the much discussed laser. At a low pumping rate the radiation is weak, proportional to the pumping, and incoherent, but at a certain threshold value the self-multiplication of photons suddenly starts. For single-mode operation the following master equation for the number n of photons has been derived.[49]

$$\dot{P}(n, t) = (\mathbf{E}^{-1} - 1)\frac{A(n + 1)}{1 + (n + 1)/\Omega} P + C(\mathbf{E} - 1)nP \tag{20.3}$$

The constant A is the linear gain, C the loss rate per phonon, Ω the volume of the cavity, in such units as to make the "saturation parameter" equal to $1/\Omega$. The stationary solution is given by (6.5):

$$P^{\mathrm{st}}(n) = \frac{\mathrm{const}}{\Gamma(n + \Omega + 1)} \left(\frac{A}{C}\,\Omega\right)^n$$

$$\cong \mathrm{const}\, \exp\left[-\Omega\left\{(s + 1) \log (s + 1) - s - s \log \frac{A}{C}\right\}\right] \quad (20.4)$$

where $s = \epsilon n = n/\Omega$. This is again of the form (19.9); the function $U(s)$ is the expression $\{\ \}$ and is sketched in Fig. 11. For $A > C$ there is a potential minimum at $s = (A - C)/C$, corresponding to a macroscopic number of photons. The variance of the photon number is

$$\langle(\Delta n)^2\rangle = \Omega^2\langle(\Delta s)^2\rangle = \Omega^2[\Omega U''(s)]^{-1} = \frac{\Omega A}{C} = \langle n\rangle + \Omega \quad (20.5)$$

which shows again that the fluctuations in s are small for large Ω.

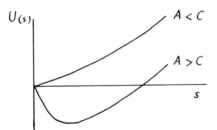

Fig. 11. The distribution given by (20.4).

As a third example remember that the Malthus–Verhulst equation in Section IX showed a phase transition at $\beta = \alpha$. We now compute the stationary solution. The general formula (6.5) yields $P^{\mathrm{st}}(n) = \delta_{n,0}$. In fact this is the sole time-independent solution, because for every other solution probability constantly leaks from $n = 1$ into $n = 0$ without returning. That is, the probability for a population to die out tends to unity, just as a gambler will always be ruined in the long run.* In practice, of course this probability is often so small that an almost stationary or *metastable* state

---

* This complication does not arise if there is an additional creation or immigration term as in (16.7). The stationary distribution for that case[50] has the same form (20.6).

prevails. Such a distribution must obey (6.4) except for small n and is therefore

$$P^{ms}(n) = \frac{g(n-1)g(n-2)\cdots g(n_0)}{r(n)r(n-1)\cdots r(n_0+1)} P^{ms}(n_0)$$

$$= \text{const} \frac{1}{n\Gamma(n+\alpha\Omega/\gamma)} \left(\frac{\beta\Omega}{\gamma}\right)^n$$

$$\cong \text{const} \exp\left[-\Omega\left\{\left(s+\frac{\alpha}{\gamma}\right)\log\left(s+\frac{\alpha}{\gamma}\right) - s - s\log\frac{\beta}{\gamma}\right\}\right] \quad (20.6)$$

The function $U(s)$ given by { } is the same as in (20.4) and as sketched in Fig. 11. For $\beta > \alpha$ it has a minimum in agreement with (9.8). The leakage into $n = 0$ is roughly of order

$$\frac{P^{ms}(1)}{P^{ms}(\text{max})} = \exp\left[-\Omega\{U(1) - U(\text{min})\}\right] \quad (20.7)$$

For $\beta < \alpha$ the minimum of U lies at $n = 1$; the leak is not small and therefore the solution (20.6) becomes meaningless.

Our last example will be discussed in more detail because it serves for computing critical fluctuations in the next section. Consider the following autocatalytic chemical reaction, studied by Schlögl.[30]

$$3X \;\overset{1}{\underset{3}{\rightleftharpoons}}\; A + 2X$$

$$X \;\overset{\beta}{\underset{\gamma}{\rightleftharpoons}}\; B \quad (20.8)$$

The macroscopic rate equation for the concentration φ of X is

$$\dot{\varphi} = -\varphi^3 + 3\varphi^2 - \beta\varphi + \gamma \equiv \alpha_1(\varphi) \quad (20.9)$$

The units of time and of Ω are chosen in such a way that the first two coefficients are 1 and 3.

For $\beta > 3$ (Fig. 12a) there is a single stationary state, which is asymptotically stable because $\alpha_1'(\varphi) < 0$.

For $\beta < 3$ and sufficiently large γ (Fig. 12b) there is still a single stationary state with $\alpha_1'(\varphi^{st}) < 0$. There is an interval where $\alpha_1'(\varphi) > 0$, but nevertheless the fluctuations cannot grow indefinitely, because φ passes through that interval in a finite time and they remain of order $\Omega^{1/2}$. A similar situation occurred in the initial stage of the population growth in Section IX.

For $\beta < 3$ and small γ (Fig. 12c) there are three stationary states, two of them stable. Hence the situation inside the reaction vessel is not uniquely determined by the externally imposed conditions. It may even happen that both situations coexist in different regions of the vessel,[30] but fluctuations in such inhomogeneous states are outside the scope of this article.

For $\beta < 3$ and one intermediate γ (Fig. 12d) there is one stationary state φ^{st} that is stable and another one φ^{us} with $\alpha'_1(\varphi^{us}) = 0$ and $\alpha''_1(\varphi^{us}) > 0$. The equation for a small deviation $\delta\varphi = \varphi - \varphi^{us}$ is

$$\frac{d}{dt}\delta\varphi = \tfrac{1}{2}\alpha''_1(\varphi^{us})(\delta\varphi)^2 \tag{20.10}$$

A negative deviation decays slowly like $1/t$ rather than exponentially. A positive deviation, however, grows. Hence the fluctuations will grow in the

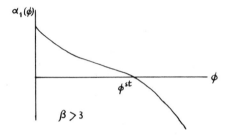

Fig. 12a. The function $\alpha_1(\varphi)$ given in (20.9) for $\beta > 3$.

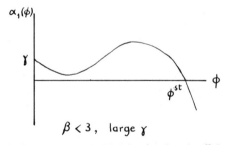

Fig. 12b. The function $\alpha_1(\varphi)$ in (20.9) for $\beta < 3$ and sufficiently large γ.

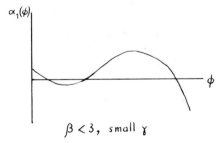

Fig. 12c. The function $\alpha_1(\varphi)$ in (20.9) for $\beta < 3$ and small γ.

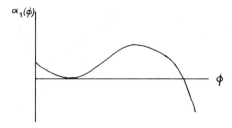

Fig. 12d.   The function $\alpha_1(\varphi)$ in (20.9) for $\beta < 3$ at one special intermediate value of γ.

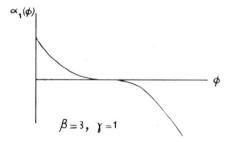

$\beta = 3, \ \gamma = 1$

Fig. 12e.   The function $\alpha_1(\varphi)$ in (20.9) for $\beta = 3$ and $\gamma = 1$.

positive direction and ultimately cause a transition from φ^{us} into φ^{st}. The Ω expansion can describe the onset of this transition, but not the entire transition, since it breaks down by the time the fluctuations have reached a macroscopic size.

For $\beta = 3, \gamma = 1$ (Fig. 12e) there is a single stationary state, φ^{cr}, and $\alpha_1'(\varphi^{cr}) = \alpha_1''(\varphi^{cr}) = 0$, which is analogous to the critical point in fluids or magnetic systems. In the next section we compute the fluctuations about φ^{cr} and φ^{us}.

XXI. CRITICAL FLUCTUATIONS

The expansion method is based on the idea that the fluctuations in n are small and remain small, that is, of order $\Omega^{1/2}$ as compared to the macroscopic part of order Ω. It is therefore necessary to verify *a posteriori* that this is so (see Section V). Whereas the diffusion term in the Fokker–Planck equation (4.8) tends to make the fluctuations grow proportionally with $\sqrt{\alpha_2 t}$, it is counteracted by the "convection term," which tries to squeeze them back towards equilibrium at the rate $|\alpha_2'| x$. Roughly speaking one has therefore[9]

$$\frac{\langle \Delta x^2 \rangle}{\Delta t} = \alpha_2 - 2|\alpha_2'|\langle x^2 \rangle \tag{21.1}$$

which leads to the estimate

$$\langle x^2 \rangle = \frac{\alpha_2}{2|\alpha_1'|} \tag{21.2}$$

In equilibrium α_1' and α_2 are constant; then this estimate is exact and expresses the connection between dissipation and fluctuations.

This argument assumes that $\alpha_1'(\varphi) < 0$, which is also the condition for the macroscopic solution φ to be asymptotically stable. In the multivariate case too it was found that asymptotic stability of the macroscopic equation guaranteed that the fluctuations remain of order $\Omega^{1/2}$. This condition is not satisfied in the case of a limit cycle (Section XVIII); as a consequence the fluctuations grow and render the formal expansion invalid. The equations indicate that this happens, but do not describe the behavior after a time $t \sim \Omega$.

An even more disastrous example is provided by the Lotka–Volterra equations for two species, one of which feeds on the other.[51] The rate equations are

$$\dot{n} = -\alpha n + \gamma \Omega^{-1} nm \tag{21.3a}$$

$$\dot{m} = \beta m - \delta \Omega^{-1} nm \tag{21.3b}$$

The same equations (with $\gamma = \delta$) also describe the chemical reaction

$$X \xrightarrow{\alpha} A$$

$$B + Y \xrightarrow{\beta} 2Y \tag{21.4}$$

$$X + Y \xrightarrow{\gamma} 2X$$

The macroscopic equations do not have a proper limit cycle, but *all* solutions are periodic.* Hence the fluctuations grow unchecked not only along the solution curves, but also perpendicular to them.

Returning to the one-variable case, one sees that difficulties arise when $\alpha_1' > 0$. The convection term then magnifies the fluctuations exponentially rather than counteracting the diffusive tendency; the expansion breaks down at $t \sim \log \Omega$. When, however, $\alpha_1' = 0$ the asymptotic stability of the macroscopic equations, and thereby the growth of the fluctuations, is determined by the higher derivatives of α_1.

Suppose that the first jump moment $\alpha_1(\varphi)$ has a zero φ^{st} where the first $q - 1$ derivatives vanish while $\alpha_1^{(q)}(\varphi^{st}) \neq 0$. In order to compute the fluctuation we start from the general master equation (4.3). Instead of the

---

* This follows from the fact that there is a constant of the motion: $-\alpha \log m - \beta \log n + \Omega^{-1}(\gamma m + \delta n) = C$. The solution curves in m, n space are determined by this equation.

transformation (4.6) we set more generally

$$n = \Omega\varphi(\tau) + \Omega^\mu x \tag{21.5}$$

The constant μ is still adjustable, subject to the condition $0 < \mu < 1$. The transformed master equation is

$$\frac{\partial \Pi}{\partial \tau} - \Omega^{1-\mu} \frac{d\varphi}{d\tau} \frac{\partial \Pi}{\partial x} = - \Omega^{1-\mu} \frac{\partial}{\partial x} \alpha_1(\varphi + \Omega^{\mu-1}x)\Pi$$

$$+ \frac{1}{2}\Omega^{1-2\mu} \frac{\partial^2}{\partial x^2} \alpha_2(\varphi + \Omega^{\mu-1}x)\Pi$$

$$- \frac{1}{3!}\Omega^{1-3\mu} \frac{\partial^3}{\partial x^3} \alpha_3(\varphi + \Omega^{\mu-1}x)\Pi + \cdots \tag{21.6}$$

The largest terms are of order $\Omega^{1-\mu}$; they are made to vanish by choosing for φ a solution of (4.7). Take the stationary solution φ^{st}; the fluctuations are then determined by

$$\frac{\partial \Pi}{\partial \tau} = - \frac{1}{q!} \alpha_1^{(q)}(\varphi^{st})\Omega^{(1-\mu)(1-q)} \frac{\partial}{\partial x} x^q\Pi$$

$$+ \frac{1}{2}\alpha_2(\varphi^{st})\Omega^{1-2\mu} \frac{\partial^2 \Pi}{\partial x^2} + \text{higher orders} \tag{21.7}$$

In order to obtain an equation that causes x to remain of order unity we determine μ by equating both powers of Ω,

$$(1 - \mu)(1 - q) = 1 - 2\mu, \qquad \mu = \frac{q}{q + 1} \tag{21.8}$$

The result is

$$\frac{\partial \Pi}{\partial \tau} = \Omega^{-(q-1)/(q+1)}\left[-\frac{1}{q!} \alpha_1^{(q)} \frac{\partial}{\partial x} x^q\Pi + \frac{1}{2}\alpha_2 \frac{\partial^2 \Pi}{\partial x^2} \right] \tag{21.9}$$

The reader will remember that $\alpha_1^{(q)}$ and α_2 are taken at φ^{st}. Observe that the fluctuations vary on a slower time scale than the macroscopic quantity φ. Another difference with the normal case is that the equation for the average

$$\frac{d}{d\tau} \langle x \rangle = \Omega^{-(q-1)/(q+1)} \frac{1}{q!} \alpha_1^{(q)}\langle x^q \rangle \tag{21.10}$$

is no longer just the variational equation associated with the macroscopic equation.

When q is even (as in Fig. 12d) it appears from (21.10) that positive fluctuations grow and negative ones decay, or vice versa, depending on the sign of

$\alpha_1^{(q)}$. According to the equation they even become infinite after a time of order $\tau \sim \Omega^{(q-1)/(q+1)}$, but of course the expansion breaks down as soon as they are of order $x \sim \Omega^{1-\mu} = \Omega^{1/(q+1)}$, because then the distinction (21.5) between a macroscopic and a mesoscopic part becomes blurred.

The first interesting value of q is 3, which occurs at the critical point in Fig. 12e. Equation (21.9) then becomes

$$\frac{\partial \Pi}{\partial \tau} = \Omega^{-1/2}\left[-\frac{1}{6}\alpha_1''' \frac{\partial}{\partial x} x^3 \Pi + \frac{1}{2}\alpha_2 \frac{\partial^2 \Pi}{\partial x^2}\right] \tag{21.11}$$

If $\alpha_1''' < 0$ the value of x remains finite, that is, the fluctuations in n remain of order $\Omega^{3/4}$. More precisely, Π tends to

$$\Pi(x, \infty) = \frac{1}{\Gamma(\frac{1}{4})}\left(\frac{4|\alpha_1'''|}{3\alpha_2}\right)^{1/4} \exp\left[-\frac{|\alpha_1'''|}{12\alpha_2} x^4\right] \tag{21.12}$$

The rate of approach, however, is of order $\Omega^{-1/2}$, which exhibits the critical slowing down.

XXII. CONCLUSIONS

The Ω expansion is the appropriate way for solving master equations of the type (1.6) when the numbers involved are so large that the fluctuations are relatively small. It dispenses with *ad hoc* assumptions about the smallness of certain terms, which gave rise to controversies and paradoxes.

The terms of order $\Omega^{1/2}$ yield the macroscopic equation, which may well be nonlinear. Terms of relative order Ω^{-1}, which are sometimes included in the macroscopic equation, are left out, as they should be in the thermodynamic limit.

The terms of order Ω^0 give rise to a linear Fokker–Planck equation with time-dependent coefficients, describing the fluctuations around the macroscopic behavior (linear noise approximation). A nonlinear Fokker–Planck equation is merely an artifact, due to making *ad hoc* approximations rather than a systematic expansion.

Fluctuations in a stationary state are in this approximation described by the familiar linear Fokker–Planck equation with constant coefficients and therefore constitute an Ornstein–Uhlenbeck process. Alternatively they may be treated by a Langevin equation, but the Langevin approach fails for fluctuations around time-dependent states.

In the next order the average value of n no longer coincides with the macroscopic value as determined by the macroscopic equation. The average does not obey an equation by itself, but can be found by solving the coupled equation (2.4) and (2.7).

In this and higher orders the Fokker–Planck equation is modified: additional powers are added in its coefficients, but at the same time higher derivatives appear. As a consequence the stationary distribution need no longer be Gaussian, and the autocorrelation function of the fluctuations is no longer a single exponential. Nevertheless the moments can be computed successively.

The multivariate master equation can be treated in the same way. The only difficulty is solving the resulting set of coupled macroscopic equations. Once this is done the fluctuations can be found (in the linear noise approximation).

With slight modifications the Ω expansion also applies to fluctuations around critical points and other special solutions of the macroscopic equations, as long as they are asymptotically stable, so that the fluctuations remain small.

Exceptions are the solutions that are unstable (as occurring in phase transitions) or merely orbitally stable (such as limit cycles). The fluctuations around such solutions grow so as to obliterate the distinction between the macroscopic value and the mesoscopic fluctuations. In these cases the Ω expansion is valid only for a limited time and can at best indicate what happens after that.

These conclusions remain true in many cases having a continuous variable q in lieu of the discrete n, but that is not the subject of this article.

APPENDIX

In order to carry out the program of Section IV one first has to solve (4.7) with given initial $\varphi(0)$. This can be done by a quadrature

$$\tau = \int_{\varphi(0)}^{\varphi(\tau)} \frac{d\varphi}{\alpha_1(\varphi)} \tag{A.1}$$

followed by an inversion to find $\varphi(\tau)$ as a function of τ. We suppose this done and write for the solution φ_τ rather than $\varphi(\tau)$.

Next one may solve (5.1) by using φ_τ as the independent variable instead of τ

$$\frac{d\langle x \rangle}{d\varphi_\tau} = \frac{\alpha_1'(\varphi_\tau)}{\alpha_1(\varphi_\tau)} \langle x \rangle \tag{A.2}$$

The solution is given by

$$\frac{\langle x \rangle_\tau}{\alpha_1(\varphi_\tau)} = \frac{\langle x \rangle_0}{\alpha_1(\varphi_0)} \tag{A.3}$$

Alternatively one may write

$$\langle x \rangle_\tau = \langle x \rangle_0 \frac{\dot{\varphi}_\tau}{\dot{\varphi}_0} \tag{A.4}$$

which shows more clearly that $\langle x \rangle \to 0$ when φ tends to a limiting value φ^{st}, so that $\dot{\varphi}_\tau$ tends to zero.

In the same way (5.2) is solved with the result

$$\frac{\langle x^2 \rangle_\tau}{\alpha_1(\varphi_\tau)^2} = \frac{\langle x^2 \rangle_0}{\alpha_1(\varphi_0)^2} + \int_{\varphi_0}^{\varphi_\tau} \frac{\alpha_2(\varphi)}{\alpha_1(\varphi)^3} \, d\varphi \qquad (A.5)$$

Again, when $\varphi_\tau \to \varphi^{st}$ one has of course $\alpha_1(\varphi^{st}) = 0$ so that

$$\langle x^2 \rangle_\tau \to \alpha_2(\varphi^{st}) \lim \alpha_1(\varphi_\tau)^2 \int^{\varphi_\tau} \frac{d\varphi}{\alpha_1(\varphi)^3}$$

$$\to \frac{\alpha_2(\varphi^{st})}{\alpha_1'(\varphi^{st})} \lim (\varphi_\tau - \varphi^{st})^2 \int^{\varphi_\tau} \frac{d\varphi}{(\varphi - \varphi^{st})^3}$$

$$\to \frac{1}{2} \frac{\alpha_2(\varphi^{st})}{|\alpha_1'(\varphi^{st})|} \qquad (A.6)$$

This expression is of course the same as the one obtained directly from (5.2) with $\varphi = \varphi^{st}$ and identifying the left-hand side with zero.

Having found $\langle x \rangle_\tau$ and $\langle x^2 \rangle_\tau$ one obtains the solution of (4.8) by taking for Π a Gaussian with these values of $\langle x \rangle$ and $\langle x^2 \rangle$. The fact that it obeys (4.8) is explicitly verified for the multivariate case in Section XIV and need not be repeated here. We only note that the next order equation (5.5) can also be solved in this way with the following result:

$$\frac{\langle x \rangle_\tau}{\alpha_1(\varphi_\tau)} = \frac{\langle x \rangle_0}{\alpha_1(\varphi_0)} + \frac{1}{2} \Omega^{-1/2}$$

$$\times \left[\frac{\alpha_1'(\varphi_\tau) - \alpha_1'(\varphi_0)}{\alpha_1(\varphi_0)^2} \langle x^2 \rangle_0 + \int_{\varphi_0}^{\varphi_\tau} \frac{\{\alpha_1'(\varphi_\tau) - \alpha_1'(\varphi)\}\alpha_2(\varphi)}{\alpha_1'(\varphi)^3} \, d\varphi \right]$$

In the case of the Malthus–Verhulst equation the solution (A.1) is given in (9.7). Substitution in (A.3) gives

$$\langle x \rangle_t = \langle x \rangle_0 e^{(\beta - \alpha)t} \left[1 + \varphi(0) \frac{\gamma}{\beta - \alpha} \{ e^{(\beta - \alpha)t} - 1 \} \right]^{-2}$$

Regardless of $\langle x \rangle_0$ this actually tends to zero both for $\beta > \alpha$ and for $\beta < \alpha$. In a similar way it is possible to compute $\langle x^2 \rangle_t$ from (9.11) using (A.5), but the resulting expression is too long to be reproduced here.

References

1. Reviewed in: A. A. Maradudin, E. W. Montroll, G. H. Weiss, and I. P. Ipatova, *Theory of Lattice Dynamics in the Harmonic Approximation*, 2nd ed., Academic Press, New York, 1971.

2. For example, L. Onsager and S. Machlup, *Phys. Rev.*, **91**, 1505 and 1512 (1953); L. D. Landau and E. M. Lifshitz, *Fluid Mechanics*, Pergamon, Oxford, 1959, Chap. 17; L. D. Landau and E. M. Lifshitz, *Electrodynamics in Continuous Media*, Pergamon, Oxford, 1960, Chap. 13; A. A. Abrikosov and I. M. Khalatnikov. *Sov. Phys. JETP*, **34** (7), 135 (1958); M. Bixon and R. Zwanzig, *Phys. Rev.*, **187**, 267 (1969); R. F. Fox and G. E. Uhlenbeck, *Phys. Fluids*, **13**, 1893, 2881 (1970); R. Zwanzig in *Proceedings of 6th IUPAP Conference on Statistical Mechanics*, S. A. Rice et al., Eds., Univ. of Chicago Press, Chicago, 1972; E. H. Hauge and A. Martin-Löf, *J. Stat. Phys.*, **7**, 259 (1973).

3. A. Nordsieck, W. E. Lamb, and G. E. Uhlenbeck, *Physica*, **7**, 344 (1940).

4. N. G. van Kampen, *Can. J. Phys.*, **39**, 551 (1961).

5. N. G. van Kampen, in *Fluctuation Phenomena in Solids*, R. E. Burgess, Ed., Academic Press, New York, 1965.

6. D. R. McNeil, *Biometrika*, **59**, 494 (1972).

7. A different (and somewhat confusing) interpretation of the transition probabilities is advanced in I. Oppenheim and K. E. Shuler, *Phys. Rev.*, **138**, B 1007 (1965); I. Oppenheim, K. E. Shuler, and G. E. Weiss, *J. Chem. Phys.*, **46**, 4100 (1967).

8. J. E. Moyal, *J. Roy. Statist. Soc.*, **B11**, 150 (1949).

9. N. G. van Kampen in *Proceedings of NUFFIC International Summer Course*, E. G. D. Cohen, Ed., North-Holland, Amsterdam, 1962.

10. R. Kubo, K. Matsuo, and K. Kitahara, *J. Stat. Phys.*, **9**, 51 (1973); R. Kubo, in *Quantum Statistical Mechanics in the Natural Sciences*, S. L. Mintz and S. M. Windmayer, Eds., Plenum, New York, 1974.

11. R. E. Burgess, *Proc. Phys. Soc. London*, **B68**, 661 (1955); **B69**, 1020 (1956); A. van der Ziel, *Noise*, Prentice, Englewood Cliffs, N.J., 1954; K. M. van Vliet and J. R. Fasset, in *Fluctuation Phenomena in Solids*, R. E. Burgess, Ed., Academic Press, New York, 1965.

12. L. M. Milne-Thomson, *The Calculus of Finite Differences*, Macmillan, London, 1951.

13. N. G. van Kampen, *Biometrika*, **60**, 494 (1972).

14. G. H. Weiss, *J. Stat. Phys.*, **6**, 179 (1972).

15. H. Mori, H. Fujisaka, and H. Shigematsu, *Progr. Theor. Phys.*, **51**, 109 (1974).

16. C. T. J. Alkemade, N. G. van Kampen, and D. K. C. MacDonald, *Proc. Roy. Soc.*, **A271**, 449 (1963).

17. For an example, see E. W. Montroll and K. E. Shuler, in *Adv. Chem. Phys.*, **1**, 361 (1958).

18. S. R. de Groot and P. Mazur, *Non-Equilibrium Thermodynamics*, North-Holland, Amsterdam, 1962.

19. J. T. Ubbink, *Physica*, **52**, 253 (1971).

20. W. Bernard and H. B. Callen, *Rev. Mod. Phys.*, **31**, 1017 (1959); *Phys. Rev.*, **118**, 1466 (1960).

21. N. G. van Kampen, *Physica Norvegica*, **5**, 279 (1971).

22. A. J. Lotka, *Elements of Mathematical Biology*, Dover, New York, 1956; E. W. Montroll, in *Quantum Theory and Statistical Physics, Boulder Lectures in Theoretical Physics*, Vol. 10A, A. O. Barut and W. E. Brittin, Eds., Gordon & Breach, New York, 1968; E. Batschelet, *Introduction to Mathematics for Life Scientists*, Springer, Berlin, 1971.

23. G. E. Uhlenbeck and L. S. Ornstein, *Phys. Rev.*, **36**, 823 (1930).

24. R. L. Stratonovich, *Topics in the Theory of Random Noise*, I, Gordon & Breach, New York, 1963.

25. R. Courant and D. Hilbert, *Methods of Mathematical Physics*, II, Interscience, New York, 1962, p. 230.
26. F. R. Gantmacher, *Matrizenrechnung* I. *Allgemeine Theorie*, 2nd ed., VEB Deutscher Verlag Wissens., Berlin, 1965, Chap. 8.
27. F. R. Gantmacher, *Applications to the Theory of Matrices*, Interscience, New York, 1959, p. 220 ff. See also M. Lax, *Rev. Mod. Phys.*, **32**, 25 (1960).
28. P. Glansdorff and I. Prigogine, *Structure, Stability and Fluctuations*, Wiley-Interscience, London, 1971.
29. D. M. McQuarrie, in *Adv. Chem. Phys.*, **15**, 149 (1969).
30. F. Schlögl, *Z. Phys.*, **253**, 147 (1972).
31. G. Nicolis, *J. Stat. Phys.*, **6**, 195 (1972); G. Nicolis, P. Allen and A. van Nypelseer, *Progr. Theor. Phys.*, **52**, 1481 (1974).
32. G. Nicolis and J. F. G. Auchmuty, *Proc Natl. Acad. Sci. USA*, **71**, 2748 (1974).
33. G. Nicolis, in *Cooperative Phenomena*, H. Haken, Ed., North-Holland, Amsterdam, 1974.
34. J. Keizer and R. F. Fox, *Proc. Natl. Acad. Sci. USA*, **71**, 192 and 2919 (1974).
35. A. Nitzan and J. Ross, *J. Stat. Phys.*, **10**, 379 (1974); Y. Kuramoto, *Progr. Theor. Phys.*, **49**, 1782 (1973); Y. Kuramoto, *Progr. Theor. Phys.*, **52**, 711 (1974); N. Saitô, preprint.
36. G. Nicolis, *Adv. Chem. Phys.*, **19**, 209 (1971).
37. R. Lefever and G. Nicolis, *J. Theor. Biol.*, **30**, 267 (1971).
38. R. J. Field and R. M. Noyes, *J. Chem. Phys.*, **60**, 1877 (1974).
39. R. L. Stratonovich, *Topics in the Theory of Random Noise*, II, Gordon & Breach, New York, 1967.
40. C. T. J. Alkemade, *Physica*, **24**, 1029 (1958); N. G. van Kampen, *Physica*, **26**, 585 (1960); R. McFee, *Amer. J. Phys.*, **39**, 814 (1971).
41. N. G. van Kampen, *J. Math. Phys.*, **2**, 592 (1961).
42. R. Graham, in "Quantum Statistics in Optics and Solid-State Physics," *Ergebn. Exact. Naturw.*, Vol. 66, Springer, Berlin, 1973.
43. R. Landauer in *Proceedings of 6th IUPAP Conference on Statistical Mechanics*, S. A. Rice et al., Eds., Univ. of Chicago Press, Chicago, 1972; R. Landauer, *J. Stat. Phys.*, **9**, 351 (1973).
44. L. Brillouin, *Phys. Rev.*, **78**, 627 (1950); D. K. C. MacDonald, *Phil. Mag.*, **45**, 63 (1954); *Phys. Rev.*, **108**, 541 (1957). D. Polder, *Phil. Mag.*, **45**, 69 (1954). R. O. Davies, *Physica*, **24**, 1055 (1958). A. Marek, *Physica*, **25**, 1358 (1959); J. B. Gunn, *J. Appl. Phys.*, **39**, 5357 (1968).
45. H. A. Kramers, *Physica*, **7**, 284 (1940); *Collected Scientific Papers*, North-Holland, Amsterdam, 1956, p. 754; R. J. Donnelly and P. H. Roberts, *Proc. Roy. Soc.*, **A312**, 519 (1969).
46. K. Huang, *Statistical Mechanics*, Wiley, New York, 1963; G. H. Wannier, *Statistical Physics*, Wiley, New York, 1966; A. Münster, *Statistical Thermodynamics*, Vol. 2, Springer, Berlin, 1974.
47. Th. W. Ruijgrok and J. A. Tjon, *Physica*, **65**, 539 (1973).
48. R. J. Glauber, *J. Math. Phys.*, **4**, 294 (1963).
49. M. O. Scully and W. E. Lamb, *Phys. Rev.*, **159**, 208 (1967); M. Scully in *Proceedings Varenna Summer School 1967*, R. J. Glauber, Ed., Academic Press, New York, 1969.
50. K. J. McNeil and D. F. Walls, *J. Stat. Phys.*, **10**, 439 (1974).
51. For literature see N. S. Goel, S. C. Maitra, and E. W. Montroll, *Rev. Mod. Phys.*, **43**, 231 (1971); E. H. Kerner, *Adv. Chem. Phys.*, **19**, 325 (1971).

AUTHOR INDEX

Numbers in parenthesis are reference numbers and show that an author's work is referred to although his name is not mentioned in the text. Numbers in *italics* indicate the pages on which the full references appear.

311

SUBJECT INDEX

321